THE GOVERNANCE OF SOLAR GEOENGINEERING

Climate change is among the world's most important problems, and solutions based on emission cuts or adapting to new climates remain elusive. One set of proposals receiving increasing attention among scientists and policymakers is "solar geoengineering" (also known as solar radiation modification), which would reflect a small portion of incoming sunlight to reduce climate change. Evidence indicates that this could be effective, inexpensive, and technologically feasible, but it also poses environmental risks and social challenges. Governance will thus be crucial. In *The Governance of Solar Geoengineering*, Jesse Reynolds draws on law, political science, and economics to show how solar geoengineering is, could, and should be governed. The book considers states' incentives and behavior, international and national law, intellectual property, compensation for possible harm, and nonstate governance. It also recommends how solar geoengineering could be responsibly researched, developed, and – if appropriate – used in ways that would improve human well-being and ensure sustainability.

Jesse L. Reynolds is an Emmett / Frankel Fellow in Environmental Law and Policy at the University of California, Los Angeles School of Law, as well as an associate researcher at Utrecht University and a research affiliate at Harvard University. He has degrees from Tilburg University; the University of California, Berkeley; and Hampshire College, and has been a US Environmental Protection Agency Science to Achieve Results Graduate Fellow and a Fulbright Scholar.

concludes that nonstate governance can play an essential role in solar geoengineering governance.

Chapters 11 and 12 concern two topics that are often invoked in the solar geoengineering governance discourse yet only infrequently explored in any depth. These are sufficiently detailed that my prescriptive suggestions are presented in these chapters instead of Chapter 13. The first of these is the roles of nonstate actors, and especially commercial ones. Chapter 11 begins by considering concerns that nonstate actors might implement solar geoengineering. However, they are likely to play roles – most likely as contractors in public procurement – in research, development, and possible deployment. A leading way in which commercial and other nonstate actors are governed in an innovative domain, such as that of solar geoengineering, is through the policies for intellectual property, particularly patents. Some challenges that intellectual property related to solar geoengineering would pose, as well as a handful of proposals for how to manage these, are discussed. I then put forth a proposal for a research commons for intellectual property related to solar geoengineering.

The second common, insufficiently researched topic is compensation for harm that could result from outdoor solar geoengineering activities. Would those who undertake or approve such activities be liable, especially for transboundary harm, and should they be? If not, should those who have been harmed be otherwise compensated, and if so, how? Chapter 13 provides overviews of the challenges that this would face and of existing international law as a vehicle for liability. I then offer initial proposals for compensation for harm from large-scale outdoor research and solar geoengineering deployment. The recommendation for the former is more specific, whereas that for the latter remains a conceptual framework.

Existing norms, rules, procedures, and institutions appear to be insufficient to effectively govern solar geoengineering in the longer term. In Chapter 13, I suggest what could be done to help ensure that solar geoengineering is researched, developed, and – if appropriate – used in ways that improve human welfare, are sustainable, and consistent with widely shared norms. These are divided into rough stages of small-scale outdoor research, small-scale research, and deployment.

Before proceeding, I wish to make some of my beliefs and assumptions clear and explicit. Normatively, I am consequentialist and welfarist. I believe that policies should be designed and implemented in ways that are expected to increase people's well-being. This should not be a mere brute summation of individuals' quantified utility, or worse, financial wealth. At the very least, there is a strong argument for equity weighting, in which those who are worse off are given disproportionate consideration. Furthermore, future people have value, as do the many nonmaterial things – including justice, security, the natural world, interpersonal relations, and personal experiences – that people consider important.

Second, I generally assume that actors, including states, pursue their diverse goals in a context of incentives and constrained by limited resources. This rationality offers substantial explanatory and predictive power and is a good – albeit imperfect – starting point for understanding interactive behavior. It is "weakly" rational in that it further assumes only reflective actors and complete, transitive, and sufficiently time-consistent preferences. I do not assume "strong" rationality, in which actors always consciously assess the expected outcomes of all options and choose, sensibly and without bias, the one with the greatest expected payoff. Furthermore, larger institutions, such as states, are arguably more rational than individuals due to collective and structured decision-making processes. At the same time, one should be vigilant for institutional failures and outcomes that appear collectively irrational.

Third, a project such as this is inherently speculative. As emphasized, climate change, the responses to it, solar geoengineering, and international relations are all uncertain, yet decision-makers must consider the long term. I try to focus on what I believe to be a reasonable range of possible scenarios, giving greater attention to those that seem more probable while not neglecting the less probable, but potentially relevant, ones.

All writers balance accuracy with brevity. I thus must explain how I use certain words and phrases, lest I either repeat disclaimers and clarifications or risk ambiguity.[2] "Climate change" and "global warming" are used largely interchangeably, with a bias toward the former due to its appropriately greater breadth. "Solar geoengineering" is the intentional modification of the Earth's radiative balance, excluding changes to greenhouse gases (see Heyward 2013; Boucher et al. 2014). This encompasses large-scale actions that would reflect or block some incoming sunlight – which is elsewhere called solar radiation management or modification (SRM), solar climate engineering, albedo modification, climate remediation, and radiation modification measures. Solar geoengineering also includes cirrus cloud thinning, which would allow more infrared radiation – that is, heat – to escape and has similar relevant characteristics. I sometimes use "geoengineering" to encompass a wider range of "deliberate large-scale intervention in the Earth's climate system" (Shepherd et al. 2009, ix), which encompasses both solar geoengineering and large-scale NETs.

The phrase "solar geoengineering" includes the entire set of actual and possible activities, such as brainstorming, serious discussion, lab work, models, small field experiments, climate responses tests, deployment, and assessment activities that would inform and enable such interventions. "Outdoor activities" are both experimental and operative, although the line between these would not be distinct. "Experiments," "tests," and "research" – often preceded by "outdoor" – are intended to be roughly synonymous, as are "implementation" and "deployment" for the use of solar geoengineering to globally counter climate change. Mundanely, "billion" and "trillion" refer to their US or short-scale meanings of 10^9 and 10^{12} respectively.

---

[2] My word choices are without prejudice to others'.

"Governance" is the goal-oriented, sustained, focused, and explicit use of authority to influence behavior. This can be done through diverse means including unwritten norms, nonbinding principles and rules, laws, administrative regulations, market instruments, procedures, institutions, funding, and international law. Furthermore, governance can be performed by governments, intergovernmental organizations, businesses, other authoritative institutions, and individuals, with others or themselves as the targets. I sometimes use "regulation" to imply governance through binding rules that are developed and enforced by authoritative institutions – especially state ones – that can punish violators. Exceptions are my use of "self-regulation," "private regulation," and "meta-regulation." "Law" is state-made governance that is embodied domestically in legislation, administrative rules, and case law, and internationally in custom, multilateral agreements, general principles, and decisions of international tribunals. "State," "country," and "nation" are also meant synonymously, except for US states, which are relevant mainly in Chapter 9. These words are often anthropomorphized to indicate their leaderships.

As this book concerns a speculative topic, I use probabilistic words such as "possible," "feasible," and "likely" to describe futures. These represent nothing more than my personal judgment. People of good faith can disagree. I try to use modal verbs to appropriately reflect various degrees of probability and permissiveness, although mistakes are likely. Please do not misinterpret an occasional "will" or "can" as implying precise predictions. Likewise, I mean "proposed" in the broad sense of being suggested by some experts, not necessarily in a pipeline toward expanded activities.

Climate change will affect humans, other species, and ecosystems, and mostly negatively so. Solar geoengineering appears able to reduce climate change, while other responses will almost certainly continue to be inadequate. It will thus receive increasing interest. Because of solar geoengineering's transboundary impacts, environmental risks, and social challenges, governance will be critical. Broad, in-depth, and long-term conversations are necessary to develop governance that is effective and consistent with widely shared values. I hope that this book offers a useful foundation for these processes.

2

Climate Change and Solar Geoengineering

Anthropogenic climate change is arguably the most important and difficult environmental problem that presently confronts global society. Solar geoengineering is being considered and researched as a potential response to it. To understand solar geoengineering, the motivation for it, and how it would, could, and should be governed, we must first explore the causes and risks of climate change as well as other responses. This is the topic of the first half of this chapter, while the second half describes the history, proposed methods, apparent effectiveness, physical risks, and social challenges of solar geoengineering.

2.1 THE CHANGING CLIMATE

Climate is average weather for a given area over a long period of time. The relevant conditions include temperature, precipitation (that is, rain and snowfall), humidity, air pressure, and wind. The climate is driven by the Sun's energy, which comes in a range of wavelengths. Incoming solar radiation is mostly in the visible range. Atmospheric gases are usually transparent to visible light – that is, they allow visible light to pass through them. About a third of this incoming energy is reflected into space by clouds, atmospheric aerosols (very small suspected particles), and the surface of the Earth, which absorbs the remaining energy. From this, the Earth warms and then reemits the energy at a different range of wavelengths, those of invisible infrared radiation. Some atmospheric gases – the greenhouse gases – absorb this infrared radiation and warm up. This greenhouse effect causes the planet to naturally be warmer than it otherwise would be. Naturally occurring greenhouse gases increase the Earth's average surface temperature by thirty-three degrees Celsius. The energy budget of the preindustrial atmosphere was in balance: the incoming solar energy from the Sun was balanced by the reflected light and outgoing infrared radiation.

We have changed the atmospheric concentrations of greenhouse gases, most importantly through burning fossil fuels. Coal, oil, and natural gas originate from subterranean preserved organic matter. Their carbon–carbon chemical bonds release heat when broken through combustion, making them useful energy sources. Humans have burned increasing amounts of fossil fuels, especially during the last two centuries. This has led to enormous improvements in our well-being. We live longer, suffer less, and are more secure than our ancestors, in large part due to industrial activities that rely on fossil fuels. But importantly, the combustion of fossil fuels releases the carbon as carbon dioxide, a greenhouse gas. Some other human activities, such as changing natural landscapes to agricultural ones, also release carbon dioxide. Although terrestrial ecosystems and the ocean have absorbed some of these emissions, the atmospheric concentration of carbon dioxide has increased at an unprecedented rate and is now more than forty percent greater than its preindustrial level. Carbon dioxide is the most important anthropogenic greenhouse gas and contributes the most to the increase in the greenhouse effect. This is because our emissions of it are great and are closely coupled with economic activity, which remains in the atmosphere for a long time. The concentrations of other greenhouse gases, including methane, nitrous oxide, ozone, and chlorofluorocarbons have also grown due to human activities.

The anthropogenic greenhouse effect has already changed the Earth's climate. Its average surface temperature has increased by 0.85 degrees Celsius, and the oceans have warmed as well. Precipitation appears to have increased, particularly at midlatitudes. Some extreme weather events, such as heat waves and heavy precipitation, seem to be more frequent. Glaciers and ice sheets (as well as sea ice) are melting, causing the sea levels to rise. Areas of snow cover and permafrost have shrunk. With varying degrees of confidence, scientists can attribute these phenomena to greenhouse gas emissions.

At the same time, some human activities have partially counteracted the anthropogenic greenhouse effect. For example, aerosols and other particulate pollution in the lower atmosphere reflect some incoming sunlight, cooling the planet by one-half to one degree Celsius and thus masking a large portion of anthropogenic warming (Samset et al. 2018). Ironically, as stronger regulations and improved technology lead to less particulate air pollution, more of this suppressed climate change will continue to manifest itself.

This anthropogenic increase in the greenhouse effect will continue to change the climate, in particular by warming it. Models indicate that business-as-usual scenarios, with no policies in place, could lead to 1.4 to 3.1 degrees Celsius warming by the end of this century, and up to 4.8 degrees in worst-case scenarios (Alexander et al. 2013, 20). This warming will not be distributed evenly: the polar regions will warm more than the tropics, and winter temperatures will rise more than those of summer. Precipitation will change too. In general, the intensity of the hydrological cycle will increase, with more evaporation and precipitation globally. Furthermore, wet regions will tend to become wetter, and dry regions will tend to become drier. Likewise, both extreme precipitation events and droughts will become more severe. Higher temperatures will lead to rising sea levels due both to the melting of glaciers on the Greenland and Antarctic ice sheets and to the thermal expansion of the oceans.

Forecasts of climate change and its impacts remain uncertain for several reasons. One source of uncertainty is natural variability. The climate changes naturally through processes of diverse timescales. Some of these – such as El Nino – are relatively well characterized, whereas others remain less so. Observed and future climate change is not purely anthropogenic because these natural processes reduce or increase it.

A second source of uncertainty is physical: we do not know precisely how much the climate will change due to a given increase in greenhouse gas concentrations. This "climate sensitivity" is usually expressed as the expected warming of global average surface temperature that would result from a doubling of the preindustrial carbon dioxide concentration.[1] Scientists estimate the climate sensitivity to be three degrees Celsius, but it could be substantially more or less. The effects of aerosols and clouds, both of which generally counteract some warming, are complex and difficult to assess and forecast. This physical uncertainty is compounded by possible feedback effects, through which Earth systems respond to anthropogenic climate by absorbing or releasing their own greenhouse gases. Some of these feedbacks could be negative, moderating atmospheric greenhouse gas concentrations. However, more of these appear to be positive, or exacerbating. Extreme positive feedback loops are possible and could cause dangerous climate change on a relatively short timescale, a phenomenon that is sometimes called the "tipping point" (Galaz 2014).

A third source of uncertainty is that of future human conditions and activities. Modelers have developed a range of scenarios of population growth, economic development, technological change, policies, and greenhouse gas emissions. The scenarios (or more accurately, pathways) used by the Intergovernmental Panel on Climate Change (IPCC), in its most recent authoritative Assessment Report, forecast annual net carbon dioxide emissions in 2100 from slightly negative to almost triple their present rates.

A final source of uncertainty is the impacts of climate change. For the most part, we are concerned not about a changing atmosphere or climate but instead about the effects on people and on the environment. The most recent authoritative IPCC Assessment Report identifies eight key risks, all determined with high confidence, quoted here as:

1. Risk of death, injury, ill health, or disrupted livelihoods in low-lying coastal zones and small island developing states and other small islands, due to storm surges, coastal flooding, and sea level rise.
2. Risk of severe ill health and disrupted livelihoods for large urban populations, due to inland flooding in some regions.

[1] There are multiple ways to express climate sensitivity, such as which feedbacks are considered and whether the warming is estimated after a specific period of time or at equilibrium.

3. Systemic risks due to extreme weather events leading to breakdown of infrastructure networks and critical services ...
4. Risk of mortality and morbidity during periods of extreme heat ...
5. Risk of food insecurity and the breakdown of food systems linked to warming, drought, flooding, and precipitation variability and extremes ...
6. Risk of loss of rural livelihoods and income due to insufficient access to drinking and irrigation water and reduced agricultural productivity ...
7. Risk of loss of marine and coastal ecosystems, biodiversity, and the ecosystem goods, functions, and services they provide for coastal livelihoods ...
8. Risk of loss of terrestrial and inland water ecosystems, biodiversity, and the ecosystem goods, functions, and services they provide for livelihoods (Field et al. 2014, 59–62).

Put together, this uncertainty causes the risks of climate change to have an asymmetrical probability distribution. Although there is some chance of minor impacts, there is also a small chance of very severe impacts, sometimes called "fat tail risks." A distribution such as this influences risk management strategies.

To some unknown degree, humans and ecosystems will adapt to climate change. However, society has never needed to cope with these large environmental changes on such a short timescale. Independent of uncertainty, poorer people and developing countries are exposed more to climate change impacts than the wealthier ones, for several reasons. For one thing, they tend to be in warm areas. Second, the activities of poorer people are more reliant upon the climate. They spend more time outdoors, and indoors without air conditioning, and their economies depend more upon agriculture. Third, poorer people and developing countries have fewer financial resources and generally weaker institutional capacity to adapt to a changing climate.

Responding to the uncertainty is exacerbated by the fact that the effects of climate change are delayed relative to the emissions that cause them. This delay is mostly because the oceans absorb much of the excess heat, and some time is needed before the climate reaches a new equilibrium. Thus, even if anthropogenic greenhouse gas emissions were – hypothetically – to stop completely, the climate would continue to warm for some decades. In other words, at any given point in time, humanity is committed to an uncertain amount of future climate change. Greenhouse gas emissions to date might have committed us to around 1.3 degrees Celsius of total warming.

Some observers assert that human's influence on the natural world is already so great that the planet has entered a new geological epoch – the Anthropocene (Zalasiewicz et al. 2017). This claim is contentious among earth scientists because, among other reasons, they disagree as to whether the influence has been great enough and, if so, when to demarcate the beginning. Environmentalists also contest it, as it could imply that we should come to terms with, if not embrace, our impacts (see Chapter 14). Regardless, a declaration that we have moved into the "epoch of the human" would likely alter how many conceptualize our relationship with the Earth.

2.2 RESPONSES TO CLIMATE CHANGE

Human activities have already changed the climate and will continue to do so. The magnitude of its impacts on humans and nature remain within our agency, at least to some degree. We are undertaking and will continue to undertake several responses to reduce climate change and its effects.

2.2.1 Emissions Abatement

The first, and perhaps most logical, organized response to climate change has been the abatement (often called "mitigation") of greenhouse gas emissions. Since the rise of concern regarding climate change in the late 1980s, abatement efforts have, relative to other responses, received most attention by policy-makers, researchers, advocates, and the media.

Abatement consists of diverse activities across the economy and society that would reduce greenhouse gas emissions. Energy supplies need the most reform. Changes are also necessary in sectors that produce significant emissions of greenhouse gases and/or consume large amounts of energy, such as agriculture, forestry, industrial processes, buildings, and transportation. Furthermore, some of the carbon dioxide produced needs to be captured at its source – such as a coal-fired electricity plant – and either reused or stored, perhaps underground or under the ocean. This "carbon capture and storage" is not yet applied at industrially significant scales.

For our purposes here, what matters is the likelihood that greenhouse gas emissions abatement could prevent dangerous climate change. Although some observers are encouraged by the progress to date, this has consisted of limited and relatively easy and inexpensive actions. Unfortunately, there are several reasons to be pessimistic that future emissions abatement will be sufficient. First is the unprecedented scale of what is necessary. In order to be confident in keeping warming within the two degrees Celsius limit established by the Paris Agreement, total future greenhouse gas emissions will need to be limited to about a thousand gigatons of carbon dioxide equivalent (United Nations Environment 2018, 17). Presently, emissions are roughly fifty gigatons per year, leaving only about twenty years in the so-called carbon budget at the current rate. Yet the rate of emissions has increased almost every year. This figure also implies that net greenhouse gas emissions should fall to zero within a few decades, a relatively short time in which the entire energy system – and more – would need to be almost completely rebuilt.

The necessary scale is exacerbated by global demographic and economic trends. The world presently has about seven billion people, of which roughly one billion enjoy consumption levels equivalent to or greater than those of industrialized countries' middle class. By the end of the century, the Earth will be home to maybe eleven billion people, most of whom will be understandably clamoring for – and many of whom will have attained – these levels of consumption. From another perspective, in a moderate scenario, the world economy is expected to be seven times as large at the end of the century as that of today (Dellink et al. 2016). For total emissions to remain less than a thousand gigatons of carbon dioxide equivalent with a world economy that will be seven times as large, greenhouse gas emissions per unit of economic activity would need to consistently decrease at around 6.5 percent annually through 2100. The average global rate for 2000–17 has been 1.6 percent, and only a few countries have decarbonized at a 3 percent rate for more than a decade (PwC 2018).

A second challenge is the problem structure. Each country would benefit from the costly abatement efforts of others, independent of whether it abated. Likewise, each country captures the rewards of its emissions while the negative consequences are distributed worldwide. Emissions abatement is thus a public good, in which all actors benefit from its provision, yet each cannot be excluded from enjoying these benefits if it refuses to contribute to the public good's cost. This presents a collective action problem, in which each actor has incentives to free ride – that is, to contribute as little as possible – in the hope that others do more. Clearly, if all actors believe this, then they will provide little or none of the public good. Although this outcome appears collectively irrational, free riding is rational for each individual actor. Due to the absence of centralized enforcement, the provision of global public goods is one of the leading challenges of international cooperation (see Chapter 4).

Some specifics of emissions abatement make it a particularly difficult collective action problem. As described, climatic change impacts are global in scope and delayed by decades relative to the emissions that cause them. This reduces the incentives, including those of political leaders, to take costly, short-term, local actions that yield long-term global benefits. Furthermore, most greenhouse gas emissions presently come from areas that are relatively less vulnerable to climate change impacts. Finally, the leading source of greenhouse gas emissions is fossil fuel combustion and other industrial processes, which are highly coupled to population growth and economic activity, with its concomitant general increase of human welfare.

The third challenge of ambitious emissions abatement is international disagreement. Developing countries point to wealthy industrialized countries, which have emitted the majority of total historical greenhouse gases despite presently representing only one-sixth of the world's population. "They can afford the expensive abatement actions," the developing states argue, "while we deserve some of the economic growth that they have experienced." The wealthier ones counter with calls for controls on the rapidly rising emissions of the developing states, whose share of annual emissions is now the majority and growing. This results in an impasse between these two groups, although the lessening of the development gap between them is perhaps reducing this barrier.

The third challenge is that meaningful abatement policies would be politically unpopular. Political scientist Roger Pielke, Jr. suggests an "iron law of climate policy," which is, "when policies focused on economic growth confront policies focused on emissions reductions, it is economic growth that will win out every time" (Pielke 2010, 46). Opinion surveys that ask respondents to either rank their policy priorities or to express a willingness-to-pay to prevent climate change reinforce this. For example, the Pew Research Center annually asks Americans to rank a set of eighteen to twenty-three policy priorities. Since its introduction as a choice in 2007, climate change has been among the bottom four in importance (see Pew Research Center n.d.). Similarly, a UN website has received almost ten million votes for what the priorities of the intergovernmental body should be. Among the sixteen choices, "action taken on climate change" is last (United Nations n.d.). Moreover, while current abatement activities are inexpensive, the costs of additional mitigation could rapidly increase as the less expensive options are undertaken. As this occurs, Pielke's "iron law" could become increasingly valid.

These challenges are evident in the greenhouse gas emissions abatement to date. The need to reduce them has been evident for about thirty years, yet actual emissions have consistently been far from what scientists and other experts have recommended, and indeed closer to more pessimistic scenarios. The first treaty on the problem – the 1992 United Nations Framework Convention on Climate Change (UNFCCC) – itself did little more than encourage research, data sharing, and vague national programs. It was supplemented by the 1997 Kyoto Protocol, which committed only the industrialized countries to specific, modest emissions caps. These were largely met due not to abatement actions but to economic downturns in the former Soviet bloc in the 1990s and throughout the world in the late 2000s. The 2015 Paris Agreement takes a different approach, in which countries are to put forth their own emissions abatement plans. The first round of voluntary commitments, which were made in the lead-up to the 2015 Conference of the Parties (COP) in Paris, is forecast to lead to 2.7 degrees Celsius warming by the end of the century – if all countries actually meet their targets (International Energy Agency 2015, 3). Parties are to submit more ambitious plans every five years. Although this bottom-up approach might be better than Kyoto's top-down one, it seems to be too little, too late if net emissions must be eliminated within a few decades.

2.2.2 Negative Emissions Technologies

Most scenarios that keep warming within two degrees Celsius assume that atmospheric greenhouse gas concentrations will surpass the corresponding level and then later be brought back down. These rely on negative emissions technologies (NETs, or sometimes "carbon dioxide removal") that would remove greenhouse gases, particularly carbon dioxide, from the atmosphere and sequester them over a long timescale. The leading NETs are direct air capture, in which carbon dioxide is captured from ambient air, and then stored; bioenergy with carbon capture and storage, in which plants are grown and burnt to produce energy, with the resulting carbon dioxide captured and stored; enhanced weathering, in which minerals are processed to accelerate natural chemical carbon dioxide sequestration; and ocean fertilization, in which nutrients are added to speed up natural marine biological carbon dioxide sequestration (see National Research Council 2015a). Although these have been researched and, to some degree, implemented, serious questions remain regarding whether any of them could be effectively, safely, and acceptably used at the very large scales that are assumed in the optimistic scenarios.

NETs are sometimes grouped with solar geoengineering in a general category of "geoengineering" or "climate engineering." Although some NETs under certain circumstances would be "the deliberate large-scale intervention in the Earth's climate system," this depends on their scale (Shepherd et al. 2009, ix). Most NETs could help reduce net greenhouse gas emissions – albeit perhaps marginally – at modest scales. Furthermore, NETs are sufficiently different from solar geoengineering – in their modes of operation, possible roles within climate policy, capacities, costs, risks, and speeds – that separating them is usually analytically justified.

NETs have gradually become part of the mainstream climate change discourse. They were implicitly recognized in 1992's UNFCCC, which calls for atmospheric greenhouse gas concentrations to be stabilized by methods that include "the conservation and enhancement, as appropriate, of sinks and reservoirs of all greenhouses gases ... including biomass, forests and oceans as well as other terrestrial, coastal and marine ecosystems" (Articles 2, 4.1(d)). Given that NETs were not well conceived at that time, the drafters of the UNFCCC likely intended this passage to reflect land and ecosystem management practices that would increase carbon storage. Nevertheless, the IPCC more recently relied on large quantities of bioenergy with carbon capture and storage in its scenarios that are expected to keep warming below two degrees Celsius. The Paris Agreement of 2015, whose drafting was informed by these IPCC scenarios, not only continues to refer to enhancement of sinks and reservoirs, but its central commitment calls on parties "to achieve a balance between anthropogenic emissions by sources and removals by sinks of greenhouse gases in the second half of this century" (Article 4.1; see also Article 5.1). The extent to which states could use NETs to satisfy their emissions targets will be important but remains unsettled.

Despite NETs' growing importance in climate change policy, scaling them up could be problematic. They are presently expensive and pose some local environmental risks that vary by the method and scale. Like emissions abatement, NETs also present an asynchronous global collective action problem, in which costs are borne locally and incurred now, whereas benefits will be globally enjoyed later.

2.2.3 Adaptation

The third primary category of responses to climate risks is to adapt society and, where possible, ecosystems to a changed climate. Adaptation is a diverse set of processes that will happen at a wide range of spatial and temporal scales. It has been part of the climate change discourse since the beginning, although in the last decade its prominence has been elevated (see Chapter 4). Some adaptation will be anticipatory, while other actions will occur only during or after the climate has changed. Furthermore, some adaption will be autonomous, undertaken by the relevant actors in response to changing conditions, while in other cases policies will play a necessary role. In fact, adaptation represents a large uncertainty in estimates of climate change's expected costs.

Most climate change risks have corresponding adaptive actions. For example, as water supplies change, such as by becoming more erratic, managing institutions could build greater storage facilities and create incentives to reduce demand. Farmers could change which crops they plant or when they plant them, perhaps assisted by experts who share best practices and other knowledge. Dikes, levees, and other infrastructure could be built in low-lying coastal areas as a response to higher sea level and storm surges. Some residents of these and other vulnerable areas will migrate.

As with greenhouse gas emissions abatement, the details of adaptation are less important than the extent to which we can expect it to reduce climate change impacts. Unlike emissions abatement, each country or other region is the primary beneficiary of its adaptive actions. It thus does not present a global collective action problem. Furthermore, in some ways – such as protecting coastal communities from sea level rise – adaptation can be cost effective. At the same time, it faces serious limitations. Because political leaders' time horizons are strongly influenced by their election cycles and careers, they may be reluctant to invest in anticipatory adaptation, which has long-term benefits, relative to other policy options that yield shorter-term benefits. Moreover, countries' vulnerabilities to climate change risks are negatively correlated with their financial resources. Adaptation that is sufficient to genuinely reduce climate change risks in vulnerable developing countries will require substantial financial support from the wealthier industrialized ones. But large cash transfers would most likely be politically unpopular in the source states. Finally, some impacts of climate change are very difficult to reduce through adaptation. An example is isolated ecosystems, for which little can be done in some cases.

Much can and should be undertaken to prevent climate change and its impacts. Emissions abatement, NETs, and adaptation each have advantages but also limitations. Their actual implementations have been insufficient thus far and will almost certainly continue to be so, in large part due to the underlying problem structures. This disheartening forecast, coupled with climate change's risks, provides motivation to consider other means to respond.

2.3 THE RISE OF SOLAR GEOENGINEERING

Facing a seemingly dire future, some scientists and others are increasingly calling for considering and researching a fundamentally different response to climate change. They suggest that the amount of incoming solar radiation that reaches or is absorbed by the Earth's surface could be reduced, or that the amount of outgoing infrared radiation could be increased. Through intentionally modifying the Earth's radiative balance, humanity may be able to cool the planet and reduce climate change. This is here called "solar geoengineering," although it is often "solar radiation management" or "modification" (SRM).

The earliest such thinking predates widespread awareness of and anthropogenic climate change.[2] In the mid-1960s, when contemporary framings of environmentalism were taking shape, a Presidential committee was formed in the United States in response to growing public worries about environmental degradation. Its 1965 report, *Restoring the Quality of Our Environment*, touches on a range of topics including the rising atmospheric concentration of carbon dioxide, the first such discussions in a government document. It suggests only one response to elevated greenhouse gas concentrations: to increase the reflectivity of the Earth. The proposed method – "spreading very small reflecting particles over large oceanic areas" – is not presently considered to be feasible and effective, and the general idea did not receive sustained interest (Revelle et al. 1965, 127).

However, the concept of solar geoengineering was not dead. Stratospheric aerosol injection, now the most widely discussed method, appears to have been first suggested by a prominent Soviet climatologist, Mikhail Budyko. He was inspired by the cooling effect of large volcanic eruptions and published his proposal in 1974, initially in Russian (Budyko 1977). Solar geoengineering surfaced again in a 1983 US National Academies report on the build-up of atmospheric carbon dioxide. Its relevant chapter, written by foreign policy expert and future Nobel Laureate Thomas Schelling, notes that states might seek to use it for their own relative gain (Schelling 1983). It also recognizes that while abatement of greenhouse gases would be expensive and pose a collective action problem, solar geoengineering might be inexpensive enough that individual states would pursue it on their own.

In the early 1990s, as climate change entered international and domestic policy debates, an entire chapter of another US National Academies climate change report was dedicated to geoengineering, a term that included NETs and reforestation in its scope.[3] Notably, it compares the feasibility, potential effectiveness, and approximate costs of several suggested methods, including stratospheric aerosol injection, mirrors and dust in space, and increased cloud abundance, and concludes "that further inquiry is appropriate" (Institute of Medicine, National Academy of Sciences, and

---

[2] For more history of geoengineering, see Keith (2000) and Morton (2015).
[3] The chapter was written by, but not explicitly attributed to, Robert A. Frosch, a former National Aeronautics and Space Administration administrator.

National Academy of Engineering 1992, 496). The chapter was controversial among the report's authors, many of whom were concerned that such investigations would weaken abatements efforts (Schneider 1996, 295). This sentiment would remain the basis of a taboo on solar geoengineering among climate scientists for years to come (see Chapter 3). A subsequent symposium at the 1994 annual meeting of the American Association for the Advancement of Science led to a special issue of *Climatic Change*, including the first published writing on geoengineering and international law (Bodansky 1996).

In the late 1990s, solar geoengineering was an occasional topic of discussion at Lawrence Livermore National Laboratory, near San Francisco, California. Lowell Wood, a sometimes controversial physicist there, published a white paper and gave a talk in Aspen, Colorado, in which he provocatively argues that it is the preferred response to rising greenhouse gas concentrations (Teller, Hyde, and Wood 1997). Climate scientists David Keith and Ken Caldeira were present and skeptical, and particularly concerned that solar geoengineering would spatially have dramatically heterogeneous impacts (Morton 2015, 148–53).[4] The latter, who worked at Lawrence Livermore separately from Wood, ran with a colleague the first model of solar geoengineering and was surprised by the resulting simulated effectiveness and consistency (Govindasamy and Caldeira 2000).

Caldeira's paper was read by Paul Crutzen, an eminent atmospheric chemist who had received a Nobel Prize for his work on the damage done by humans to the stratospheric ozone layer and is now one of the leading proponents of the Anthropocene categorization. He subsequently drafted, circulated, and published an essay, arguing for solar geoengineering research – and particularly stratospheric aerosol injection – because emissions abatement efforts "have been grossly unsuccessful" and look "like a pious wish" (Crutzen 2006, 211–12, 7). Crutzen notes that our sulfate aerosol pollution in the lower atmosphere already offsets some climate change; reducing it contributes to warming. In principle, he asserts, we could transfer a small portion of this pollution to the upper atmosphere, where it would be less harmful and have a stronger cooling effect. Although solar geoengineering remained controversial, such an argument coming from a scientist with Crutzen's reputation cracked the taboo.

In 2007, the Royal Society formed a committee to investigate geoengineering. The resulting report, *Geoengineering the Climate: Science, Governance and Uncertainty*, presents the state-of-the-art thinking, calls for publicly funded research, and frames issues in ways that continue to hold sway. Notably, it asserts that "The acceptability of geoengineering will be determined as much by social, legal and political issues as by scientific and technical factors" (Shepherd et al. 2009, ix).

After the Royal Society report, research accelerated but remains modest (Necheles, Burns, and Keith 2018). The European Commission and a handful

---

[4] Keith had previously written on solar geoengineering (Keith and Dowlatabadi 1992).

of national governments have funded projects, which now total a few million US dollars annually. International conferences focused on geoengineering were held at the Asilomar Conference Grounds near Monterey, California in 2010 and in Berlin in 2014 and 2017. The US National Academies released a pair of reports on geoengineering, one on NETs and the other on solar geoengineering, in 2015, calling for the government to support research (National Research Council 2015a, b).

Solar geoengineering research has progressed, although not to the extent to which the US and UK reports call. Two outdoor experiments have taken place, although their relevance was limited, and one of these was not explicitly described as solar geoengineering until after the fact (Izrael et al. 2009; Russell et al. 2013). An English research team planned an outdoor test of aerosol injection equipment but cancelled it (Stilgoe, Watson, and Kuo 2013). Scientists are currently planning a few more outdoor experiments, described in Section 2.4. Meanwhile, scholarly interest in solar geoengineering within the humanities and social sciences has grown, and their volume of output now eclipses that of the natural sciences. Popular media coverage is likewise common. Some environmental advocacy organizations are now engaged, with a few taking positions that range from opposition to cautious support of research. There are at least three nongovernmental organizations dedicated to geoengineering. The university-affiliated Forum for Climate Engineering Assessment assesses the social implications to ensure that "justice, equity, agency, and inclusion" remain prominent within the dialogue (Forum for Climate Engineering Assessment, n.d.). The Solar Radiation Management Governance Initiative (SRMGI) expands the solar geoengineering conversation to include voices from developing countries. And the Carnegie Climate Geoengineering Governance Initiative (C2G2), led by veteran UN diplomat Janos Pasztor, works toward international governance of geoengineering, primarily through international organizations. Notably, governments remain mostly absent from the solar geoengineering discourse.

A small group of people believe that solar geoengineering is already happening. They argue that some coalition of powerful actors is secretly spraying material into the sky, leaving observable "chemtrails" in the sky behind aircraft. Their suggested reasons for this are inconsistent and include covert research, influencing the weather, and mind control. These conspiratorial beliefs are false but appear to have grown and influence public opinion on solar geoengineering (Cairns 2016; Tingley and Wagner 2017).

2.4 PROPOSED METHODS

A handful of solar geoengineering methods have been proposed during recent decades. Two of these – marine cloud brightening and especially stratospheric aerosol injection – appear to be potentially effective, technologically feasible, inexpensive in their direct costs, fast acting, and reversible in their climatic effects, and have thus received the most attention. A range of solar geoengineering methods is introduced here, including cirrus cloud thinning, which aims to increase outgoing infrared radiation rather than reduce incoming sunlight.

2.4.1 Stratospheric Aerosol Injection

The solar geoengineering method that presently holds the most potential is stratospheric aerosol injection. Its inspiration is large volcanic eruptions, whose erupted material can reach the stratosphere, the atmospheric layer eight to twenty kilometers above the Earth's surface. Within this, volcanic sulfate compounds form tiny sulfuric acid droplets. This aerosol layer reflects some of the incoming sunlight back into space, cooling the planet. Because the stratosphere is stable and lacks rain that would remove the aerosols, these linger there for longer than they would in the lower atmosphere, giving them relatively more cooling power.

Some scientists have consequently suggested creating an anthropogenic analog of this natural phenomenon. Sulfates are the logical leading contenders for the injected precursor substance, because their behavior and effects are familiar, although other materials are under consideration and may offer advantages. To counteract the warming effect of a doubling of preindustrial atmospheric carbon dioxide – which serves as something of a standard of measurement in climate change conversations – about five to ten megatons of sulfur would be injected annually. The aerosols would gradually leave the stratosphere and fall to the surface in rain over a few years, Therefore, injection would need to be continuous, or at least repetitious, to maintain a consistent concentration. If it were to be reduced or stopped, a new lower equilibrium of atmospheric sulfate concentration would be reached within a year or two, potentially causing dangerously rapid climate change.

Researchers have considered various delivery mechanisms. These include new or modified high flying aircraft, artillery, and a hose or pipe held up by a balloon (McClellan, Keith, and Apt 2012). The basic necessarily technologies of stratospheric aerosol injection already exist (although not for tethered balloons), and it could thus be implemented relatively rapidly, at least in principle.

Studies have estimated the delivery costs of stratospheric aerosol injection to be on the order of a few to tens of billion dollars annually (Crutzen 2006; Robock et al. 2009; McClellan, Keith, and Apt 2012; Moriyama et al. 2016; Smith and Wagner 2018).[5] Monitoring systems and redundant delivery infrastructure would substantially increase these costs.

Research of stratospheric aerosol injection is more advanced than that of other solar geoengineering methods. Most evidence comes from modeling and – to a lesser

---

[5] The expected costs of climate change damages and of emissions abatement are each on the order of trillions of dollars annually.

extent – observations of natural analogs, primarily volcanoes. There was an outdoor experiment of aerosols in the lower atmosphere, although this has been of limited utility (Izrael et al. 2009). A UK team planned, as part of the Stratospheric Particle Injection for Climate Engineering (SPICE) project, an outdoor test of a delivery system that would spray seawater into the air at one kilometer altitude via a lofted hose (SPICE n.d.). However, this was cancelled after concerns were raised regarding potential conflicts of interest and inadequate public engagement (Stilgoe, Watson, and Kuo 2013; see Chapter 10). A US group is planning an experiment in which a small amount of various aerosols would be injected into the stratosphere to monitor impacts on atmospheric chemistry. This Stratospheric Controlled Perturbation Experiment (SCoPEx) "will only proceed ... if it passes independent risk assessment" (Dykema et al. 2014, 15).

2.4.2 Marine Cloud Brightening

Clouds substantially influence the Earth's reflectivity, or albedo, mostly by increasing it, which results in cooling. Their shade varies from light to dark. Low clouds, particularly those over dark ocean areas, have the greatest cooling effect. Marine cloud brightening aims to increase the reflectivity of these low marine clouds. The water droplets that constitute clouds typically require aerosol particles, called cloud condensation nuclei, on which to aggregate. Clouds that form in the presence of a greater concentration of cloud condensation nuclei have more, smaller cloud droplets. They consequently reflect light more strongly than those with fewer, larger droplets. Marine cloud brightening would involve spraying sea salt aerosols into the air as a source of additional cloud condensation nuclei for marine clouds, brightening them and in turn cooling the planet.

As with stratospheric aerosol injection, there is an existing phenomenon that offers some empirical evidence. Most large ships' exhaust emissions include small particles, causing brightened "ship tracks" in the clouds behind them. One outdoor experiment – the second one of solar geoengineering in general – released smoke into the atmosphere and examined aerosols' role in cloud droplet characteristics and brightness. Although its primary stated purpose was to understand aerosol–climate interactions generally, the researchers also noted its implications for solar geoengineering (Russell et al. 2013). Scientists have proposed further outdoor experiments, and some of them formed the Marine Cloud Brightening Project to advance this (Wood et al. 2017; Chhetri et al. 2018, 14). Another Australian team plans to research the method to protect the Great Barrier Reef from coral bleaching (Marine Cloud Brightening for the Great Barrier Reef, n.d.).

Significant hurdles remain to marine cloud brightening. The interactions between clouds and aerosols are one of the greatest uncertainties in climate science. Evidence specific to marine cloud brightening is still scant. In fact, because clouds also retain heat, the technique could cause net warming. The degree to which injected particles, such as salt aerosols from sprayed seawater, would spread vertically and horizontally is also unclear. A delivery system would be needed, including high-performance nozzles that can spray sufficiently fine yet not become clogged. Furthermore, costs are largely unknown, although the US National Academies report estimated the cost of implementing marine cloud brightening at roughly five billion dollars (National Research Council 2015b, 101).

2.4.3 Space-Based Solar Geoengineering

In principle, objects could be placed in space, blocking some portion of incoming solar radiation. These could be put in orbit around the Earth, or at a special point – the L1 Lagrange Point – directly between the Sun and the Earth where their two gravitational pulls are equal. The objects could be mirrors, dust particles, a disk, a sail, or something else. An advantage of space-based solar geoengineering is that it would not directly intervene in Earth systems, but instead simply reduce incoming sunlight. However, studies consistently indicate that the implementation of space-based solar geoengineering would be very expensive and time-consuming, requiring either large amounts of material to be launched into space or the construction of objects while in space. These factors might change through technological innovation. Space-based solar geoengineering receives much attention in the popular media, as well as occasional scientific work.

2.4.4 Surface-Based Solar Geoengineering

The surface of the Earth could be changed to reflect more incoming sunlight. Roofs of human-made structures could be painted white. Crops or natural plants could be made more reflective, perhaps through genetic modification. Tiny air bubbles could be injected into the ocean, causing it to have a lighter color. However, models indicate that the terrestrial surface methods would have limited global efficacy and be expensive. They would also require a long time to implement and – if needed – to reverse. However, these techniques could have significant local cooling benefits (Seneviratne et al. 2018). Indeed, some Australian scientists are researching whether a film on the ocean's surface could help protect the Great Barrier Reef from warming (Great Barrier Reef Foundation, 2018).

2.4.5 Cirrus Cloud Thinning

The final suggested method of solar geoengineering discussed here is distinct from the others in that, instead of blocking or reflecting some incoming visible light, it would increase the amount of outgoing infrared radiation. Like other clouds, cirrus clouds – those that occur at relatively high altitudes and appear feathery – both cool the planet due to their albedo effect and warm it by absorbing outgoing infrared radiation. Unlike low-lying marine clouds, their warming effect seems to be greater than their cooling.

Some scientists are researching how to reduce the frequency, size, or lifetime of cirrus clouds, and what the expected climatic effects would be. These reductions could be done by injecting ice nuclei, such as bismuth triiodide, into the areas where cirrus clouds are likely to form. These nuclei would promote the growth of fewer, larger ice crystals that would be less effective at absorbing infrared radiation and would fall more quickly. Notably, because cirrus cloud thinning functions by allowing more heat-causing infrared radiation to escape the planet, it would cool even in the absence of sunlight and thus possibly counter anthropogenic global warming more homogeneously than other solar geoengineering methods.

Research of cirrus cloud thinning remains at a very early stage, and some of the core conclusions are still contested. For example, it is unclear what fraction of cirrus clouds normally form in the absence of ice nuclei and would thus be susceptible to seeding. The technique could result in net warming. There have been no cost estimates.

Although cirrus clouds thinning might not be strictly *solar* geoengineering, as usually defined elsewhere, it shares several key characteristics. It would reduce climate change by a large-scale technological intervention in a natural system, most likely by injecting a substance into higher altitudes of the atmosphere. It would leave atmospheric greenhouse gas concentrations unchanged and would have transboundary environmental impacts. Many of the applicable legal instruments and governance needs are congruent between cirrus cloud thinning and the other suggested techniques. Therefore, the term "solar geoengineering" is defined in this book to include cirrus cloud thinning.

2.5 EFFECTIVENESS

Evidence consistently indicates that some solar geoengineering techniques could effectively counter anthropogenic changes in temperature and – to a lesser extent – the hydrological changes as well as the consequent risks (see Irvine et al. 2016; Keith and Irvine 2016). For example, the most recent Assessment Report of the IPCC states that "Models consistently suggest that SRM would generally reduce climate differences compared to a world with elevated greenhouse gas concentrations and no SRM" (Boucher et al. 2013, 575). Given that atmospheric concentrations of greenhouse gases and global warming are likely to overshoot internationally agreed-upon limits, solar geoengineering appears at least temporarily necessary to stay within these limits and to manage short-term climate change risks.

The proposed methods' capacities to counteract climate change vary. Space-based solar geoengineering and stratospheric aerosol injection would be essentially unlimited. Regarding the latter, IPCC concludes that stratospheric aerosol injection "is the most researched SRM method with *high agreement* that it could limit warming to below 1.5°C" (de Coninck et al. 2018, 350; italics in original).

In contrast, models indicate that marine cloud brightening might be able to cool the planet by a couple degrees Celsius (National Research Council 2015b). Cirrus cloud thinning might counteract the warming from a fifty to eighty percent increase of preindustrial atmospheric carbon dioxide, which implies around 1.5 to 2.4 degrees Celsius (Lohmann and Gasparini 2017).

Solar geoengineering would imperfectly reduce climate change. One way would be through its spatial and temporal heterogeneity. The warming effect of elevated greenhouse gas concentrations occurs in all places and at all times, whereas solar geoengineering (except for cirrus cloud thinning) would cool only in proportion to incoming sunlight. A spatially uniform deployment would thus compensate climate change relatively more at low latitudes and during the summer and daytime and less at high latitudes and during the winter and night. But solar geoengineering could be implemented in ways to reduce the spatial heterogeneity of its effects. Greater amounts of stratospheric aerosol could be injected away from the equator, and cirrus cloud thinning could act as an important complement (MacMartin and Kravitz 2019).

Furthermore, a climate resulting from elevated atmospheric greenhouse gas concentrations and solar geoengineering would differ substantially from the preindustrial one. At the very least, the greater retention of energy from the greenhouse gases coupled with the greater reflection of energy from the solar geoengineering would increase climatic uncertainty. An analogy is an object that experiences counteracting forces on its opposing sides, such as a door being pushed by a person on each side. If both of those counteracting forces increase, then the object might remain generally motionless or it could move unexpectedly due to, for example, slight fluctuations in the forces. This is the main reason that the research, development, and possible use of solar geoengineering does not remove the need to reduce and – ideally – to eliminate net greenhouse gas emissions.

Solar geoengineering would counter the changes in temperature and in precipitation differently. Anthropogenic climate change will likely intensify the global hydrological cycle, increasing global average precipitation, making wet regions wetter and dry regions drier, and strengthening floods and droughts. Solar geoengineering (again except for cirrus cloud thinning) would somewhat overcompensate these changes in precipitation relative to its effect on temperatures. In other words, if solar geoengineering were used to restore global mean temperature, then global average precipitation would decrease to below its preindustrial amount. This suggests that a moderate solar geoengineering deployment would be preferable to one that aims to restore global average temperatures. One possibility is to "shave the peak" of global warming, limiting it to some amount, until greenhouse gases concentrations are lower. Another is to reduce the rate of warming by some proportion, such as by half. Furthermore, evaporation would also be reduced in a solar geoengineered world, partially offsetting the impact on surface runoff and soil moisture. Finally, models indicate that extreme heat and precipitation events would be less common (Curry et al. 2014).

The climatic effects of the solar geoengineering would occur rapidly – on a timescale of a few years – once implemented and would respond rapidly if the intervention were reduced or stopped. Therefore, it could reduce climate change risks in ways that carbon dioxide emissions abatement and NETs – whose effects are not experienced until decades later – could not. Indeed, solar geoengineering is one of the few ways to manage the short-term climate risks that are caused by extant greenhouse gas emissions.

Plants form the basis for both human food systems and ecosystems, and their productivity depends upon factors that include temperature, moisture, and carbon dioxide concentrations. Climate change and solar geoengineering would each ease some of these constraints in some ways but put pressure on others. The net effects remain unclear and would vary by species and local conditions. For example, solar geoengineering might lead to greater net water runoff and sometimes greater water availability (Glienke, Irvine, and Lawrence 2015). The elevated atmospheric carbon dioxide concentration will fertilize plants and reduce transpiration, which will boost their productivity, especially in dry areas. Solar geoengineering could reduce climate change impacts on crops, if not actually improve yields, although these effects could be cancelled out by the reduced incoming sunlight (Parkes, Challinor, and Nicklin 2015; Xia et al. 2016; Proctor et al. 2018).

Solar geoengineering appears to necessarily have global effects. As described, the timing and latitude of stratospheric aerosol injections could be adjusted. Regionalization is difficult but might not be out of the question (Quaas et al. 2016). For example, marine cloud brightening might be able to be somewhat regionally tailored, and surface-based methods might be cost-effective in mitigating local climate risks.

The value of solar geoengineering's effectiveness can be expressed in economic terms.[6] Without delving into specifics, climate change damages might be about twenty-three trillion dollars in the absence of abatement policies (Nordhaus 2008, 204).[7] Optimal emissions abatement policy has been estimated to cost two trillion dollars and to reduce damages by five trillion dollars (Nordhaus 2008, 15). Importantly, note that even with optimal policy, approximately seventeen trillion dollars of damages would remain. Various economic models have placed the value of using solar geoengineering on the order of ten to twenty-five trillion dollars (Nordhaus 2008, 19; Bickel and Lane 2010; Heutel, Moreno-Cruz, and Shayegh 2018). Furthermore, the *option* to use solar geoengineering could provide a sort of

---

[6] For reviews, see Klepper and Rickels (2014); Harding and Moreno-Cruz (2016); Heutel, Moreno-Cruz, and Ricke (2016).

[7] These values are expressed in present value and discounted for the future, and thus can include the indefinite future.

insurance against the low probability "fat tail risks" of future dangerous climate change.

A few final qualifications regarding solar geoengineering's effectiveness are necessary. First, most evidence regarding their expected impacts is based upon models that were developed to forecast the climatic effects of elevated greenhouse gas concentrations. These models' results are uncertain. Indeed, the leading proposed techniques rely upon clouds and aerosols, two of the greatest uncertainties in climate models. Some of this uncertainty could be resolved through research, and some through learning and correction during implementation. Some uncertainty will remain unresolvable. Furthermore, models are based upon assumptions about the natural and social world, and their validity is usually not immediately clear. Second, solar geoengineering's impacts will depend upon, among other things, social and political choices regarding implementation. For example, observers sometimes claim that deploying it would result in a dryer world. Yet this assumes that solar geoengineering would be used to fully offset changes in global average temperature, whereas a more moderate implementation would bring both variables closer to their preindustrial values. Likewise, some models assume optimized solar geoengineering, but its actual use might not be optimal for numerous reasons.

2.6 PHYSICAL RISKS

Although solar geoengineering has the potential to reduce climate change, it would pose problems of its own. These are divided here into physical risks and social challenges, although closer examination reveals that this distinction is imperfect. Indeed, many of the social challenge, left unaddressed, would result in physical risks.

Solar geoengineering's most important physical risk is that it would imperfectly compensate climatic changes, as described in Section 2.5. This would result in regions that are too warm or cool, and too wet or too dry, when compared with a climate without elevated greenhouse gas concentrations. These residual climatic anomalies could put people – especially through agriculture – and ecosystems at risk. Current evidence from models indicates that under moderate deployment scenarios the anomalies would be much less intense than those from climate change. For example, one paper modeled using solar geoengineering to offset half of the change in the Earth's energy balance arising from doubling the preindustrial carbon dioxide concentration (Irvine et al 2019; see also Kravitz et al. 2014; Tilmes, Sanderson, and O'Neill 2016; MacMartin, Ricke, and Keith 2018). On ice-free land, no region and only 0.4% of the model's high-resolution grid cells were worse off, relative to a climate changed world, in terms of average temperature, extreme temperature, water availability, or extreme precipitation. Most of the anthropogenic increase in tropical cyclone intensity was also offset.

2.6 Physical Risks

Other physical risks would arise because the atmosphere and climate are complex systems that are not fully understood. Additional consequences of solar geoengineering probably remain unknown. For example, the climate is subject to several long-term cycles such as the Arctic, quasibiennial, and El Niño-Southern oscillations. These and other larger-scale phenomena could be affected in unpredictable ways.

However, this does not imply that solar geoengineering's uncertainty would compound that of climate change. In fact, some of the most uncertain factors, such as the impacts of clouds and aerosols, that could cause climate sensitivity to be greater (or less) would likewise cause solar geoengineering to be more (or less) effective. Likewise, most uncertainties of climate change's impacts increase with a greater anthropogenic change in the Earth's radiative balance, which a judicious use of solar geoengineering would reduce. Because of these relations, some important characteristics in a world of climate change and solar geoengineering would be less uncertain than one of climate change alone.

Solar geoengineering would not counteract all negative impacts of elevated greenhouse gas concentrations. Most importantly, atmospheric carbon dioxide dissolves in sea water as an acid. Because of this, the oceans are becoming more acidic, threatening marine species and ecosystems. Solar geoengineering would do little to reduce this trend. However, it would indirectly and significantly reduce atmospheric carbon dioxide concentrations by decreasing ecosystem respiration, increasing primary productivity, and increasing the oceans' carbon dioxide dissolution (Keith, Wagner, and Zabel 2017).

Stratospheric aerosol injection would carry some specific physical risks, the first of which involves stratospheric ozone. This natural gas – a form of oxygen – blocks much of the Sun's dangerous incoming ultraviolet radiation. Certain persistent anthropogenic chemicals find their way to the upper atmosphere, where they have depleted some stratospheric ozone, raising concerns about increased rates of cancer. Fortunately, the problem was identified, and effective legal and technical solutions were implemented (see Chapter 6). Although stratospheric ozone is now recovering, particles injected into the stratosphere could slow this recovery by providing surfaces on which ozone-depleting reactions could take place. This is one of the primary motivations for the proposed SCoPEx field experiment and for research into alternative materials for stratospheric injection, some of which could even accelerate stratospheric ozone recovery (Dykema et al. 2014; Keith et al. 2016). Overall, impacts on ozone would probably be "a relatively small effect that would not pose substantial risks" (Irvine et al. 2016, 823; see also Eastham et al. 2018).

Another possible physical risk of stratospheric aerosol injection is acid rain. Because sulfates – the leading proposed material for stratospheric aerosol injection – are acidic, a logical concern is that their injection would increase acid rain. However, this would not be significant. Sulfate pollution in the lower atmosphere acidifies rain because the releases are large and concentrated in a few areas, such as the eastern US, Europe, and East Asia. In contrast, estimates of the quantity of sulfate needed for stratospheric aerosol injection are roughly one-tenth of that of the current lower atmospheric sulfate pollution, and this would be dispersed over a much larger area (Eastham et al. 2018).

2.7 SOCIAL CHALLENGES

Another set of challenges associated with solar geoengineering are of a social character. The IPCC concluded that these methods face "substantial risks and institutional and social constraints to deployment related to governance, ethics, and impacts on sustainable development" (Allen et al. 2018, 15). These are more speculative than the physical risks because they rely also on assumptions regarding how social systems will operate. Although some of these assumptions may be reasonable, social systems cannot be as easily modeled as the physical world can.

Who decides whether, when, and how solar geoengineering might be implemented is arguably its leading social challenge. Because it, as presently understood, would affect the entire planet, international disagreements are possible. Reaching agreement – in whatever forum – might be difficult, and disagreements could create or exacerbate international tensions or hostilities (Schellnhuber 2011; Maas and Scheffran 2012, 196; see Chapter 4). Some writers have claimed that solar geoengineering could be used for hostile purposes or even militarized, or perhaps deployed covertly (Adger et al. 2014, 777; Lin 2016, 2545–6). On the other hand, if international relations are already strained due to climate change, reducing it through solar geoengineering could lessen tensions (Maas and Scheffran 2012, 196; Dalby 2017, 234). Independent of international tensions, appropriately integrating the voices of the already vulnerable – who have the most to gain and lose from solar geoengineering – into decision-making presents a challenge of procedural justice.

International relations might be further complicated by solar geoengineering's apparent low costs and technical simplicity. A single country or a handful of them could deploy it, at least in principle. It would also raise serious questions of legitimacy. On the one hand, this could be an advantage, as solar geoengineering would not require the international cooperation that emissions abatement and NETs do. On the other hand, such uni- or minilateral implementation could run against the desires of other states or of the broader international community. At the very least, widely distributed capacity among countries to implement could lead to uncoordinated solar geoengineering, muddling results of its climatic impacts, and a greater-than-optimal magnitude.

Solar geoengineering might strain international relations also after implementation. Some countries may perceive themselves, rightly or wrongly, to have suffered negative climatic effects, and their leaders might blame the implementing states, particularly if their countries had experienced extreme weather events (Scheffran

2013, 338; Dalby 2015, 196–7). It is possible that a state that perceives itself to be the victim of others' solar geoengineering could demand compensation (see Chapter 12) or even use extreme greenhouse gases to internationally warm the planet as a form of counter–solar geoengineering. This could engender a sort of solar geoengineering arms race or act as a counterbalance to problematic uni- or minilateral implementation (Maas and Comardicea 2013, 43–4; Parker, Horton, and Keith 2018).

The social challenge that has received the most attention is the concern that consideration, research, and development of solar geoengineering would undermine efforts to reduce greenhouse gas emissions (see Chapter 3). Indeed, this is the leading reason that the topic was for years a taboo and that solar geoengineering remains controversial. As noted, a climate in which the anthropogenic greenhouse effect is substantially suppressed through solar geoengineering would differ from the preindustrial one and would, in particular, be more uncertain. This is the central reason why solar geoengineering does not eliminate the need for aggressive emissions abatement.

Another risk of solar geoengineering is that of sudden and sustained cessation. Because its effects would be short lived, solar geoengineering would need to be continuously or regularly maintained to sustain them. If it were to be implemented under conditions of elevated greenhouse gas concentrations and then stopped – for whatever reason – in a sudden and sustained way, then the theretofore suppressed climate change would quickly manifest. Such rapid climate change could be dangerous to humans and ecosystems, as there would be less time to adapt. Furthermore, if such "termination" is to be avoided, then the necessity of maintaining deployment over the long term could be seen as an unjust burden on future generations.

The consideration, research, and development of solar geoengineering also might bias future decision-making, causing its implementation to be unduly probable (Cairns 2014; Lin 2016; Chhetri et al. 2018; McKinnon 2019; but see Bellamy and Healey 2018; Callies 2018). Writers sometimes refer to this, in general, as a "slippery slope" or as lock-in. Although the idea appears logical, definitions are difficult (see Volokh 2003). There are two general theoretical bases for this phenomenon. The first is a rational but socially suboptimal one, in which society adopts a technology at a time when it is indeed optimal, but a later superior alternative is not adopted due to the original choice's lock-in effect. Possible causes of this include economies of scale, learning effects, barriers to entry, network effects, and the undue influence of vested interests (Liebowitz and Margolis 2013). This latter cause of "rational" lock-in – that of vested interests – is the most frequently cited by scholars in the case of solar geoengineering and could be especially problematic if many relevant experts have substantial professional, personal, or economic interests (see Chapter 11; Long and Scott 2013). The second general basis is cognitive or epistemic lock-in, in which the early framing of a technology or a social challenge privileges particular responses. In turn, this creates a mutually reinforcing dynamic in which the technology or response at hand "unreflectively" becomes part of the dominant social paradigm (Jamieson 1996, 333). In the case of solar geoengineering, these possibilities might be exacerbated by the fact that the line between its large-scale research and global implementation is indistinct (Bellamy et al. 2012). At the same time, cognitive and epistemic lock-in remains largely theoretical. Examples that clearly differ from both rational lock-in mechanisms and general historical causality, in which earlier conditions influence later ones, are difficult to identify (Mahoney 2000; Vergne and Durand 2010).

There would be limits to knowledge before outdoor solar geoengineering activities that could pose risks would be undertaken. As noted, the line between research and deployment is hazy, the climate is complex, and some uncertainty might be irreducible (Reynolds 2011). Solar geoengineering would require decision-making under uncertainty and perhaps even ignorance. Surprises are possible (Lempert and Prosnitz 2011). There are reasonable scenarios in which countries or other actors hastily implement solar geoengineering with inadequate prior research, perhaps due to perceptions of a climate change emergency.

Finally, solar geoengineering is ethically challenging. For example, would the doctrine of double effect – which suggests that sometimes it is acceptable to cause an expected but unintended harm in the course of an intended good or beneficial action – be applicable (Morrow 2014c)? Solar geoengineering would change people's relative welfares, raising questions of distributive justice. Furthermore, responses to climate change are inexorably linked to which actors have caused it. If solar geoengineering reduced the burden of emissions abatement, either optimal or actual, would it allow them to inappropriately escape their obligation to pay for their emissions' harm? To some ethicists, solar geoengineering could pose a genuine moral dilemma that threatens to corrupt us (Gardiner 2010). Others consider it through the lens of nonideal theory (Svoboda 2017). Some scholars argue that it would constitute an inappropriate, hubristic intervention in the natural world (Stilgoe 2018).

2.8 SUMMARY AND CONCLUSION

Anthropogenic climate change poses serious risks to humans, other species, and ecosystems, especially the most vulnerable. It is arguably the most important and difficult environmental problem that global society presently confronts. Greenhouse gas emissions abatement, NETs, and adaptation have been insufficient in reducing these risks, and there are good reasons to believe that this will continue to be the case. Solar geoengineering is a set of proposed technologies to reduce climate change by deliberately manipulating the Earth's temperature without altering the atmospheric concentration of greenhouse gases. The leading technique is stratospheric aerosol injection. Evidence from models and natural analogues indicate that its moderate deployment could globally and regionally reduce climate change. It also appears to be technologically feasible, rapid in its effects, inexpensive in its direct deployment costs, and reversible. Solar geoengineering poses several physical risks and social challenges of its own, some of which appear serious, suggesting that governance would be beneficial.

3

Solar Geoengineering and Emissions Abatement

Discussions of solar geoengineering are intertwined with other responses to climate change, especially greenhouse gas emissions reductions.[1] The most important relationship is the concern that solar geoengineering would undermine emissions abatement efforts and that this would be harmful. This has loomed over the discourse since its beginning and has been the most widespread basis for resistance to solar geoengineering.

I suggest that a two-part concern is at stake: first, that the consideration, research, development, and possible implementation of solar geoengineering would cause greenhouse gas emissions to be greater than they otherwise would be; and second, that this outcome would be undesirable. Here, this is called the "emissions abatement displacement concern." If this holds, policy-makers should presumably try to avoid or minimize the phenomenon. However, what they could do is unclear.

This chapter first provides a background of this concern in the climate change context in Section 3.1 and then examines this by asking three questions. First, how likely is it that solar geoengineering would reduce emissions abatement? Addressing this requires an introduction to two similar phenomena and a review of public opinion studies. Second, if so, would this be a problem? The consideration of this question uses an extended thought exercise in ethics, economics, and consequences. Third, what could policy-makers do? The chapter's final substantive part, Section 3.6, considers why the abatement displacement concern might be so widespread and seemingly strong

3.1 BACKGROUND

Understanding the concern that solar geoengineering might undermine emissions abatement benefits from a brief examination of the relationship between abatement and the other responses to climate change, that is, negative emissions technologies

[1] This chapter draws from Reynolds (2015a).

(NETs) and especially adaptation. When climate change first arose as an international scientific and policy matter in the 1980s, the responses of emissions reduction and adaptation were presented on roughly the same footing (Jaeger 1988; Schipper 2006). However, the first few Intergovernmental Panel on Climate Change (IPCC) Assessment Reports, the UN Framework Convention on Climate Change (UNFCCC), and the Kyoto Protocol each gave primacy to abatement over adaptation. For example, the UNFCCC's objective is the "stabilization of greenhouse gas concentrations" (Article 2). In 1992, the then-US Vice President Al Gore called adaptation "a kind of laziness, an arrogant faith in our ability to react in time to save our skin" (Gore 1992, 240).

There were several reasons for the early resistance to adaptation. Some observers believed that adaptation would be an admission of defeat. Others were concerned that it was "an unacceptable, even politically incorrect idea [that] could make a speaker or a country sound soft" on emissions abatement (Burton 1994, 14). As a third reason, in a related vein, there was concern that considering adaptation would divert political support, limited financial resources, and cognitive attention from abatement, resulting in greater long-term harm from climate change (Anderson 1997, 13; Kates 1997, 31–2; Kane and Shogren 2000, 94). Finally, in the eyes of some, greenhouse gas emissions were deeply and inherently problematic. Thus, if adaptation were to displace efforts toward their reduction, then it was undesirable independent of net changes in risk. Steve Rayner observes that "discussion of adaptation to climate change is viewed with the same distaste that the religious right reserves for sex education in schools. Both [adaptation and sex education] are seen as ethical compromises that will in any case only encourage dangerous experimentation with the undesired behavior" (Rayner 1991, 265).

The acceptance and position of adaptation in the international climate change discourse changed rapidly. In the late 1990s and early 2000s, some voices called for greater attention to adaptation, and much of this advocacy came from those countries that face the greatest risks of climate change (Pielke 1998; Pielke et al. 2007).[2] At the 2010 UNFCCC Conference of Parties (COP), countries agreed that adaptation must be given roughly the same priority as greenhouse gas emissions cuts (UNFCCC COP Decision 1/CP.16). The Paris Agreement arguably does so. Gore later admitted that he was "wrong in not immediately grasping the moral imperative of pursuing both policies [abatement and adaptation] simultaneously, in spite of the difficulty that poses" (Lind 2013). No serious arguments are presently put forth that adaptation efforts are undercutting those for greenhouse gas emissions abatement. In fact, some experiments have shown that people who are exposed to information about adaptation exhibit slightly stronger preferences for emissions abatement (Evans, Milfont, and Lawrence 2014; Carrico et al. 2015). Regardless, it is impossible

---

[2] For example, the Alliance of Small Island States was instrumental in advancing the adaptation agenda at the Conferences of the Parties (COPs) (see International Institute for Sustainable Development 1998, 5).

to determine whether adaptation has reduced abatement, as no counterfactual history without it is available.

More recently, a similar shift regarding NETs appears to be underway. Originally, these were frequently lumped together with solar geoengineering as "geoengineering" or "climate engineering," and together these have raised concerns of undermining abatement. Yet NETs are steadily becoming part of the mainstream climate change discourse. In both the academic and popular media, they are increasingly presented as distinct from solar geoengineering and as necessary to prevent dangerous climate change. In 2011, the more optimistic of the new scenarios used by the IPCC assumed the large-scale implementation of NETs (van Vuuren et al. 2011). Furthermore, the Paris Agreement implicitly endorses them in its call "to achieve a balance between anthropogenic emissions by sources and removals by sinks of greenhouse gases" (Article 4.1; see also Article 5.1). The international climate change community is now struggling with the apparent need for NETs to limit global warming to two degrees Celsius, considering the technologies' present limitations and uncertainties.

Given abatement's primacy within the climate change discourse and the initial resistance to both adaptation and NETs, it is unsurprising that the same concern served as the basis for solar geoengineering's taboo. This was the case as early as 1992, when the authors of a US National Academies climate change report

> were worried that even the very thought that we could offset some aspects of inadvertent climate modification by deliberate climate modification schemes could be used as an excuse by those who would be negatively affected by controls on the human appetite to continue polluting and using the atmosphere as a free sewer. (Schneider 1996, 295)

This was again the case in a rejected 2001 proposal within the US White House to explore geoengineering (MacCracken 2006, 239). Around this time, climate researcher David Keith labeled this concern "moral hazard," a term borrowed from the insurance literature that, despite a weak analogy, has generally remained in use (Keith 2000, 276–67). He noted that solar geoengineering accompanied by reduced emissions abatement could still improve welfare through reduced net climate change risks. Since then, almost all major publications on solar geoengineering have addressed the abatement displacement concern, often by emphasizing the primary importance of abatement.

In recent years, a handful of scholars have more closely considered the ethics of and possible policy responses to the concern. For example, philosopher Benjamin Hale unpacks the concept. He proposes that "moral hazard has become a sort of catch-all used to refer to a suite of objections and hazards . . . [It] functions as a falsely concrete straw man, and therefore is both easy to offer as a criticism and equally easy to dismiss" (Hale 2012, 114). Hale concludes that moral hazard arguments fail because they remain ambiguous and vague. That is, they are unclear whether the concern is that solar geoengineering would make it possible for humans to continue to emit greenhouse gases, would itself become "business as usual," or would incentivize us to behave in ways that increase risk. Furthermore, Hale asks, are the objectives at hand those of efficiently reducing risk, changing actors' motivations, or encouraging moral behavior? Since his chapter's publication, some scholars have asserted that emissions abatement displacement will or would likely be a genuine problem (Lin 2013; Baatz 2016; McLaren 2016), whereas others – including me – are less concerned (Morrow 2014b; Reynolds 2015a; Lockley and Coffman 2016; Halstead 2018).

3.2 MORAL HAZARD AND RISK COMPENSATION

Asking whether solar geoengineering would likely lessen abatement requires a discussion of two related phenomena: moral hazard and risk compensation. In general, moral hazard occurs when one party increases her risk-taking once another party has assumed some of the resulting negative consequences of the first party's risk-taking. Typically, the second party does not fully know of this increased risk-taking by the first one. The term arose, and is still most often used, in the insurance context. Because this originally referred to a conscious increase in risk-taking by an insuree, the name was intended to indicate "unscrupulous" insurees (Black 1910, 563). Its meaning broadened over the twentieth century to include any increase in risk-taking, conscious or not, as well as contexts outside insurance. For example, government bailouts of failed banks could pose a moral hazard that bankers will invest too riskily in the future. Moral hazard behavior can be rational for the "insuree" (broadly speaking, including situations outside of the insurance context). However, it is unlikely to be socially optimal because information asymmetry will cause the insuree to take excessive risk. *If* the "insurer" (again, broadly speaking) *does* know of the behavior, then he would increase his rates accordingly. Aware of prospective higher rates, the insuree might then increase her risk-taking to a lesser degree or not at all.

Moral hazard can be a problem, but its magnitudes in specific contexts are difficult to measure. After all, it is caused by information asymmetry. More precisely, there are three distinct behaviors that each result in insurees taking more and greater risks. The first is that already described – more precisely called ex ante moral hazard – in which insurees take greater risks after obtaining or increasing insurance but before any accident or other loss has occurred. For example, after I get automobile insurance, I might drive faster. This resembles the emissions abatement displacement concern, albeit imperfectly. Beyond insurance, ex ante moral hazard can exist in contexts as diverse as mutual defense treaties, foreign aid, humanitarian intervention, and financial investments. The second behavior is ex post moral hazard, in which the insuree who increases his insurance coverage subsequently files more and larger claims after accidents, regardless of whether he has increased risk-taking. Here, after I increase my automobile insurance, I would likely request reimbursement in cases where I had not before. The third behavior is adverse selection, in which an actor who knows that she is riskier chooses to obtain more insurance. Continuing the example, knowing that I like to drive fast, I purchase more automobile insurance.

Furthermore, there are confounding variables that make empirical evidence of moral hazard difficult to acquire. For example, obtaining insurance can expose insurees to information – such as their health insurer's pamphlet espousing the benefits of exercise – that cause them to alter their behavior, including by acting less riskily. Also, most insurers adopt policies, such as deductibles and copayments, to reduce moral hazard and adverse selection by keeping some burden of the risk on the insurees.

Many studies confirm that insurees take more and greater risks than noninsurees, but few can disentangle the three behaviors as well as the confounding variables. One review of insurance concluded, "This literature identifies a moral hazard effect in some contexts but not in others," (Cohen and Siegelman 2010, 72) and another that "there are theoretical reasons to believe that health insurance coverage may cause a reduction in prevention activities, but empirical studies have yet to provide sufficient evidence to support this prediction" (Dave and Kaestner 2009, 369).

As noted, moral hazard is an imperfect characterization of the emissions abatement displacement concern. A better – but still imperfect – analogy is risk compensation, in which an actor alters her risk-taking in response to perceiving that her risk exposure has exogenously changed. If she believes that her risk exposure has lowered, the actor will take greater risks, and vice versa. The perception of changed risk exposure is often due to the introduction or awareness of a safety regulation or technology. For example, a car driver might drive faster after the introduction of air bags. Notably, the scholarly literature does not indicate that the introduction of a safety regulation or equipment followed by risk compensation causes a net increase in harm to the risk-taker, but only that the decrease in harm is less than that which would be expected from the regulation or equipment with constant risk-taking. Because risk compensation is not a transfer of risk's negative consequences onto another party under conditions of information asymmetry, it can be welfare improving for the risk-taker and thus possibly for society in total, assuming that the risk-taker's perception is accurate. What can decrease total, or social, welfare are possible secondary victims – such as pedestrians and bicyclists in the case of faster driving – that might suffer from the risk compensation. Likewise, there could be secondary beneficiaries, such as sports spectators who prefer more aggressive play by athletes who now wear safety equipment.

Risk compensation has been observed in numerous settings. Automobile safety is the best studied. Initial research found that seatbelt laws have led to more car accidents with pedestrians and bicyclists, presumably due to more aggressive driving (Peltzman 1975). However, as with moral hazard, the empirical evidence remains inconclusive. Research encounters barriers, such as confounding variables and information effects, which are similar to those for moral hazard but not as problematic. For example, the presence of safety equipment might cause the risk to be more salient in the minds of potential risk-takers, who might respond by being more – not less – cautious. A recent paper concluded that "If anything, these [seatbelt] laws and the accompanying increase in belt use result in safer driving behavior" (Houston and Richardson 2007, 933). There are similar debates regarding risk compensation in, for example, sports protective equipment, hypertension drugs, and vaccines and condoms to prevent AIDS and sexually transmitted diseases. For example, the controversy about the effectiveness of so-called harm reduction measures for drug use – such as providing clean needles and opioid substitutes to users – ultimately concerns a conflict of goals in a context of risk compensation. These measures are contested not only because of the measures' potential effect on assessable harm, but also because some people object to the original drug use, independent of the behavior's risks.

Compared to ex ante moral hazard, risk compensation is a closer analogy to the emissions abatement displacement concern.[3] In this concern, society's exposure to the risks from greenhouse gas emissions is lowered by the prospect of solar geoengineering, a potential technological change. In principle, an increase in emissions might not only be rational but could also improve social welfare, at least for the emitting society. After all, we obtain benefits, such as economic development and the reduction of poverty, from fossil fuel combustion, and an increase in their use in response to a decrease in their risks would be rational and welfare-improving. However, this does not rule out negative impacts on other parties. Just as the driver who drives more riskily after putting on a seatbelt puts pedestrians and bicyclists at greater risk, so too might a decrease in net risk by a society engaged in solar geoengineering put another one – elsewhere or in the future – at greater risk.

3.3 PUBLIC OPINION STUDIES

Although reduced effort to abate greenhouse gas emissions in the face of solar geoengineering appears logical, it might not necessarily be the case. Legal scholar Albert Lin notes that "Fundamentally, the claim that geoengineering presents a moral hazard is an empirical claim about attitudes and behavior" (Lin 2013, 692). But as described, the empirical evidence for the similar phenomena of ex ante moral hazard and risk compensation remains less than conclusive. A handful of projects to gauge how people might react to the development of solar geoengineering provides some guidance. Although several other studies have explored

---

[3] Ex ante moral hazard would require the risk to be consensually shifted to some other society that has inferior information. Although greater greenhouse gas emissions in response to solar geoengineering might increase future generations' risk, this shift would neither be consensual and nor involve information asymmetry.

participants' concerns that solar geoengineering would undermine emissions cuts, the ten described here explore – to varying degrees – how the respondents claim that they themselves would behave.[4] These are reviewed here, with some important caveats. These studies report either what the respondents say that they would do or what they did under limited experimental conditions. What ultimately matters is what people – especially decision-makers – actually will do under uncertain future conditions. Furthermore, we do not know how future people will perceive climate change, solar geoengineering, their potential benefits and risks, risk in general, the natural world, and technology. Finally, each of these studies has its own methodological limitations, perhaps the most important of which is that the answers of respondents – who are mostly ignorant about solar geoengineering – are highly sensitive to how the researchers introduce the topic to them.

1. As part of the preparation of its influential report, the Royal Society convened four focus groups. In these, "Although participants were generally cautious, or even hostile, towards geoengineering proposals, several agreed that they would actually be more motivated to undertake mitigation actions themselves (such as reducing energy consumption) if they saw government and industry investing in geoengineering research or deployment" (Shepherd et al. 2009, 43).
2. The UK's Natural Environment Research Council organized a public dialogue to inform its future decision-making regarding geoengineering. Participants in this dialogue wished to link geoengineering with abatement efforts. The report from the dialogue described this outcome as "contrary to the 'moral hazard' argument" (Ipsos MORI 2010, 1–2).
3. The first large-scale international survey of the public perception of solar geoengineering queried approximately three thousand people in the United States, Canada, and the UK. Participants were asked to what extent they agreed with various statements, including "Solar Radiation Management should be used so we can continue to use oil, coal and natural gas." The average answer was close to "somewhat disagree" (Mercer, Keith, and Sharp 2011, 5).
4. An experimental survey exposed some US subjects, but not others, to information about solar geoengineering. Those who were exposed subsequently showed statistically significant – albeit slight – greater concern about climate change than those who did not (Kahan et al. 2015, 203–4).
5. The Integrated Assessment of Geoengineering Proposals project led a series of public discussion groups in the UK. It reported that "[Some] people were not as concerned about climate change when they first started the group discussion, but by the time they'd spent the day talking about geoengineering were a lot more worried" (Integrated Assessment of Geoengineering Proposals 2014, 3).
6. In semi-structured focus groups, Swedish participants responded to several open-ended questions. When they were exposed to a "climate emergency"

---

[4] For a review of public opinions of solar geoengineering more generally, see Burns (2016).

scenario, in which solar geoengineering could be used as a response to sudden and dangerous climate change, the participants became more skeptical of solar geoengineering and turned to arguments for emissions reductions (Wibeck, Hansson, and Anshelm 2015, 29).
7. When the researchers from the previous study extended their work to groups in Japan, New Zealand, and the United States, "climate engineering seemed to have a 'reverse moral hazard' effect, that is, increased willingness to consider and even advocate behavioral changes, rather than being deterred by the introduction of new technologies for large-scale climate control" (Wibeck et al. 2017, 8).
8. A survey of almost one thousand British residents provided half with information about solar geoengineering and gave the other half of them no such information. The information significantly affected neither their support for a tax on greenhouse gas emissions nor on their trust in climate science (Fairbrother 2016).
9. In a field survey, German research subjects were each given a small amount of money that they could choose to either spend on offsets, which would fund emissions reductions, or keep. Some of the subjects were exposed to information about solar geoengineering, while others were not. Those who were informed of solar geoengineering purchased more emissions offsets. The researchers inferred that this was not because these subjects became more concerned about climate change, but instead because some of them saw solar geoengineering as a threat that could be prevented by enhanced emissions cuts. Notably, this is the only one of these studies that relies on respondents' revealed preferences (Merk, Pönitzsch, and Rehdanz 2016).
10. A survey of diverse international experts in abatement, adaptation, NETs, and solar geoengineering asked them to allocate a hypothetical research budget among the four responses. Although they allocated more to the response of their own expertise, the solar geoengineering (and NETs) experts did not allocate relatively less to abatement. However, it is possible that they all would have allocated more to abatement had the solar geoengineering option been unavailable (Merk, Pönitzsch, and Rehdanz 2019).

In none of these public opinion studies did the participants indicate that they would be less concerned about climate change or less willing to reduce greenhouse gas emissions in response to the prospect of solar geoengineering. In fact, in almost all of them, they indicated the opposite, in which learning of solar geoengineering increased their concern about climate change and/or their support for emissions abatement. Although it is premature to draw firm conclusions, this at least implies the reverse of emissions abatement displacement (Victor 2011, 190). This could be caused by some mix of climate change risks seeming more salient and solar geoengineering seeming riskier and more uncertain than abatement. Although these results cast some doubt on the claims that solar geoengineering would lead to greater greenhouse gas emissions, we do not know what would be likely to happen. As stated, future actions, especially those of decision-makers, are what matter, not present survey public opinion studies.

3.4 ETHICS, ECONOMICS, AND CONSEQUENCES

Assume for the moment that the consideration, research, and development of solar geoengineering would indeed displace abatement and cause greenhouse gas emissions to be greater than they otherwise would be. Would that be a problem? After all, few people are directly concerned about the changes in the atmosphere's composition. Instead, the risks to humans, other species, and ecosystems are the most common reasons to abate emissions. Yet if solar geoengineering were to reduce climate change – as models consistently indicate that it could – then the net effect of it and an increase in greenhouse gas emissions due to abatement displacement could be either greater or lesser risk.

Although some writers simply assume that emissions abatement displacement would be bad (Robock 2008, 17), others are explicit as to why. Christian Baatz's leading reason for emissions cuts' primacy is the risk of sudden and sustained termination of solar geoengineering in a world with elevated atmospheric greenhouse gas concentrations (Baatz 2016, 32). He further argues that if solar geoengineering implementation were to cause unexpected serious negative environmental effects, then future decision-makers would face the dilemma of either maintaining solar geoengineering with its concomitant negative effects or suddenly terminating it. Additionally, Baatz notes that solar geoengineering would not address ocean acidification and, through emissions abatement displacement, would actually worsen it (Baatz 2016, 43). Duncan McLaren concurs, and adds that emissions displacement would result in spatially imperfect compensation of climate change and would prevent ecological modernization, restraint of market capitalism, "desirable new behaviors and values," and punishment of countries and corporations that have emitted greenhouse gases (McLaren 2016, 597–8).

Although some of Baatz's and McLaren's concerns are legitimate, they do not conclusively demonstrate that emissions abatement displacement would be bad, only that it might be. The risk of termination might not be so great (Parker and Irvine 2018; Rabitz 2019). Unexpected negative environmental effects might be detected early, and solar geoengineering could be ended slowly instead of suddenly. The net impact on ocean acidification would depend on the magnitude of emissions abatement displacement and the decrease in atmospheric carbon dioxide concentrations due to solar geoengineering's secondary effects of lowering atmospheric carbon dioxide concentrations (Keith, Wagner, and Zabel 2017). A moderate implementation of solar geoengineering would, according to models, bring all regions closer to their preindustrial climatic conditions. The value of ecological modernization, restraint of capitalism, "desirable new behaviors and values," and punishment are unclear and contested, and might by outweighed by the reduction in climate change by solar geoengineering.

A simple qualitative economic model helps clarify the relationship among emissions abatement, solar geoengineering, and other responses to climate change.[5] Some readers might find this tangential and may skip it. Suppose that society must choose the size of its investments to reduce climate change risks. At the margin, these investments' costs increase, and their benefits decrease. Benefits and costs include secondary positive effects (such as reduced ocean acidification from emissions abatement), negative effects (such as higher energy prices from emissions abatement), and deontological preferences (such as McLaren's restraint of market capitalism, "desirable new behaviors and values," and punishment of emitters) (see Colyvan, Cox, and Steele 2010).[6] They furthermore include both present and future benefits and costs, in which the future can be appropriately discounted due to opportunity costs and uncertainty.[7] The values can also be weighted for risk aversion and equity. Assume at first that emissions abatement is the only available means to reduce climate risks; that decisions are made by a single omniscient, intergenerationally considerate, and rational decision-maker; and that preferences are consistent across space and time. Under these assumptions, society invests in an optimal quantity of emissions abatement, and net benefits are maximized.

These assumptions can now be removed stepwise, beginning by adding other responses to climate change. These responses are imperfect substitutes for each other, in that they satisfy the same desire for lower climate change risks but do so in slightly different ways. If a new means becomes available, then society invests some in it and less in the existing one(s), in turn increasing social welfare. If adaptation is the first to be added, society invests some in it and less in emissions abatement decreases, and social welfare increases. This increase in welfare could be due to greater protection from climate change impacts, despite the higher emissions. Alternatively, the introduction of adaptation might increase welfare yet result in less net protection against climate change due to the greater importance of some secondary effects or deontological preferences. For example, it is possible that society would invest heavily in adaptation due to its secondary economic benefits – such as improved infrastructure – and would reduce its emissions abatement,

---

[5] See also Chapter 4 for a discussion of game theoretic models of state decision-making.
[6] I discovered this via Morrow (2014b). The implementation of nonconsequentialist ethical frameworks in decision-making, particularly for those that impact large numbers of people, is challenging. Norms, reasoning, and conclusions vary within these frameworks and are contested. Adopting what Morrow calls "fairness-adjusted utilitarianism" requires decision-makers to confront how much additional climate risk, if any, they are willing to accept to make the deontologically preferred or virtuous decision.
[7] Discounting is adjusting future benefits and costs downward to reflect opportunity costs and uncertainty. Although discounting is mostly unproblematic for short time spans, its intergenerational application is controversial. Both high and low intergenerational discount rates can lead to absurd conclusions.

resulting in higher welfare yet greater net climate change risks. The same process occurs as NETs and solar geoengineering are added. In the model thus far, the addition of solar geoengineering would reduce investments in greenhouse gas emissions abatement but would increase social welfare. Its effect on net climate risks could go either way, depending on society's relative valuation of other benefits and costs. Nevertheless, investments in each of these four responses to climate change are each socially optimal, given the availability of the other responses.

The second assumption to remove is that of a single decision-maker. Let us disaggregate the world into numerous decision-makers – such as the leaderships of self-interested states – with diverse preferences. Because decision-makers face different incentives when considering the diverse means to reduce climate change risks, the results of this disaggregation vary among the means. First, emissions abatement and NETs present collective action problems, and the investments in them in a world of many decision-makers are substantially lower than optimal, as is presently the case. In principle, these can be brought closer to optimal through cooperation among decision-makers, although actual international cooperation to date has been disappointing. Second, the benefits of adaptation are mostly local. In the simple economic model presented here, its quantities purchased vary, and could approach optimality at the country scale. Unfortunately, poor developing countries are unable to sufficiently invest in adaptation and will still face substantial climate risks. Total welfare could be increased through transfers from wealthy to poor countries, although the evidence thus far is again disappointing. In other words, investments in adaptation could be optimal in some locales but globally suboptimal. Third, solar geoengineering is complicated because countries have divergent preferences and capabilities. How divergent these preferences are, which states have the technical and financial capacities and the international political capital to implement, and the extent of international cooperation all remain uncertain. If the preferences are widely divergent, if many states can and are willing to deploy solar geoengineering themselves, and if international cooperation is weak, then investments in solar geoengineering would be substantially greater than optimal.[8] This is because, as it is presently understood, solar geoengineering would be additive: states could increase its intensity (or technically, the magnitude of its negative radiative forcing) but not decrease it.[9] On the other hand, if preferences are relatively homogeneous (for example, all states might prefer the preindustrial climate), if few states are able and willing to implement, and if international cooperation is effective, then investments in solar geoengineering would be closer to optimal.[10]

---

[8] One model estimated that noncooperative solar geoengineering would be at four times the intensity as under full cooperation (Emmerling and Tavoni 2017a).

[9] This is the "free driver" problem. This would change if counter–solar geoengineering were available. See Chapter 4.

[10] See Ricke, Moreno-Cruz, and Caldeira (2013) for a model of region's preferred intensities of solar geoengineering when they each prefer a climate as close as possible to preindustrial conditions.

Based upon this logic, actual multiple decision-makers cause investments in emissions abatement, NETs, and adaptation to be below optimal, even in the absence of solar geoengineering. The availability of solar geoengineering further lowers these through imperfect substitution, as described in the previous paragraph. However, it also decreases their optimal amounts. It is unclear whether this would increase or decrease the gap between the actual and the optimal investments. Solar geoengineering itself would range from optimal to greater than optimal.

Now we can remove the third assumption, that of omniscience, and allow for uncertainty. This has three implications for the relationship between greenhouse gas emissions abatement and solar geoengineering. First, decision-makers do not and will not know the precise effects of climate change and the various means to reduce its risks. Some of the uncertainty can be reduced through research, while some cannot be. Eventually, the actual effects of climate change and the chosen policies will be revealed. The sensitivity of the climate to greenhouse gas emissions; humans' and ecosystems' vulnerability, adaptive capacity, and resilience to climate change; the speed of technological development; and the actual amounts of emissions abatement, adaptation, NETs, and solar geoengineering might each turn out to be substantially greater or lesser than expected. Assuming that decision-makers are risk averse and that solar geoengineering has greater uncertainty than other responses, they might rationally prefer emissions abatement and NETs over solar geoengineering, even if the latter is expected to result in lower climate change damage.

The second implication of decision-makers' imperfect knowledge is that solar geoengineering's greater uncertainty leads to the first of four genuine hazards related to the relationships among responses to climate change: *the mix of responses that appears optimal at one point in time can later be revealed to be suboptimal.* That is, one generation might invest in emissions cuts, NETs, adaptation, and solar geoengineering at some ratios based upon the then-extant knowledge, but the next generation subsequently might learn that the benefits and costs of these responses and of climate change itself were substantially different than previously thought, leading to a suboptimal outcome. At the very least, this implies the benefits of further research to reduce the uncertainty of all responses – especially solar geoengineering, with its greater uncertainty – as well as trying to maintain alignment between evidence and perception, particularly among decision-makers.

The third implication of imperfect knowledge involves how decision-makers can respond to learning. Because climate change occurs decades after the emissions that cause it, emissions reductions and NETs have delayed effects. Adaptation is also a slow process, as it requires changing institutions and behaviors. In contrast, the climate would respond relatively rapidly to solar geoengineering. Thus, it has a high option value, in that developing the knowledge and capacity to implement it provides a sort of insurance. On the other hand, its secondary effects – some of which might be negative – might not be observed and attributed until after they have existed for quite some time.

The fourth assumption to remove is that of consistent preferences within society and across time. Regarding the former, decision-makers and the general population might differentially bear the costs or reap the benefits of climate change and the various responses, especially if the two groups live in different locations. This points to the second genuine hazard related to the relationships among responses to climate change: *the preferences of decision-makers and those of the general population can differ systematically*. In the real world, those states with high per capita greenhouse gas emissions are on average less vulnerable to climate change and have more international influence than the relative low emitters (although these correlations are weakening). This difference causes emissions abatement to be even more suboptimal because vulnerable, low emitting states have relatively less capacity to reduce their emissions. In contrast, some of these vulnerable states could possibly deploy solar geoengineering, even contrary to the high emitting states' preferences.[11] If they can leverage their threat to do so, then the high emitting countries might increase their abatement (Millard-Ball 2012; Urpelainen 2012; Moreno-Cruz 2015).

Preferences can also vary across time. Generations can differ in their prioritization of the various benefits and costs of the responses to climate change, risk aversion, equity, and deontological preferences. The current generation might invest in various responses to climate change at certain ratios based on its preferences and assumptions of future preferences. If preferences subsequently change, future generations would wish that the current one had made different investments. This leads to the third general hazard: *preferences are likely to change among generations*. These changes are difficult to predict.

The fifth, penultimate assumption to remove is that of intergenerational consideration, in which the present generation cares about future ones. Ensuring that present decision-makers bear the future in mind is a long-standing problem in political science. Rewards, such as re-election and promotion, for politicians and administrators to act consistently with the public's preferences operate on short timescales. Yet besides a desire to leave a positive legacy, there are few effective mechanisms to ensure that they take the longer term into account. Removing the assumption of intergenerational consideration reduces investments in those responses that have short-term costs and long-term benefits (that is, emissions abatement, NETs, and adaptation). This gives rise to the fourth and final genuine hazard related to the relationships among responses to climate change: *decision-*

---

[11] A relevant related question is whether certain classes of countries are disproportionately exposed to the risks of solar geoengineering. In general, current models do not indicate that the regions that are dominated by vulnerable, low emitting countries would experience less positive or more negative environmental effects. However, when environmental effects are equal, countries that are more vulnerable to climate change are also vulnerable to solar geoengineering's negative effects. This is for similar reasons: they tend to be poor, and poor countries rely more on the environment for their well-being and lack resources to adapt to major environmental changes.

3.4 Ethics, Economics, and Consequences 45

makers might lack intergenerational consideration. This appears to me to be what many scholarly and popular commentators have in mind when they speak of emissions abatement displacement.

The final assumption to remove is that of rationality. People exhibit biases and rely upon heuristics, many of which are somewhat predictable. Lin correctly notes that "the availability heuristic, optimism bias, hyperbolic discounting, and lack of outrage undermine public concern about climate change and support for any policy response" (Lin 2013, 696). Then he claims that these phenomena "might foster unduly favorable public perceptions of specific geoengineering options," later adding the affect heuristic to his list (Lin 2013, 696–9). However, only one of Lin's five behavioral phenomena – discounting – implies emissions abatement displacement. In order to be problematic, discounting future benefits and costs need not be hyperbolic, only excessive.[12] The results of this would be like those of limited intergenerational consideration, described in the previous paragraph. Lin's availability heuristic, optimism bias, and lack of outrage indicate only that people are less likely to act to prevent climate change and reveal nothing specific about the relationship between solar geoengineering and greenhouse gas emissions abatement. The availability heuristic occurs when situations that are easier to bring to mind shape one's thinking. Outrage is a bias toward action when a problem's cause is clearly identifiable.[13] Optimism bias is an unjustified conscious or subconscious belief that negative consequences will not arise and that problems will be resolved.[14] Lin's final phenomenon, the affect heuristic, is people's reliance on emotions to guide decision-making. Contrary to Lin, empirical research indicates that people favor emissions abatement relative to solar geoengineering, driven in part by affect (Pidgeon et al. 2013; Wright, Teagle, and Feetham 2014; Merk and Pönitzsch 2017; Sütterlin and Siegrist 2017; Wibeck et al. 2017). This is consistent with, and might be related to, positive perceptions of activities that seem more natural relative to those that are seen as "tampering" with nature, a tendency that is evident in the case of solar geoengineering (Sjöberg

---

[12] In fact, hyperbolic discounting could increase the relative importance of longer-term benefits and costs and consequently make solar geoengineering less appealing (Quaas et al. 2017). The difference between appropriate and excessive discounting is unclear. A reasonable case can be made that, when designing policy with intergenerational effects, discounting should reflect actual opportunity costs and uncertainty about the future, but not pure time preference.

[13] Solar geoengineering is consistent with many of outrage's components, including being or seeming involuntary, unnatural, unfamiliar, memorable, dreadful, seemingly catastrophic, controlled by others, potentially unfair, morally relevant, and impossible to eliminate (Sandman 1993). Thus, if anything, outrage could lead to relatively less investment in it.

[14] I believe that, if anything, the solar geoengineering discourse presently suffers from pessimism bias, in which popular, scholarly, and government writings focus on its risks while regularly downplaying its apparent effectiveness in reducing climate change. For example, in one review, double the percentage of policy documents expressed concerns about geoengineering as the portion that expressed hopes (Huttunen, Skytén, and Hildén 2015; see also Buck 2013, 176).

2000; Mercer et al. 2011; Corner and Pidgeon 2014; Merk et al. 2015; Visschers et al. 2017).

This section offered a simple qualitative economic model for how society might invest in various responses to climate change with particular attention to the relationship between emissions abatement and solar geoengineering. Throughout this, the availability of solar geoengineering, in several places, reduces the investment in emissions abatement but increases social welfare. The removal of some assumptions, like those of a single decision-maker and of omniscience, lead in the model to emissions abatement deviating further from its optimum than solar geoengineering would. The process identified four genuine hazards regarding the relationships among responses to climate change. First, the mix of responses that appears optimal at one point in time can later be revealed to be suboptimal. Second, the preferences of decision-makers and those of the general population can differ systematically. Third, preferences are likely to change among generations. Fourth, decision-makers might lack intergenerational consideration. Of these, only the final one is necessarily consistent with the emissions abatement displacement concern. The other three could favor any of the responses to climate change relative to the others. Notably, all four are challenges to governance in general and are not limited to climate change policy.

My intention is not to belittle or dismiss the emissions abatement displacement concern. Indeed, there are feasible scenarios in which the displacement would harmfully manifest. Instead, my goal is to encourage careful and explicit thinking about decisions where much is at stake so that they can be made in ways that are expected to improve human well-being and foster sustainability.

3.5 POSSIBLE POLICIES

Suppose that policy-makers wish to minimize any emissions abatement displacement. We can approach this question in three ways. The first is to consider how existing policies address the analogous phenomena of ex ante moral hazard and risk compensation. In the case of the former, insurers reduce information asymmetry (for example, by monitoring insurees) and share risk (for example, through deductibles). However, the emissions abatement displacement does not involve information asymmetry. To some degree, liability for harm from solar geoengineering would resemble risk sharing, in that the actor who creates a risk bears more of its negative consequences (see Chapter 12). Another way to share risk would be to limit the use of solar geoengineering by reducing climate change by some fraction, such as half. The resulting impacts would provide incentives to future actors to abate emissions. However, this might result in more harm than good. Ultimately, ex ante moral hazard is a weak analogy to the emissions displacement concern and is therefore not helpful here.

3.5 Possible Policies

Risk compensation is more like the emissions abatement displacement concern. In this, the underlying causes are generally promoted. After all, consistent with a simple economic model of imperfect substitution, the introduction of a safety rule or technology followed by risk compensation decreases net harm, albeit less than one might initially expect. For example, automobile seat belts are often mandated, even though people drive more dangerously with them. This is because the primary purpose of automobile safety policy is not to promote cautious driving but to reduce injuries and fatalities while allowing efficient transportation. Two economists jokingly proposed that, if the goal were indeed cautious driving, then a spike in the steering wheel pointed at the driver would an effective mechanism (McKenzie and Tullock 1981, 40). There are numerous other examples of policies promoting the underlying causes of risk compensation: sports safety equipment, gun storage, public health measures, street lighting, and treatments for medical conditions that result from personal choices. In the case of climate change, this implies that solar geoengineering should be researched, even if it causes greenhouse gas emissions to be higher than they otherwise would be.

A second way to approach this question is to consider the genuine hazards that were identified in Section 3.4. Reviewing them shows that they are specific to neither solar geoengineering nor climate change, but instead are challenges to governance in general. How can we make the best-informed decisions in the face of uncertainty? How can we maintain alignment between the preferences of decision-makers and those of the general population? How can policy with long-term impacts be crafted when future generations will have different preferences? How can we ensure that decision-makers take the longer term into account? These difficult questions extend well beyond climate change and solar geoengineering. Overcoming them points toward structural legal and political reform.

Nevertheless, possible solar geoengineering policies for each hazard can be considered. The first hazard would arise if the responses to climate change were chosen, and we later learned that the various responses' benefits and costs are substantially different. This indicates a need to reduce uncertainty through the research of all the responses, including solar geoengineering. This should strive to produce and refine estimates of their environmental and social benefits and risks, at the present and into the future. In addition, the results of research should be transparent, independently assessed, made compatible with other relevant sets of information, and effectively communicated to decision-makers so that they can develop policy and update it based upon learning.

The second hazard regarding the responses to climate change would occur if decision-makers' preferences differ substantially from those of the general population. This suggests that understanding public preferences and integrating them into decision-making is essential, although low levels of awareness and knowledge present challenges. Importantly, developing countries should be particularly engaged, including through international cooperation in research, as they are poised to gain or lose the most (see Liu and Chen 2013; Lin 2014; Winickoff, Flegal, and Asrat 2015; Visschers et al. 2017; Lefale and Anderson 2018; Rahman et al. 2018; Weili and Ying 2018). In addition, the possible influence of concentrated interests on decision-making should also be proactively minimized (Long and Scott 2013).

The third hazard would manifest if preferences substantially change over the course of time. This is difficult to address, because future generations do not yet exist and there is no basis to predict how their preferences will differ from ours. Regardless, this hazard highlights that decision-makers should prefer policies that would allow future generations some flexibility to alter those policies. To a substantial degree, the actions to take in this regard resemble those ensuring that policy-makers can learn from and respond to new information, such as public engagement processes described in the previous paragraphs. To complicate matters, the chosen mix of responses to climate change can itself influence future preferences. For example, a generation born and raised with adaptation or solar geoengineering might be less averse to them than the current one. This seems to be a sort of "slippery slope." If this might be true, it remains unclear whether the current generation has an obligation to work to prevent such a change in preferences, or whether it would be unethical to do so (see Bykvist 2009). After all, previous generations had different preferences than we do, and we would likely bristle at the suggestion that they should have taken steps to prevent the subsequent intergenerational changes in preferences.

The final hazard regarding the responses to climate change is that decision-makers undervalue the future. Recall that this is the only one of the four that necessarily points toward emissions abatement displacement. Such undervaluing is evident throughout policy-making, including in funding pensions, investing in infrastructure, limiting public debt, and emissions abatement itself. How to ensure that decision-makers consider longer timescales is one of the central challenges of political science, especially in democracies. Plenty of politicians have campaigned on platforms to tackle some of these problems, but even the successful ones often do not implement such reforms. Nevertheless, policy-makers could limit the freedom of future ones. This is a common practice and is among the purposes of constitutions and treaties. Thus, current decision-makers could try to make it easier for future ones to reduce emissions and more difficult for them to develop and deploy solar geoengineering (Lloyd and Oppenheimer 2014, 52; but see Barrett 2014, 255). For example, policies that limit the growth and influence of actors who have concentrated interests in solar geoengineering might be warranted. Furthermore, the suggestion to cap solar geoengineering to reducing climate change by only half or less would require restricting choices in the future. Note, though, that this set of suggestions is somewhat contrary to those in the previous paragraph, which call for policies that would allow future decision-makers the flexibility to adapt.

Lin and fellow legal scholar Edward Parson also make policy recommendations to reduce emissions abatement displacement that would limit the freedom of future policy-makers. They each suggest that states agree to implement solar geoengineering only if they have met emissions abatement targets (Lin 2013, 710; Parson 2014). However, Parson critiques his own proposals as a sort of noncredible intertemporal threat by the present to the future. If future decision-makers failed to meet their targets and faced serious harm from climate change, then they would likely renege on the old agreement to withhold solar geoengineering. In addition, such an agreement would incentivize states to set unambitious emissions abatement targets in order to retain the option of using solar geoengineering.

Parson thus puts forth a more elaborate option, in which only those states that meet greenhouse gas emissions abatement targets would be allowed to participate in international decisions regarding solar geoengineering. States' motivations to "have a seat" at the decision-making table would depend upon how widely their preferences regarding the parameters – such as timing, location, intensity, and form of solar geoengineering diverge. If these preferences closely align, then countries would have little additional incentive to abate emissions. Furthermore, hegemonic states or those that are weakly integrated into the international community might not greatly value being part of the decision-making club, as they might be willing to act contrary to the international agreement (see Reynolds 2015a, 185).

Nevertheless, policy linkages between emissions abatement and solar geoengineering might hold potential to reduce abatement displacement, particularly if the linkage were to disaggregate heterogeneous states. For example, states with the desire and capacity to deploy solar engineering could promise to refrain from doing so, provided that other states aggressively abate emissions. This lacks the shortcomings of the original intertemporal linkage proposal, those of possible little reason to participate, perverse incentives, and noncredible intertemporal threats. However, this linkage would constitute a threat by some states to use solar geoengineering, which might not be well-received in the international community.

The third and final way to approach developing policies for multiple responses to climate change is to consider their complementarity.[15] That is, greenhouse gas emissions abatement, NETs, adaptation, and solar geoengineering could each lower climate change risks in different ways. One can conceive of them as components of a climate risk management portfolio (see Wigley 2006; Shepherd et al. 2009, 56; McNutt 2016). This depends on their characteristics, some of which are given in Table 3.1. These reinforce the fact that the research, development, and possible use of solar geoengineering would not eliminate the need to aggressively abate emissions.

These responses could constitute a portfolio, presented schematically in Figure 3.1 (based on Long and Shepherd 2014, 765). Assume that atmospheric greenhouse gas concentrations, climate change, and climate risks are all roughly

[15] This is not meant in the economic sense of a price decrease in one causing a demand increase in the other. Economically, they remain imperfect substitutes. Instead, I mean that they can fill different roles.

TABLE 3.1 *Some relevant characteristics of the responses to climate change.*

	Abatement	NETs	Adaptation	Solar Geoengineering
Rapidly effective and able to respond to new information?	No	No	Possibly, but probably not	Yes
International cooperation required?	Yes	Yes	No, but helpful for developing countries	No, but helpful
Able to reduce risks from past emissions?	No	Yes	Yes	Yes
Reduces climate change?	Yes	Yes	No	Yes, but imperfectly

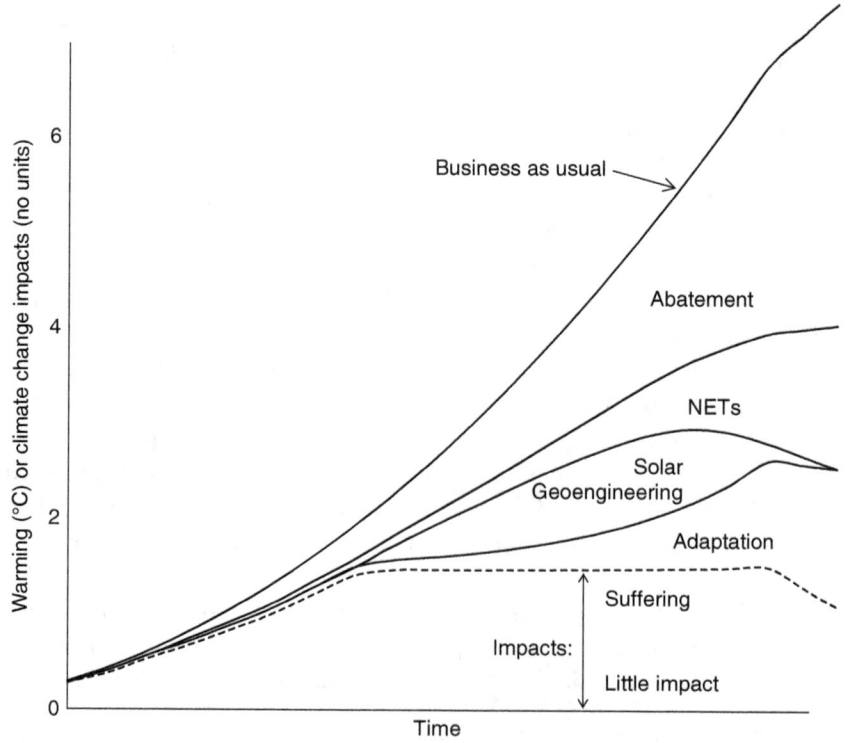

FIGURE 3.1. Possible complementary roles of responses to climate change.

proportional (although the latter two will be delayed relative to the first). On the timescale of a couple of centuries, emissions abatement cannot reduce the atmospheric carbon dioxide concentration but instead only slow its rate of increase. In contrast, NETs could do so but would require years of research and development before they might (or might not) be able to be deployed at large scales. Even then, it would be decades before the decrease in atmospheric concentration would be substantial. Adaptation could be carried out soon, although it reduces only climate risks, not actual climate change. Solar geoengineering could be a temporary measure to reduce the short-term climate change risks resulting from previous emissions until the other responses would become sufficient. If benign and acceptable solar geoengineering were available, then decision-makers could adjust its deployment parameters to new information and conditions more rapidly than the other three means. The remaining area represents the climate change that people and ecosystems would experience. Some impacts will be negligible or possibly beneficial, while the rest would cause suffering. Importantly, lowering the quantity of one response necessitates the increase of another one or – as a default – the impacts.

3.6 POLITICS, CULTURE, AND PERCEPTION

There is more to climate change and the responses to it than mere risk management. Geographer Mike Hulme writes that "depending on who one is and where one stands – the idea of climate change carries quite different meanings and seems to imply quite different courses of action ... [C]limate change possesses a certain plasticity ... allow[ing it] easily to be appropriated in support of a wide range of ideological projects" (Hulme 2009, xvi-xviii). People's understanding of the responses to climate changes is shaped by, among other things, their culturally conditioned views of humans, society, risk, technology, nature, and our role in nature.

An important related question is, why do diverse political actors support action to prevent climate change? I suggest that there are three roughly identifiable supportive groups. The first of these consists of those whose primary goal is minimizing risks to people and the environment. They are pragmatic. If a technology or other means reduces harm, then they are usually open to it. The second group is "green" environmentalists, who wish for humanity to consume less and to be less intrusive in nature. Much contemporary environmentalism often calls for humility and has been skeptical – if not hostile toward – technological interventions into nature, especially those that are large-scale and centralized. The final group wishes to change the dominant political and economic order and to redistribute wealth and power, and its members see climate change action as a means to do so.

These three groups react differently to the possible responses to climate change. Greenhouse gas emissions abatement works toward all three groups' primary goals: it lowers risks; it reduces human's environmental impacts; and it might restrict the economic power of rich individuals, countries, and corporations. Adaptation likewise furthers the objectives of at least two of these groups. For the pragmatists, it reduces risk, and for redistributionists, serious adaptation through international cooperation would move wealth from rich countries to poor ones. Green environmentalists might be resistant – at least initially – to adaptation, because it does not contribute to deindustrialization or because it might undermine abatement efforts, as described in Section 3.1. However, they might also be agnostic. In contrast, solar geoengineering clearly furthers only the first group's goal, the pragmatic reduction of risk. It would contribute neither to minimizing human's intervention in nature (in fact, it would amplify interventions) nor to redistributing power and wealth (Virgoe 2009). Therefore, solar geoengineering is not only counter to the goals of two groups that support action to prevent and reduce climate change, it threatens the political coalition for climate change action.

This political landscape has several implications for the relationship between emissions abatement and solar geoengineering. First, the "moral hazard" concern assumes a genuinely moral justification for some observers. Hale describes three views of moral hazard, one of which he calls the "vice view" in which solar geoengineering

> encourages parties to engage in overindulgent (e.g. undesirable, negative, naughty) behavior, thus suggesting that temperance and prudence have fallen by the wayside. My suspicion, in fact, is that this is something like the commonsense view of the moral hazard. If this is the case, the value element implied by the moral hazard is a matter separate from the assessment of its alleged wrong. (Hale 2012, 118)

Second, solar geoengineering might be a type of forbidden knowledge (Kempner, Merz, and Bosk 2011). In this, some topics domains are too sensitive, consequential, or taboo for researchers to pursue. These are off-limits – often only implicitly so – to researchers through regulation, social norms, or self-restraints. Steve Rayner sees the rhetoric of opposition to solar geoengineering utilizing calls for ignorance in two ways: "Ignorance is a binding constraint – 'we simply cannot know' – and ignorance is a source of virtue – 'it saves us from folly'" (Rayner 2015, 313).

Third, the implicit goals of climate policy are shown to be plural. This chapter describes how policy usually encourages the underlying causes of risk compensating behaviors. However, if the underlying cause is seen by some as immoral, then such policies are contested. For example, distributing clean needles to drug users is aptly called "harm reduction" because it reduces the harm of disease transmission, despite potentially increasing drug use. However, if one sees drug use itself as the problem, instead of diseases, then these measures should be opposed. Models indicate that solar geoengineering could reduce climate change and its risks, although greenhouse gas emissions might consequently increase. Yet if some actors' goal with respect to climate change policy is abating emissions, instead of lowering climate risks, then solar geoengineering should be forsaken.

The final implication regards harm from climate change. If the introduction of solar geoengineering into a mix of climate policies would reduce climate change impacts, then inversely its exclusion from the mix would likely increase these impacts. Science writer Oliver Morton says, "I think there are people who, though they would not necessarily put it this way, would rather see greater suffering due to climate change than have human empire dominate the previously natural world to the extent geoengineering would seem to entail" (Morton 2015, 124–5). Given the forecasted harms from climate change, particularly those to the most vulnerable people and ecosystems, removing a means to reduce it would have serious consequences.

3.7 SUMMARY AND CONCLUSION

The most frequently voiced – and arguably influential – concern about solar geoengineering is that its consideration, research, and development would undermine emissions abatement efforts, and that this would be harmful. I call this the "emissions abatement displacement concern." There was a similar objection to adaptation, but one no longer hears it. Ex ante moral hazard and risk compensation are two well-studied phenomena that are imperfect analogies to abatement displacement, although the empirical evidence for their magnitudes in various contexts is mixed. Public opinion studies that ask people how they would respond to solar geoengineering consistently do not indicate abatement displacement. In fact, they often point toward its reverse in which solar geoengineering could cause greater emissions abatement. An extended but simple qualitative economic model identifies four genuine hazards regarding the relationships among responses to climate change. Notably, all four are challenges to governance in general, and are not limited to climate change policy. It may be that there is little that policy-makers could effectively do if they wished to reduce emissions abatement displacement, although international policy linkages may have some potential. I suggest that the emissions abatement displacement concern is widespread for reasons unrelated to reducing climate change and its negative impacts, but instead is grounded in issues of political coalitions and wider worldviews.

4

International Relations

The pioneering political scientist Harold Lasswell called politics the process of deciding "who gets what, when, and how" (Lasswell 1936). Solar geoengineering would have distributional implications, in that some groups would benefit more than others and some might experience net harm. (In this way, it is like climate change and the other responses to it.) Decisions regarding its research, development, and potential implementation will concern, to some degree, who gets what, when, and how, and will thus be political. Furthermore, states will insist on being the primary actors in much – but not all – of this process. They interact with each other through international relations, international law, informal cooperation, and transboundary impacts. States also offer incentives for nonstate actors through funding, intellectual property law, and other domestic laws and administrative regulations. With respect to solar geoengineering, states will likely cooperate internationally, at least to some degree; make decisions regarding whether, how, and when it might be deployed; and respond to potential and actual international conflict. Therefore, the governance of solar geoengineering is an issue of international relations.

Surprisingly, actual scholars of international relations are weakly represented in the solar geoengineering discourse (Horton and Reynolds 2016, 441). Instead, most of the possible scenarios of state interactions have been proposed and discussed – in some cases exclusively so – by commentators outside of political science. Furthermore, a large portion of publications by international relations scholars remains in the grey literature.

Like other disciplines, the scholarly field of international relations has various schools of thought. As noted in the Chapter 1, I am a welfare-oriented consequentialist and assume actors' rationality, at least as a starting point. I thus find my thinking to be most closely aligned with Institutionalism. In this school, states are the relevant actors in the international politics, they rationally pursue their diverse interests, they have varying and limited capacities, and the international arena is anarchic. States can and do cooperate to satisfy their preferences and achieve absolute gains. I do not necessarily claim any superiority on Institutionalism's part, only that it has some useful explanatory and predictive powers. The other schools of thought within international relations, such as Realism, Liberalism, and Constructivism, have much to offer and can improve our understanding of how solar geoengineering might play out in the international arena (see Horton and Reynolds 2016; Corry 2017).

This chapter considers some leading matters of solar geoengineering and international relations. Section 4.1 describes the problem structures of global public goods. Section 4.2 examines concerns regarding the ability and likelihood of a single or small number of states to deploy solar geoengineering, possibly contrary to the international community's wishes. Section 4.3 takes on other ways in which solar geoengineering could cause problematic international relations. Section 4.4 reviews some economic models of rational state behavior.

4.1 GLOBAL PUBLIC GOODS

To help explain how states have and will likely carry out solar geoengineering and other responses to climate change, we need to understand the relevant incentives that they face. Some general concepts, some of which were encountered in Chapter 2, should first be introduced. Actors such as individuals, businesses, or states have limited resources and generally act – including through voluntary exchanges and cooperation – in pursuit of their diverse interests, which improves their individual welfare and consequently the total social welfare. However, under some circumstances, potentially welfare-increasing actions may not be able to occur, and those that do take place may reduce others' well-being. One problem is externalities, which are positive or negative effects on someone who does not voluntarily participate in an activity. An emblematic negative externality is air pollution. A second problem is collective action, in which the actors would collectively be better off if they all undertook some costly action, but are individually better off if they each do not do so. In other words, they lack incentives to contribute to a welfare-increasing cooperative effort and assurance that, if they do contribute, that others will likewise. Notably, individually rational behavior leads to a suboptimal outcome that appears collectively irrational. Not contributing is called "free riding." The provision of a beneficial public good is a collective action problem. A public good is a good or service whose effects are nonexcludable.[1] Because the provider cannot exclude those who do not pay for the public good's provision, she will lack incentives to provide it. National defense is the quintessential public good: all people benefit from protection, but each would hesitate to voluntarily contribute to it. Other barriers to welfare-increasing actions include information asymmetry and nonrationality (touched upon in Chapter 3), principal-agent problems (see Chapter 12), as well as high transaction costs and monopoly. These final two are not discussed in this book.

---

[1] In addition to being nonexcludable, a public good must also be nonrivalrous, in that its use by one does not diminish others' ability to use the public good. However, nonrivalry is not important here.

Note that externalities and public goods are here presented as distinct ideals, when in fact they vary along gradations. One person's positive externality may be another's negative one due to different impacts, values, and uncertainties. Likewise, public goods can have positive effects for some people and negative ones for others who cannot exclude themselves. Public goods are thus not always "good" in the normative sense. Instead, the word "good" is used in the economic sense of something that some people want. Moreover, a positive or negative externality increasingly resembles a beneficial or harmful public good as the number of affected people increases. A public good is a sort of universal externality. The reduction of a harmful one can often be thought of as a beneficial public good.

In the case of climate change, greenhouse gas emissions are often called a negative externality, although harmful public good might be more accurate. Likewise, emissions abatement is a beneficial public good. The implementation of solar geoengineering would also be a public good, in that its effects would be nonexcludable. Models presently indicate that, under conditions of elevated atmospheric greenhouse gas concentrations, its judicious use would reduce climate change and it would thus be a beneficial public good, at least for most regions. Likewise, reckless solar geoengineering would be a harmful public good. Furthermore, the research and development of solar geoengineering would be, like most other instances of information generation, a beneficial public good, assuming that information is shared (Foray 2004, 113–29). For the sake of brevity, I henceforth use "public goods" to mean generally beneficial ones.

Externalities, collective action, and other barriers to greater social welfare can sometimes be addressed through governance, including by the state. Governance can facilitate the generation of positive externalities, hinder that of negative ones, and provide public goods. The state can do so, sometimes effectively, in large part because of its power to collect taxes backed by the (usually implicit) threat of force. However, the situation is much different in the international arena, where there is no supra-authority. The provision of global public goods is further challenging due to the large number of states involved. In fact, some writers propose that global public goods provide both a purpose and basis for legitimacy for international law and global governance (Sandler 1997; Bodansky 2012).

Global public goods vary, including in their problem structures and the incentives that they present to states. These are essential to understanding the responses to climate change, including solar geoengineering. Economist Scott Barrett has developed a useful taxonomy, from which three forms are described here (Barrett 2007; see also Bodansky 2012; Sandler 2017). The first form is an aggregate effort global public good, whose provision depends upon the sum of all countries' efforts. Although a contribution yields net global benefits, these are distributed among all countries while the acting country bears all the costs. This is the classic collective action problem. No single state can ensure the success of the endeavor, although the major countries are usually necessary for success. These global public goods require treaties to determine what is to be done and who is to contribute how much, to encourage participation and compliance with agreements, to assure complying participants that others will also contribute, to punish free riders, and often to reallocate costs. Enforcement of these treaties is challenging, both because of the temptation to free ride and the general difficulty in enforcing international law. An important example is greenhouse gas emissions abatement and negative emissions technologies (NETs). Their aggregate effort structure is the primary reason for their continued insufficiency.

Second, a single best effort global public good is one whose provision depends on the single largest contribution. Furthermore, the benefits to a single country might be greater than the costs of providing the public good. Although a single best effort need not necessarily be done by one state, it could be. In either case, the acting country would rather not pay all the costs when others also benefit and may seek financial contributions for its effort. That is, free riding is possible but – unlike in the case of aggregate efforts – will often not prevent the provision of the single best effort. Some degree of international cooperation is often helpful and sometimes needed. In particular, cooperation can allocate the costs and in some cases institutionalize decision-making, especially among the financial contributors. States sometimes rely upon treaties for this, although less legalized means of cooperation often suffice. Regardless, single best effort global public goods are easier to govern than aggregate effort ones because they require only the providing country's active contribution, although support from others can help (Weitzman 2015, 1066–7). Barrett categorizes the implementation of solar geoengineering as a single best effort global public good (see Morrow 2014a).

Although single best efforts are likely to be provided, they can be problematic. Some states might prefer not to be affected by the public good or might prefer a lower quantity of it. That is, single best effort global public goods can be prone to oversupply, particularly if states' preferences for it diverge substantially. Economists Martin Weitzman and Gernot Wagner call this a "free driver problem," to contrast it with free riding (Wagner and Weitzman 2012).[2] Consequently, some states may offer to contribute to the good's provision, in terms of finances, effort, or other resources, in order to have input into decision-making. Alternatively, a state that could provide a single best effort global public good might not want to bear all the burdens of doing so. It could insist on international cost sharing, and if other states could also provide the single best effort, then they could each resist providing it in an attempt to force the other(s) to do so (Sandler 2017). Thus, international cooperation can be helpful for single best efforts. Specifically, international law and organizations, as well as less legalized process, can help coordinate their provision, such as by detailing the when, where, how, and by whom of provision and by managing the sharing of costs and other burdens.

[2] Weitzman's economic model that inspired this term is described in Section 4.4.

This distinction between emissions abatement and NETs as aggregate efforts and solar geoengineering as a single best effort is very important. Game theorist and Nobel laureate Thomas Schelling wrote that solar geoengineering "totally transforms the greenhouse issue from an exceedingly complicated regulatory regime to a simple – not necessarily easy but simple – problem in international cost sharing" (Schelling 1996, 305). This was also the bases of Barrett's description of solar geoengineering's economics as "incredible" (Barrett 2008) and of political scientist David Victor's claim that it "turns the politics of climate protection upside down" (Victor 2008, 323). Solar geoengineering's single best effort structure and its free driver problem cause the concern of uni- or minilateral implementation and some other potentially problematic scenarios, discussed more in Section 4.2.

Third, a mutual restraint global public good is one whose provision depends on some or all countries *not* doing something that they each are tempted to do. It is in their interest to restrain each other from undertaking the action. If only one state succumbs to the temptation and does the action, then the public good of mutual restraint is not "provided." This is thus something of a mirror image of aggregate efforts, in that each country often needs assurance that others will also restrain themselves or contribute, respectively. International cooperation is often necessary to provide assurance, establish the standards for nonaction, and facilitate communication. Treaties can sometimes accomplish this but are not always the best instruments because participation in them is voluntary. At the same time, some states might need to be given incentives to refrain from acting. This can be done through technical and institutional assistance, side payments such as financial aid or concessions on an unrelated issue, rhetorical persuasion and the resulting reputational consequences, or coercion. Regardless, establishing such cooperation and maintaining compliance can be difficult, as participants will be tempted to defect. As an example, Barrett provides the nonuse and nonproliferation of nuclear weapons.

Solar geoengineering could also be considered a problem of mutual restraint, in addition to one of single best effort. It is expected to be relatively inexpensive and technologically feasible, but states might vary widely in when and to what degree they want it to be developed and implemented. Some of them may be tempted to use it before there is sufficient international consensus. Moreover, if solar geoengineering continues to be controversial, its large-scale outdoor research or implementation could be perceived by decision-makers and the wider public as a hasty and illegitimate act. The resulting political backlash could delay or entirely prevent subsequent solar geoengineering activities, even those that would be widely beneficial and otherwise supported.

4.2 UNI- OR MINILATERAL DEPLOYMENT

From early on, the scholarly, grey, and popular literatures identified several possible problematic international dynamics that solar geoengineering could engender. This

4.2 Uni- or Minilateral Deployment

section and Section 4.3 review these, with the caveat that the distinctions among them are not always clear.

Many of these challenges to international relations derive from solar geoengineering's problem structure as a single best effort global public good (for example, Schelling 1983, 470). That is, one or a small number of states, or even a nonstate actor, could implement it globally, including in the absence of an international decision-making process and perhaps contrary to the wishes of the international community. This possibility is due to solar geoengineering's low financial costs of implementation and apparent technical feasibility. For example, a state that is threatened by climate change's impacts might implement it (Cascio 2009, 61–3). This appears more probable in the cases of a so-called rogue state whose leadership is not well-integrated into the international community or one whose leadership has a weak hold on power. The prospect that one or a handful of countries could alter the world's climate – and by extension, all other states' environments and their residents' well-being – in the absence of consensus runs contrary to core values of legitimate decision-making within the contemporary international legal order, and would likely increase international tensions.

This is often described as unilateral or, sometimes, minilateral implementation, which is a leading concern with respect to solar geoengineering and international relations. Another, less binary way to think of this is that states' preferences for the presence (or absence) and intensity (technically, the magnitude of radiative forcing) of solar geoengineering vary. Somewhere near the middle of the distribution of intensity is the optimum, where social welfare would be maximized. The challenge is that solar geoengineering, as presently understood, would be additive, in that deployers could add to solar geoengineering but not stop or subtract from it.[3] This is the "free driver" problem. The country with the preference for the greatest intensity of solar geoengineering that can deploy it and is willing to act contrary to other states would get its preference. This would likely be sooner and at a greater intensity than socially optimal.

Uni- or minilateral deployment of solar geoengineering would be undesirable for several other reasons. At the very least, it runs contrary to states' territorial sovereignty and the sense that their domestic environments should be free from such intentional external effects. Furthermore, uni- or minilateral implementation could cause or exacerbate international tensions (Chalecki and Ferrari 2018). If other countries experienced negative impacts, or perceived that they did, then they might try to persuade or compel the implementer to stop. The use of force might not be out of the question, depending on the circumstances. Some observers have even suggested that war could result (Keith 2013, x–xi; Maas and Comardicea 2013, 43). Moreover, if

---

[3] This would apply only in the absence of counter-solar geoengineering, which remains only theoretical (see Parker, Horton, and Keith 2018).

the deployer was a powerful state, then others might perceive its actions as an inappropriate way to leverage and to secure its position in the international arena.

At the same time, potential uni- or minilateral implementation may not be as likely or problematic as some writers have claimed, for four reasons. First, the number of actors who have both the interest and the financial, technical, and political capabilities to deploy and maintain solar geoengineering would be small. The direct implementation costs of tens of billions of dollars annually are not sustainable for nonstate actors and for most countries (for more on implementation by nonstate actors, see Chapter 11). In fact, bearing the implementation costs would be rational, in a narrow sense, for only those few countries whose expected damages from climate change would be greater than these costs. If one assumes globally uniform climate change damages of two percent of economic activity, then this would justify annual deployment costs of thirty billion dollars for the fourteen largest economies.[4] Barrett suggests that India – a country that has a sufficiently large economy, nuclear weapons, and fairly robust technical capacity; is vulnerable to climate change; and must respond to domestic democratic pressures for action – as a candidate for possible unilateral implementation (Barrett 2014, 256; see also Emmerling and Tavoni 2017b). Regardless, public condemnation, international political pressure, issue linkage, threats by powerful countries, and other traditional tools of statecraft would likely end unwelcome solar geoengineering by middle and weak powers (Parson 2014, 98–103; Keohane 2015, 23; Rabitz 2016; Halstead 2018). This seems to leave only great powers such as the United States and China, and perhaps Russia as feasible unilateral deployers of solar geoengineering, and even this would raise concerns among other states of abuse of power.

Second, the international system is arguably dominated by a growing logic of multilateralism (Keohane and Nye 2011). In general, a world of enormous international economic and cultural exchange is one in which both the value of cooperation and the cost of conflict are high. International relations scholar Joshua Horton identifies a handful of reasons that are specific to solar geoengineering that would further encourage cooperation (Horton 2011). At the very least, a potentially deploying state would prefer to not bear the entire burden of an endeavor that costs tens of billions of dollars yet could benefit the entire world. Instead, it would likely seek others to share these costs, and participation in decision-making would be an important way to attract partners. It would also rather not raise the ire of other countries. On this point, law professor Daniel Bodansky argues that even though unilateral action is permitted under international law unless otherwise proscribed, the international political order would default to a restrictive position in solar geoengineering negotiations: "The absence of an effective process for making international decisions is far more likely to frustrate proposals for climate engineering. Countries would be unwilling to incur the political costs of proceeding without

---

[4] Climate change impacts could be greater or less and will be spatially heterogeneous.

international approval; but there would be no effective way of obtaining this approval" (Bodansky 1996, 310).[5] Furthermore, a state considering unilateral deployment would know that any such action would set a precedent and that other countries could subsequently undertake solar geoengineering as well. If the state was considering implementation as a means to gain relative power, this precedent-setting and possible subsequent deployment by other countries – which could include rivals – might undermine any possible gain. Finally, countries that prefer less or no solar geoengineering might, in principle, be able to threaten or engage in counter–solar geoengineering, by releasing a short-lived, powerful greenhouse gas (Parker, Horton and Keith 2018). However, such methods remain only hypothetical. In the face of possible neutralization of its own efforts, plus environmental risks, the original state would likely either refrain or engage with the international community.

The third reason that uni- or minilateral implementation may not be so problematic is that such action could be beneficial, in that it could reduce climate change without requiring arguably unlikely global cooperation (Michaelson 1998, 121; Virgoe 2009, 116). Suppose that greenhouse gas emissions abatement remains insufficient, perhaps due to its collective action problem, and severe climate change impacts manifest. Then one or a few states deploy solar geoengineering without global consensus. If the climate change impacts were substantially reduced, as models indicate as possible, then this uni- or minilateral implementation would not necessarily be problematic and could improve human well-being and foster sustainability. Such unilateral action may not be as undesirable as it first appears:

> [D]espite the growth of multilateral decision making, international cooperation often remains unachievable or illusory. In such cases, where there is no real prospect for effective international action, unilateral action may be the only means of promoting and enforcing shared values. Although such actions do not comport with how we think decisions should ideally be made, they may nevertheless further important substantive goals, such as the protection of human rights or the environment. Rather than rejecting them outright, we should evaluate each particular unilateral action (or inaction) to determine whether, on balance, it advances or detracts from desired ends. (Bodansky 2000, 347)

Oliver Morton describes a future in which a handful of developing countries that are particularly vulnerable to climate change reveal that they have secretly been injecting low levels of aerosols into the stratosphere (Morton 2015, 347–59). Although they initially frame their efforts as a type of civil disobedience to catalyze abatement by others, they also invite states to join their initially small consortium, which intends to limit global warming to 0.1 degree Celsius per decade. These other states might

---

[5] Consider the example of the Large Hadron Collider, which was proposed and built by the European Organization for Nuclear Research (CERN). Some observers were concerned that its operation would cause a very small chance of catastrophe, including destruction of the Earth. CERN commissioned research into this but did not consult with other states. It built the collider (see Johnson 2008).

realize that the amount of future solar geoengineering will depend, in part, on their own emissions and might increase their abatement efforts.[6] In this way, solar geoengineering could dramatically redistribute international power, at least with respect to climate change. Nevertheless, and to be clear, uni- or minilateral implementation that would increase social welfare could still have normatively undesirable distributional and procedural aspects.

Finally, the nature of unilateral implementation may not be as unprecedented as it initially appears. Individual and small groups of countries regularly influence others without consent. Often this is – at least purportedly – simply for the welfare of the acting countries, not to negatively affect others. For example, the leaderships of countries (or the quasi-federal European Union) with large economies make influential policy decisions, such as central banks' setting of interest rates. In the case of the US Federal Reserve, the most affected countries are those that use the US dollar as their official or unofficial currency, hold large amounts of dollars as a reserve currency, or rely heavily on the United States for trade or investment. A decision regarding a key interest rate influences their inflation, asset prices, unemployment, wages and income, economic security, and more. Yet the consent of these affected countries is not required, although consultations are increasingly common, especially through intergovernmental forums.

4.3 OTHER CHALLENGES TO INTERNATIONAL RELATIONS

Solar geoengineering could foster challenges in international relations other than uni- and minilateral implementation. This section explores several feasible scenarios in which international tensions heighten, physical risks increase, and widely shared norms are challenged.

Decision-making regarding large-scale outdoor experiments and deployment of solar geoengineering could be the outcome of a multilateral decision-making process. Whether and how such decisions would be seen as legitimate is central to its governance. Because solar geoengineering as presently understood would affect the entire planet, a global body might appear to be an appropriate locus of decision-making. However, there is no existing body whose decision-making is both seen as legitimate as well as effective (see Chapters 5, 6, and 13). The UN General Assembly offers a deliberative forum for all countries, but cannot make decisions that legally bind states. Indeed, its consensus-seeking culture means that a few small states can hold up a decision that otherwise has broad support. It seems unlikely, although not out of the question, that the world's powerful countries would concede that such consequential decisions should be made in a consensus-oriented deliberative forum in which each state has one vote. The UN Framework Convention on Climate Change (UNFCCC) also has universal participation and operates on a consensus

---

[6] This is like linkage between solar geoengineering and abatement policy, discussed in Chapter 3.

basis. However, its objective is limited to the stabilization of atmospheric greenhouse gas concentration, and solar geoengineering deployment is outside of this scope. Furthermore, its Conference of the Parties (COP), which facilitates the implementation of the UNFCCC, does not take action with broader consequences.[7] Some scholars have asserted that, if the topic of solar geoengineering were to be introduced into global forums such as the UN General Assembly or the UNFCCC COP, then the likely result would be an international prohibition (Bodansky 1996, 319–20; Victor 2008, 331; Davis 2009, 921–36). A stalemate also seems feasible in these consensus-based forums.

In contrast, the small UN Security Council can issue legally binding resolutions but only "for the maintenance of international peace and security" (Charter of the UN, Article 24.1). It has expanded its scope in recent decades and has recently considered climate change. In a nonbinding statement, it indicated that climate change impacts might "in the long run, aggravate certain existing threats to international peace and security," and at least two of its resolutions cite climate change as factors that affect security (United Nations Security Council S/RES/2349, S/RES/2408; see also S/PRST/2011/15). Under some circumstances, solar geoengineering could fall within the Security Council's purview. At the same time, its five permanent members' veto power means that they must all consent, which might be difficult to achieve. Furthermore, in a world in which power and population are shifting, the unrepresentativeness of the permanent members is increasingly seen as undermining the Security Council's legitimacy.

Even if decisions regarding large-scale outdoor activities were made multilaterally, states would likely still disagree about solar geoengineering. Although preindustrial conditions or the internationally agreed-upon two-degree warming limit could offer focal points for negotiations concerning implementation, countries may not necessarily strive to return to a pre–climate change world (Heyen, Wiertz, and Irvine 2015). Instead, they may have other preferences, due to risk aversion, a belief that they would be better off or rival states would be worse off with a novel climate, or cultural reasons, such as the importance of a more "natural" world and the role of technology therein. Edward Parson writes that

> the discussion thus far may understate the prospects for opposition, because it assumes some rational process of forming nationally aggregated interests, based on realized or projected climate effects, with each region viewing its recent climate as ideal; but any of these assumptions might not hold. State interests could be driven by smaller-scale patchiness of climate effects within countries and resultant domestic political conflict. Alternatively, climate preferences might shift in response to realized climate changes or to recognition of the possibility of intentional climate control, such that a region's present climate is no longer judged to be ideal. State interests in

---

[7] It may "Exercise such other functions as are required for the achievement of the objective of the Convention as well as all other functions assigned to it under the Convention" (UNFCCC, Article 7.2(m)).

CE [climate engineering] might also be dominated by non-consequential or non-rational processes, such as religious or symbolic commitments, general technological optimism or pessimism, or generalized suspicion about other states' intentions. To the extent that these other processes show strong regional variation, they could further increase the possibility of interstate conflict over CE. (Parson 2014, 101)

At the same time, solar geoengineering deployment would likely be neither binary – "yes" or "no" – nor even a single scalar variable – "how much." Instead, it would have multiple parameters, including how much, which techniques, which materials, at what latitudes, and at what times of year (MacMartin and Kravitz 2019). Although a greater number of adjustable parameters might offer bases for more international disagreements, it could allow a greater diversity of states' preferences to be satisfied, or at least more closely approximated.

An additional source of disagreement regarding solar geoengineering could be how to share the costs. Although the direct financial costs of implementation are presently thought to be low, they could be greater if they include monitoring, assessment, redundant systems, and especially compensation for harm.

Independent of whether decision-making was global, multi-, mini-, or unilateral, solar geoengineering could affect international relations and security. On the one hand, if climate change impacts were great enough to threaten international security, then the use of solar geoengineering to reduce them would lessen this threat, as a first approximation. On the other hand, the mere presence of solar geoengineering could destabilize relationships, exacerbate tensions, and disrupt the international order through several mechanisms (Kellogg and Schneider 1974; Schelling 1983, 470; Corry 2017, 302–4). First, it could simply constitute an additional source of strain where relations were already tense. Countries that are not deploying solar geoengineering, perhaps because they are unable, could perceive it as a threatening, dual-use technology, one that can be used for both beneficial and intentionally harmful purposes (Scheffran 2013, 338). For example, some current signs point toward a future bipolar world dominated by the United States and China. To the extent that they will have a tense relationship, solar geoengineering could offer an additional point of contention. Second, once used, solar geoengineering's direct or indirect environmental effects could fuel existing conflicts (Maas and Scheffran 2012, 198). Countries might respond with force to solar geoengineering that they consider to be deeply contrary to their interests (Cascio 2009, 56–9; Schellnhuber 2011; Chalecki and Ferrari 2018). A third path to greater international tensions is that, as noted, some states might perceive others' capacity for solar geoengineering as threatening. This could propel a race to develop solar geoengineering – as well as possible counter-solar geoengineering – more rapidly than their rivals (Maas and Comardicea 2013, 43). Fourth, and similarly, countries could be suspicious that their rivals are actually using solar geoengineering – maybe secretly – in order to gain relative advantage, perhaps even intentionally harming rivals (Kellogg and

Schneider 1974, 186; Adger et al. 2014, 777; Dalby 2015, 195–6; Lin 2016, 2546–7). A subset of these writers who make such assertions goes further, claiming that solar geoengineering could be militarized. Others are skeptical of militarization, because solar geoengineering would be difficult to control, have indirect and delayed effects, and could risk drawing additional parties into a conflict (Maas and Scheffran 2012, 196; Briggs 2018; Halstead 2018).[8] In either case, the fearful country could take preemptive measures, including risky or destabilizing ones. Fourth and finally, in the wake of a harmful extreme weather event, some leaders could blame this on a state that has been researching or implementing solar geoengineering. In fact, they might do so independent of actual evidence and beliefs; blaming external forces is a common tactic among politicians. The "victim" state might demand compensation for harm, pursue claims in international legal forums, or retaliate (Scheffran 2013, 338; Halstead 2018; see Chapter 12). Blame might be more likely to originate from a relatively weak state as an attempt to reduce a hegemon's stature.

Independent of their precise cause, solar geoengineering might also reproduce and widen existing international fault lines between cohorts of states, such as between developing and industrialized countries (Zürn and Schäfer 2013). The former might be concerned that the latter's solar geoengineering activities are displacing emissions abatement and adaptation efforts (Virgoe 2009, 113; see Chapter 3). Alternatively, developing countries might take the lead in solar geoengineering due to their generally greater vulnerability to climate change and to the techniques' relative low costs and technical barriers. If so, then industrialized countries could protest and even retaliate for several reasons. To the extent that aggressive emissions abatement and financial transfers for adaptation hold the potential to fundamentally change international power relations, such a perception of abatement displacement could lead to a belief that solar geoengineering was perpetuating an unjust international order (Dalby 2017, 235).[9]

Another possible fault could be between powerful and weak countries. For example, a weak state that is vulnerable to climate change might wish to implement solar engineering but refrain from doing so under pressure from powerful ones. It might perceive this as coercion. Alternatively, a powerful country that is not gravely threatened by climate change with the capacity for solar geoengineering

---

[8] A related claim is that powerful countries' militaries would inevitably assume control of solar geoengineering (Fleming 2007; Cairns and Nightingale 2014).

[9] Indeed, a North–South split might be emerging: the only three identifiable national positions on solar geoengineering to date are the UK's and Germany's cautious support for research, and Bolivia's opposition, which was based in part on the emissions abatement displacement concern (Government Response to the House of Commons Science and Technology Committee 5th Report of Session 2009–10: The Regulation of Geoengineering, 2010; Estado Plurinacional de Bolivia, Submission to Joint Workshop of Experts on Geoengineering, Lima, Peru, June 20–22, 2011; Schütte 2014). Meanwhile, research is moving forward most rapidly in the United States (see Necheles, Burns, and Keith 2018).

might not deploy it, even though some countries that are vulnerable to climate change implore it to do so.

Throughout issues of international relations, states' perceptions are highly salient. Indeed, some international relations scholars assert that perceptions are more important to states' decision-making than facts. Solar geoengineering might qualitatively alter how countries' leaders, voters, and other political actors see the climate. It could cease to be understood as a mix of natural and unintentionally anthropogenically altered and change to a vision of being controlled, mostly by others. If widespread, such a perception could catalyze the so-called securitization of climate, in which it is seen as an object of contestation in the international arena (Corry 2017, 305–7; McDonald 2018). Notably, perceptions' importance highlights the limitations of treating states as "black boxes," as Institutionalist international relations scholars often do. In reality, decisions are made by leaders, bureaucrats, voters, and other actors, all of whom are humans with the concomitant bounded rationality. Furthermore, perceptions are generated in complex and often unpredictable interactive social processes of ideas, identity, understandings of history, assumptions, and interpretations, further highlighting the limitations of the modeling of states as rational actors.

There is a scenario in which solar geoengineering is not used until climate severe change impacts manifest relatively fast, at least by climatic standards, possibly through positive climatic feedback. Some scholars and other observers have written of deployment in response to a "climate emergency." In this, decisions would likely be made in haste and under conditions of low, if not insufficient, knowledge. Furthermore, leaders could use such an emergency – legitimately or otherwise – to expand their domestic and international power in exceptional ways (Markusson et al. 2014). Some writers have recently argued that effective governance can anticipate and avoid such situations (Horton 2015; Lederer and Kreuter 2018).

Independent of how decision-making regarding deployment had occurred, if solar geoengineering were to be used to substantially reduce climate change and then stop suddenly and for an extended time, then the climate would change suddenly. This "termination" would have strong negative impacts and is a frequent concern regarding solar geoengineering. Indeed, some scholars have questioned whether the stratospheric aerosol injection could ever be consistent with international law for this reason (Reichwein et al. 2015, 169–70). However, for termination to occur and to pose substantial risks, solar geoengineering would need to be implemented at a high intensity, stopped suddenly, and not resumed for several years. If cessation would pose such great risks, then states could and most likely would take steps to ensure solar geoengineering's continuity. For example, redundant systems could resume implementation responsibilities in case of failure, accident, or attack. Andy Parker and Pete Irvine write that "In a world where multiple parties were capable of deployment, SRM [solar geoengineering] could not be terminated unilaterally" (Parker and Irvine 2018; see also Rabitz 2019). The possibility remains that a global catastrophe of some sort could make solar geoengineering's continuation impossible (Baum, Maher, and Haqq-Misra 2012). However, "in such a case the global community would likely have more serious problems to address than the unmanaged termination of geoengineering" (Barrett 2014, 255).

More generally, the development of international governance of solar geoengineering might affect the larger course of global governance. From one side, some scholars have expressed concern that implementation would undermine democratic or legitimate governance. For example, sociologist Bronislaw Szerszynski and colleagues argue that solar geoengineering will "necessitate autocratic governance," perhaps at the global scale:

> the social constitution of [large-scale] SRM geoengineering through stratospheric aerosol injection would be strongly compatible with a centralised, autocratic, command-and-control world governing structure, in tension with the current, broadly Westphalian, international system based on national self-determination. (Szerszynski et al. 2013; see also Hulme 2014; Owen 2014; Stirling 2014, 21; Eckersley 2017, 6)

A group of scholars, including myself, counter that solar geoengineering would not necessarily require authoritarianism (Horton et al. 2018). In fact, its governance might, like deepening economic interdependence and cultural exchange, foster and deepen complex interdependence among states. Legal scholar Gareth Davies further writes that

> There is at least a good argument that the lack of a global environmental governance system is a rather frightening absence and that, from a strategic environmental point of view, any action which can help generate such institutions is to be welcomed, at least on these grounds. From this point of view, the fact that SRM entails a long-term commitment, and that all but the most suicidal state will recognize that once begun they cannot simply walk away, is an added strength. A decision to start SRM forces states to commit to collective long-term responsibility for sustaining the global environment. There may be other downsides, but that in itself looks rather like a plus. (Davies 2010, 279)

The future of global governance, including of solar geoengineering is indeed unclear.

4.4 MODELS OF RATIONAL STATE BEHAVIOR

Sections 4.1 to 4.3 reveal the uncertainty of the international relations of solar geoengineering. One way to understand, explain, and predict states' behavior is to explicitly model them as rational actors. In this, they are usually abstractly represented as unitary and having diverse exogenous preferences, constraints, and options. Their behavior is then methodically analyzed in a manner akin to that of traditional welfare-maximizing economic actors. This often relies upon game theory, in which each actor's welfare (or other outcome) depends on both its own decisions and those of others. This approach has a substantial history in understanding international relations generally and cooperation in environmental issues specifically. By the 1980s, this method supported Institutionalists' arguments that, contra the Realists, cooperation is indeed possible and that, contra the Liberals, it could be driven by rational self-interest instead of norms with their own compliance pull (Keohane 1984). Scholars soon applied the theories to multilateral environmental agreements (Maler 1989; Barrett 1994). This section presents a brief review of the literature regarding such modeling in the context of solar geoengineering.[10] Although this work can be useful, given the approach's limitations, these results should be interpreted with caution.

The first category of models is those that consider the world as a single actor. In these, the actor usually optimizes its climate policy under various conditions to maximize its welfare. In one study, the addition of solar geoengineering reduces the optimal level of greenhouse gas emissions abatement by eight percent, reduces average global warming from four to three degrees Celsius, and improves social welfare by four percent of global economic activity (Heutel, Moreno-Cruz, and Shayegh 2018). The model's sensitivity analysis confirms that solar geoengineering would have additional value due to its speed, in that it can act as a sort of insurance policy against possible future insufficient emissions abatement and high climate change impacts.

Some single-actor models use multiple stages to consider uncertainty and learning. These often show that solar geoengineering would be a partial substitute for emissions abatement and that knowledge regarding it would have substantial value as a sort of insurance (for example, Moreno-Cruz and Keith 2013; Heyen 2016). In one of these, Martin Quaas and colleagues find conditions under which the decision-maker at the first of three stages would prefer to not learn about solar geoengineering's effectiveness (Quaas et al. 2017). This could occur if discounting is hyperbolic and the absolute prudence of expected climate damage is less than absolute risk aversion. In another, Timo Goeschl and colleagues model generations in a sort of intergenerational strategic game (Goeschl, Heyen, and Moreno-Cruz 2013). For example, if the current generation were to expect future ones to more strongly favor solar geoengineering, then it would undertake greater emissions abatement, even while investing in solar geoengineering research and development.

A second category of models investigates how multiple states would interact. A few of these describe two-state models that have similar implications.[11] In these, if the two states confront the same costs, benefits, and options – that is, they are

---

[10] For longer reviews, see Klepper and Rickels (2014); Harding and Moreno-Cruz (2016); Heutel, Moreno-Cruz, and Ricke (2016). Models in the first two categories described in this section relate to Chapter 3's extended thought exercise.

[11] See also Manoussi and Xepapadeas (2015) and Qu and Delfino Silva (2015) for elaborations on the two-state models.

symmetrical – then the introduction of solar geoengineering would cause them to reduce emissions abatement due to the partial substitutability of the two response options. However, if one of the states would experience substantially greater negative effects from solar geoengineering than the other one – that is, they are asymmetrical – then it would rationally increase abatement in an attempt to prevent deployment by the other state. Adam Millard-Ball notes that a country that is greatly exposed to climate risks could thus threaten to implement solar geoengineering as a means to increase emissions abatement elsewhere (Millard-Ball 2012; see also Urpelainen 2012). Juan Moreno-Cruz expands his model to consider multiple countries and concludes that the resulting strategic dynamic could strengthen developing countries' bargaining position in international climate negotiations (Moreno-Cruz 2015).

The third and final category of models considers the impact of solar geoengineering's international governance. One of these is a calibrated model of international clubs for decision-making, developed by Katherine Ricke and colleagues (Ricke, Moreno-Cruz, and Caldeira 2013). They divide the world into twenty-two regions that each prefer preindustrial climate and value reducing anomalies in temperature and in precipitation equally. In their model, an international coalition determines solar geoengineering's intensity of implementation based upon an average of the coalition members' preferences. The researchers compare an open club, in which all regions may join, with an exclusive one, which can exclude additional ones. The latter consists of a group of regions whose preference regarding solar geoengineering are roughly similar, because a more diverse club would cause its decisions to differ more from its members' preferences. Ricke and her colleagues find that coalition members would experience greater reductions of their climate change damages than nonmembers, although the difference is relatively small. Likewise, the differences in outcomes between open and exclusive clubs are minor.

Another model for international governance is one in which Martin Weitzman supposes that a group of countries have diverse preferences for the intensity of solar geoengineering. In the absence of restraint, the country with the preference for the greatest amount would implement it to that amount, a condition that Weitzman calls a "free driver." This would cause states' social welfare to be suboptimal. Based upon this, he suggests an international body that would require a supermajority for decision-making, with states' votes weighted by their population. Weitzman admits that this requires "heroic abstraction" and is naïve, but that he wishes to offer "a theory-based point of departure for further discussion" regarding international governance of solar geoengineering (Weitzman 2015, 1060).

To a degree, some of the results from these economic models are intuitive and predictable. Rational states take actions that reduce their risks and uncertainty and increase their welfares. As those conditions change, so do the countries' decisions. At the same time, these can provide surprising insights, such as possibly greater abatement in response to the threat of solar geoengineering as well as a foundation for contemplating the governance of solar geoengineering.

4.5 SUMMARY AND CONCLUSION

Because of its transboundary if not likely global effects, solar geoengineering and its governance are matters of international relations. The divergent problem structures among the responses to climate change help explain action in this domain thus far, as well as what we should expect in the future. While emissions abatement and NETs are aggregate effort global public goods and will be undersupplied, solar geoengineering would be both a single best effort and a mutual restrained global public good, implying that, if anything, it might be oversupplied. Its problem structure also indicates that uni- or minilateral deployment could be a problem, such as by being done too soon or contrary to the international community's consensus. Although uni- or minilateral implementation may not be as likely or unwanted as some writers have claimed, solar geoengineering could pose other challenges to international relations such as legitimate decision-making, potential disagreements, cost-sharing, and impacts on security and international tensions. The sudden and sustained termination of solar geoengineering would have strong negative environmental impacts, but its probability is uncertain and may be low. Some economists and others have explicitly modeled states as rational actors to understand, explain, and predict their behavior. Although this work can be useful, given the approach's limitations, these results should not be interpreted as conclusive.

5

International Law: Legal Norms, Principles, Custom, and Organizations

Like anthropogenic climate change itself, solar geoengineering, as it is presently understood, would alter the entire world's climate. Scholars quickly recognized that these unencapsulated phenomena are matters of international law, which includes a robust set of legal rules, procedures, and institutions to govern states' rights and obligations with respect to, among other things, affecting each other's environment (Zaelke and Cameron 1989; Bodansky 1996). Chapters 5 to 8 describe how existing international law would and could contribute to the governance of solar geoengineering.[1] Although there are presently no international instruments in force that are legally binding and specific to solar geoengineering, and despite occasional claims to the contrary, international law is applicable and adaptable to the topics, with varying degrees of clarity. Indeed, existing international law already provides substantial governance for solar geoengineering and a foundation upon which future governance can be built (for example, see Zelli, Möller, and Asselt 2017, 686).

These chapters' scope is mostly limited to the large-scale outdoor research or implementation of solar geoengineering that would pose a significant risk of environmental harm that would be transboundary, that is, occurring in another state or area beyond national jurisdiction. At the same time, other tenets of international law are pertinent even in the absence of transboundary risk, such as the prevention of international conflicts; the creation of intergovernmental organizations for deliberation; cooperation in research, knowledge sharing, and technology transfer; expressions of states' priorities; and human rights.

Chapters 5 to 8 are largely an exercise in the interpretation of international agreements, which is variously textual, historical, and teleological in approach. This is not simple. For one thing, the interpretation and application of international law is not clear-cut but instead occurs in a political context of state and nonstate actors that have diverse interests and levels of power (see Chapter 4). These actors have interpreted, applied, and enforced international law inconsistently, and will

[1] Chapters 5, 6, and 8 are largely derived from Reynolds (2018c). See also Armeni and Redgwell (2015b) and Du (2017).

continue to do so. Second, we should be cautious about drawing firm conclusions from extant international law that was not developed with solar geoengineering in mind. Moreover, solar geoengineering would pose both potential benefits and risks, and is being considered and researched in response to anthropogenic climate change. Both solar geoengineering and the climate change (or global warming, or elevated atmospheric greenhouse gas concentrations) that it could reduce, satisfy the definition of pollution or similar terms that some international environmental legal instruments aim to minimize. This definitional scope presents a challenging tension between climate change and solar geoengineering. Because of this, in some circumstances, the net expected effects of climate change and solar geoengineering should be considered. Which factors are considered, the relative weights given to these factors, and the applicable thresholds of harm's magnitude and probability will depend on, among other things, the legal environment as well as on actual and anticipated impacts. Given the high levels of uncertainty and the limitations of early research, any such consideration should be a dynamic process that can respond to new information.

This chapter offers a foundation for understanding how international environmental law could apply to solar geoengineering. It begins by introducing how international law operates. Section 5.2 discusses several general international legal principles that would guide the interpretation and development of international law. Section 5.3 describes one source of binding international law – that of states' customary behavior – and what it might mean for solar geoengineering. The chapter closes with a review of some relevant nonbinding multilateral environmental agreements and activities of intergovernmental organizations.

5.1 INTERNATIONAL LAW

Under international law, countries – often called states or nations – are sovereign in that their leadership has supreme authority to manage their internal affairs and to carry out relations with other sovereign countries as they see fit, free from unwanted external interference. Legally, all states are equally sovereign, and there are no international institutions that can exert binding authority over sovereign states without their consent.[2] Sovereignty includes the right for countries to exploit their own natural resources as they deem appropriate, except when this poses transboundary risks. Notably, some areas – such as the high seas, Antarctica, and outer space – are not within sovereign countries' legal control, or jurisdiction.[3]

When states interact, they face many of the same problems as individuals in society do. Countries and individuals benefit by having clear expectations regarding others' behavior, and consequently make explicit and implicit mutual promises.

---

[2] States sometimes consent to external authorities, such as the European Union and the UN Security Council. They can withdraw such consent, albeit with consequences.
[3] There are jurisdictional claims in Antarctica, but these are not widely recognized.

Through this, states try to fill internationally some of the same governance functions that national law does domestically: prevent and resolve conflicts; provide beneficial public goods; reduce negative externalities and promote positive ones; offer forums for discussion, coordination, and information sharing; develop normative and evaluative principles; encourage behavior consistent with those principles; promulgate procedural rules of decision-making and substantive rules of conduct; and establish institutions and procedures to make collective decisions, enforce agreements, and resolve disputes. Although these international promises have coalesced into something that resembles law, the important difference is that people and other legal persons are domestically subject to binding national law that is enforced with the (usually implicit) threat of socially sanctioned appropriation of property, freedom, and sometimes even life, whereas sovereign countries are not. This lack of centralized or hierarchical enforcement means that the subjects of international law – that is, the states – both create and enforce it.

International law is typically described as having three primary sources. First, states make explicit agreements with one another in which they promise to do, to not do, to try, or to try not to do specific activities. These treaties, agreements, or conventions are considered legally binding (inasmuch as international law is binding) on those countries – often called parties – that ratify them.[4] Prior to ratification, states might sign a treaty, signaling their intention to ratify. In the meantime, such nonparty signatories may not act contrary to the agreement's core objective but are not bound by the agreement's provisions. Most treaties are between only two countries, but some count many more as parties. Parties generally may withdraw from treaties. Some treaties establish institutions, such as regular meetings of parties, secretariats, and scientific advisory bodies, which perform various international governance functions. Second, over the course of centuries of interaction, countries have developed customary behavior among themselves. Once such custom is widely practiced and there is evidence that this conduct arose from a sense of legal obligation, then it is considered binding. This customary international law is not codified, although some authoritative institutions suggest language. States may explicitly object to and be exempt from tenets of customary international law. Third, general principles guide the interpretation and further development of international law. These are not binding in themselves but can be operationalized in a treaty or in custom. Beyond these three primary sources, other sources such as nonbinding agreements among countries, rulings of international tribunals and domestic courts, statements of intergovernmental organizations, and scholarly writing contribute to the international law.

Despite the absence of a centralized or hierarchical power, international law can be and is enforced (Guzman 2008). First, states can sometimes enforce international law through reciprocation. In this, violators are punished by others' equivalent violations,

---

[4] In some cases, international organizations, such as the European Union, may also be a party to a treaty.

which are often internationally sanctioned. Second, states can punish violators through retaliation, either within the issue area at hand or, more often, in an unrelated one. However, retaliation – such as economic sanctions or military action – is generally costly to the punisher. Thus, enforcement by retaliation is itself a type of global public good and will consequently be undersupplied due to collective action problems. Third, violators will suffer reputational damage and will find it more difficult to reap the benefits of international cooperation in the future. Notably, states can experience reputational damage also for acting contrarily to nonbinding international law or even to nonlegal expectations, blurring the definition of binding. Finally, ex post renegotiation can be an additional mechanism to reduce noncompliance (Posner and Sykes 2013, 32). Here, other countries offer to directly or (more often) indirectly pay the violator to cease the breach, sometimes under the guise of assistance with compliance. Because these four enforcement mechanisms are of limited effectiveness, international law emphasizes preventing violations and conflicts in the first place.

The boundaries of international law with other forms of international governance are unclear. In recent decades there has been growing recognition of diverse arrangements by authoritative state, substate, intergovernmental, and nonstate institutions that seek to intentionally and explicitly influence various actors' actions. Such "transnational law" or "global governance" has advantages over narrower international law in some circumstances, including when the governed actors operate in transboundary manners, when the conditions are highly dynamic, and when political leaders have insufficient incentives or opportunities to adopt and enforce relevant policy.

International law typically governs the actions of states, not of individual people, corporations, or other nonstate actors. Although there are a few exceptions, such as international criminal law, for the most part these nonstate entities are governed indirectly. That is, countries might ratify a treaty in which they promise to require, prohibit, encourage, or regulate certain behaviors from nonstate actors that are within their jurisdiction or otherwise under their control. The participating states then usually implement the agreement through domestic law and administrative policy. If they do so consistently with the agreement and other relevant international legal rules, then they typically remain in compliance, even if a nonstate actor within one of their jurisdiction or under their control acts contrary to the agreement.

Finally, solar geoengineering is firstly an environmental issue, and international environmental law is generally not oriented toward protecting ecosystems and Earth systems for their own sake. Instead, "almost all justifications for international environmental protection are predominantly and in some sense anthropocentric" (Birnie, Boyle, and Redgwell 2009, 7).

5.2 PRINCIPLES OF INTERNATIONAL LAW

General principles are described first because they structure discussions as well as provide guidance for the interpretation and the further development of international law. They are not directly binding, but instead can be operationalized in a treaty or custom. There is no definite inventory of general principles of international law, and scholars and lawmakers do not agree upon their precise identity and substance. Principles are evident in treaties' nonobligatory text; nonbinding international agreements; decisions of international organizations, treaty bodies, and tribunals; state behavior; and national law.

5.2.1 Cooperation

One of the cornerstones of international law, cooperation in good faith among, states is necessary for good neighborliness. It was embodied in the Charter of the UN (Articles 1.3, 2.2, 33, 55, 74) and has since been explicitly stated or relied upon in most multilateral environmental agreements and in customary international law. Cooperation does not require agreement, only that states act in good faith, with due diligence, and with consideration for other states' interests and well-being.

Because of its widespread application, the principle of cooperation is already legally binding in most contexts applicable to solar geoengineering, including the UN Framework Convention on Climate Change (UNFCCC) and the customary rule of preventing transboundary harm. Any future new legal instrument that would be applicable to solar geoengineering or the interpretation of an existing one should, at the very least, call upon states to cooperate in research and to share the results thereof. Moreover, states engaging in solar geoengineering activities, especially those that would pose transboundary risks, should notify, share information with, and consult with others.

5.2.2 Equity

Equity is the consideration of a fair and reasonable expected outcome. It is particularly relevant when a potentially affected party is somehow disenfranchised or unable to effectively consult and negotiate with the decision-maker. The impacts of both climate change and solar geoengineering on populations that lack a voice in governance will require attention. Equity can be divided into that within and that between generations.

Intragenerational equity can guide the division of rights and obligations among states. For example, the implementation of solar geoengineering could reduce climate change risks for many states yet might change precipitation in others, possibly having unwanted impacts on their agriculture. Intragenerational equity is reflected in the related principles of cooperation and common but differentiated responsibilities and in the customary rule of the prevention of transboundary harm. However, intragenerational equity can take a global perspective and consider indirect effects, while the prevention of transboundary harm focuses primarily on bilateral relations and on direct impacts. Equity is especially important in the management of shared resources as seen, for example, in the first Principle of the UNFCCC: "The Parties should protect the climate system for the benefit of present and future generations of humankind, on the basis of equity and in accordance with their common but differentiated responsibilities and respective capabilities" (Article 3.1; see also Paris Agreement, Articles 2.2, 4.1, 14.1).

The second form of equity – intergenerational equity – concerns future populations, who inherently lack self-representation in the present. It is central to climate policy due to the delay between greenhouse gas emissions and their consequent climatic impacts, the long length of time before elevated atmospheric carbon dioxide concentrations will naturally lower, and the intergenerational opportunity costs of aggressive responses to climate change. Intergenerational equity implies that solar geoengineering implementation should be subject to a high degree of scrutiny, especially in the absence of widespread consensus in the international community. This and other implications are complicated by the fact that, under conditions of elevated atmospheric greenhouse gas concentrations, solar geoengineering would need to be maintained across generations. On the one hand, transferring such a burden could be contrary to the principle (Burns 2011). At the same time, all other things being equal and assuming that solar geoengineering would function as presently indicated by models, it also appears contrary to intergenerational equity to not research and develop a potential additional means to reduce climate risks that will be borne almost entirely by future generations.

5.2.3 The Environment as a Common Concern of Humankind

Some multilateral agreements explicitly designate certain aspects of the environment – including the conservation of biological diversity (Convention on Biological Diversity, Preamble recital 3) and climate change (UNFCCC, Preamble recital 1; Paris Agreement, Preamble recital 11) – as the common concerns of humankind. International environmental law also implicitly treats other components such as stratospheric ozone, Antarctica, and the marine environment, as well as the global environment in general, as common concerns. This designation as a sort of global public good does not have clear legal implications, but has generally resulted in all states having individual and collective interests in and obligations for the maintenance of the common concern, independent of any direct harm to other parties. Legally speaking, states' obligations toward common concerns might be *erga omnes*, that is, owed to the international community as a whole.

Solar geoengineering would affect common concerns of humankind, given "that change in the Earth's climate and its adverse effects are a common concern of humankind" per the universally ratified UNFCCC. Thus, uni- or minilateral implementation in the absence of notification and consultation with other states, whether or not they would face a risk of significant transboundary harm, would be contrary to this principle of international environmental law. Furthermore, the

5.2 Principles of International Law 77

probable impacts of solar geoengineering on the recognized common concerns of humanity should guide the future development of any international law specific to it.

5.2.4 Precaution

Precaution, expressed as a principle or an approach, has been an increasingly common feature of international environmental law. It is a legal tool to manage risk and uncertainty that is sometimes cited when confronting issues of emerging technologies. Its formulations vary slightly in the Rio Declaration on Environment and Development (Principle 15), the Convention on Biological Diversity (CBD) (Preamble recital 9), the UNFCCC (Article 3.3), the Oslo Protocol on Further Reductions of Sulphur Emissions to the Convention on Long-Range Transboundary Air Pollution (CLRTAP) (Preamble recitals 3–4), the London Protocol to the Convention on the Prevention of Marine Pollution by Dumping of Wastes and Other Matter (Article 3.1), and the Kiev Protocol on Pollutant Release and Transfer Registers to the Aarhus Convention on Access to Information, Public Participation in Decision-Making and Access to Justice in Environmental Matters (Article 3.4). That in the UNFCCC is typical and most relevant here:

> The Parties should take precautionary measures to anticipate, prevent or minimize the causes of climate change and mitigate its adverse effects. Where there are threats of serious or irreversible damage, lack of full scientific certainty should not be used as a reason for postponing such measures, taking into account that policies and measures to deal with climate change should be cost-effective so as to ensure global benefits at the lowest possible cost. To achieve this, such policies and measures should take into account different socio-economic contexts, be comprehensive, cover all relevant sources, sinks and reservoirs of greenhouse gases and adaptation, and comprise all economic sectors. (Article 3.3)

Scholars have debated precaution's meaning and utility. In general, some assert that it is incoherent and should be applied only narrowly (Sunstein 2005). Solar geoengineering is an especially difficult issue, given that it might, and climate change does, pose "threats of serious or irreversible damage" while demonstrating some "lack of full scientific certainty." As with the other instances where the risks and benefits of these two climatic phenomena seem to be in tension, application of this principle will depend on the specific technique, circumstances, and evidence at hand. Floor Fleurke and I assert that precaution, as embodied in the UNFCCC, calls for at least the consideration, such as through research, of all means to reduce climate change risks, including that of solar geoengineering (Reynolds and Fleurke 2013; see also House of Commons (UK), Science and Technology Committee, The Regulation of Geoengineering, 34). This argument relies upon the UNFCCC's calls for precautionary measures to mitigate climate change's adverse effects, for measures and policies to be cost effective, and for comprehensive consideration in developing response measures. Furthermore, "precautionary measures" is neither defined nor restricted, and the term could include solar geoengineering.

5.2.5 Polluter Pays

A fourth relevant principle of international environmental law is that the polluter, not the victim, should pay for environmental damage. Despite its normative appeal of requiring the party who caused the harm to bear the burden, its operationalization in international environmental law has been inconsistent, been codified mainly in regional treaties and nonbinding agreements, and utilized highly qualified language. Among the multilateral agreements considered there, the principle is in the Rio Declaration (Principle 16) and the London Protocol (Article 3.2), both of which call upon states to require polluters to bear the costs within domestic regulatory frameworks. The principle is noticeably absent in the UNFCCC, despite this convention being drafted at the same time as the Rio Declaration.

To the extent that this principle applies to solar geoengineering, it could do so through two modalities. The primary cause of climate change is greenhouse gases – a type of pollution – and solar geoengineering offers a means to reduce their effects. If so, then those actors that have contributed most to elevated greenhouse gas concentrations should bear the costs of the means to reduce the risks from climate change, including any solar geoengineering research, development, and implementation (Buck 2012, 263). The secondary source of "pollution" could be solar geoengineering itself, which could cause deleterious effects. It is less clear who should under the "polluter pays principle" pay for any resulting damage, such as through environmental restoration and compensation for victims of harm. The actors that implemented or undertook the activities are the most proximal source, yet the greenhouse gas emitters are the ultimate source of most climate change and the presumptive reason for sole geoengineering. This second modality is related to liability or compensation for harm from solar geoengineering (see Chapter 10).

5.2.6 Common but Differentiated Responsibilities

In contrast to most international legal obligations, states are not to share some environmental responsibilities equally. Instead, restoring stratospheric ozone (Montreal Protocol, Article 5), preventing climate change (UNFCCC, Preamble recital 6, Articles 3.1, 4.1; Kyoto Protocol, Article 10; Paris Agreement, Preamble recital 3, Articles 2.2, 4.3, 4.19), the conservation of biological diversity (CBD, Articles 3.1, 4.1), the preservation of the marine environment (UN Convention on the Law of the Sea (UNCLOS) Article 194(1)), and environmental protection in general (Rio Declaration, Principle 7) are explicitly or implicitly treated as common but differentiated responsibilities. This recognizes that, despite the sovereign equality of states under international law, their environmental responsibilities need not be uniform. Although all parties have some responsibilities toward the goals of the specific agreement at hand, those with greater historical contribution to the problem bear disproportionate responsibility to satisfy the commitments therein.

The climate change regime uses a slightly different approach. In the UNFCCC, the principle is given as "common but differentiated responsibilities and respective capabilities," indicating that a country's capabilities – roughly equivalent to its level of economic development – also shapes its obligations. The Paris Agreement further adds "in the light of different national circumstances" (Article 2.2), emphasizing adapting commitments to states' changing conditions. The principle implies that wealthier countries and those with greater historical greenhouse gas emissions should carry most of the burdens of researching, developing, and (if appropriate) implementing solar geoengineering (Virgoe 2009, 111–12).

5.3 NONBINDING MULTILATERAL ENVIRONMENTAL AGREEMENTS

Although the phrase "international law" generally refers to those instruments, either explicit agreements or custom, which are intended to be legally binding, nonbinding multilateral agreements and intergovernmental organizations' decisions also create expectations regarding states' behavior. These can influence the behavior of countries that wish to avoid reputational costs that would result from acting contrary to the international community's wishes. Furthermore, nonbinding agreements and the decisions of intergovernmental organizations contribute toward the further development of international law in multiple ways: first, by providing a sense of the preferences, aspirations, and interests of states and of the international community as a whole; second, by gradually crafting and clarifying norms; and third, by establishing initial terms for future negotiations toward binding agreements. Some nonbinding agreements and intergovernmental organizations' decisions can trigger obligations under domestic laws, such as reporting requirements. International organizations can also play essential coordinating roles. Their authority derives both from their international character and from the expertise – often scientific or technical – of their decision-makers and administrators.

A handful of nonbinding multilateral legal instruments and intergovernmental organizations' decisions and other outputs might have a bearing on the future governance of solar geoengineering. The former is discussed here, and the latter in Section 5.4. Attention is particularly given here to those nonbinding agreements that have been globally endorsed or nearly so. Given that these, by their nature, commonly use vague and hortatory language, inferring what they might mean for solar geoengineering is unclear.

A series of UN-organized global summits on the environment and development have substantially shaped international environmental law. In fact, its contemporary era arguably began with the 1972 UN Conference on the Human Environment, which 113 states attended in Stockholm. The Declaration that arose from the event contains proclamations and principles that sound decidedly anthropocentric (and androcentric) to twenty-first century ears and that regularly refer to the benefits of improving nature. Therefore, to the extent that the Stockholm Declaration still reflects states' preferences regarding international environmental law, it generally tilts favorably toward interventions in the natural world – implicitly including solar geoengineering – to improve humanity's well-being. For example, its purpose is to "inspire and guide the peoples of the world in the preservation and enhancement of the human environment," and it proclaims that

> In our time, man's capability to transform his surroundings, if used wisely, can bring to all peoples the benefits of development and the opportunity to enhance the quality of life ... For the purpose of attaining freedom in the world of nature, man must use knowledge to build, in collaboration with nature, a better environment. (Preamble, Proclamations 3, 6)

The anthropocentric tone continues in the principles, the first of which includes the assertion that humanity "bears a solemn responsibility to protect and improve the environment for present and future generations" (Principle 1). Consistent with this, science and technology are to be "applied to the identification, avoidance and control of environmental risks and the solution of environmental problems and for the common good of mankind" (Principle 18; see also Principle 20). The Stockholm Declaration further calls for the minimization of transboundary harm and for international cooperation in protecting and improving the environment (Principles 21, 24).

Twenty years later, the UN hosted the Conference on Environment and Development in Rio de Janeiro. The resulting Rio Declaration, which the UN General Assembly later endorsed, reflects how the international community's priorities have changed since Stockholm. It gives greater, albeit qualified, emphasis to the protection of the natural world and it elevates the importance of poorer countries' economic development. These environmental and economic goals are often folded together under the rubric of sustainable development. At the same time, some anthropocentrism remains. For example, the first sentence of the first principle is "Human beings are at the centre of concerns for sustainable development." Furthermore, the Rio Declaration refers to only two rights: that of states' "sovereign right to exploit their own resources pursuant to their own environmental and developmental policies" (coupled with their duty to prevent transboundary harm) and a right to development, which remains undefined (Principles 2, 3). More specifically, the Rio Declaration calls for "improving scientific understanding" and for creating "new and innovative technologies" (Principle 9). In one of its principles, states are to avoid transferring sources of environmental harm to other states (Principle 14). Finally, the Rio Declaration reiterates several procedural duties, such as states' obligations to conduct environmental impact assessments, notify potentially affected states, and provide public access to information, as well as several principles of international law, including common but differentiated responsibilities, precaution, polluter pays, intergenerational equity, and cooperation (Principles 3, 7, 10, 15–17, 19, 27).

The meaning of the Rio Declaration for solar geoengineering is less clear than for the Stockholm Declaration. Solar geoengineering could help protect the natural world from climate change and enable greater economic development, and it could also cause significant harm to the environment and to vulnerable populations. Furthermore, to the extent that a right to development indeed exists in international law, then solar geoengineering might help countries fulfill it, given that climate change risks will be disproportionately borne by poor states and that greenhouse gas emissions and economic development remain coupled (Buck 2012). If solar geoengineering merely moved the location of climate risks, it might be contrary to its principle regarding the transfer of environmental harm to other states.

In 2015, the UN Member States agreed upon a set of seventeen Sustainable Development Goals for 2030. Combating climate change not only has its own goal but is also integrated into several others and is emphasized in the Preamble. Three months later, the Paris Agreement incorporated much of the Sustainable Development Goals' climate-specific language. The Goals' accompanying Declaration does "note with grave concern the significant gap between the aggregate effect of parties' mitigation pledges ... and aggregate emission pathways consistent with having a likely chance of holding the increase in global average temperature below 2 degrees Celsius, or 1.5 degrees Celsius above pre-industrial levels" (Transforming Our World, Declaration paragraph 31). Furthermore, human and institutional capacity is to be improved on, among other things, reducing the impacts of climate change (Transforming Our World, target 13.3).

Few implications specific to solar geoengineering can be gleaned from the Sustainable Development Goals. A report on geoengineering and the Sustainable Development Goals can conclude only vaguely:

> Delivery of at least three-quarters of all SDGs [Sustainable Development Goals] ... is expected to be affected in some way if Solar Geoengineering ... were deployed. These implications could be positive or negative in how they help attenuate climate change impacts or result in unwanted physical, socioeconomic or political outcomes. (Honegger et al. 2018, 6)

5.4 INTERGOVERNMENTAL ORGANIZATIONS

5.4.1 UN Environment

Since its launch soon after the Stockholm Conference, UN Environment (until 2016 known as the UN Environment Programme, or UNEP) has played an important role in coordinating other international bodies and national governments and in laying the foundation for the development of international environmental law. In 1980 it approved, with the World Meteorological Organization (WMO), Provisions for Cooperation between States in Weather Modification. Although the original implicit conception was of activities with regional impacts, the term "weather modification" is defined, in a footnote, in a manner that would include solar geoengineering. The document generally supports states' use of weather modification when it is for "the benefit of mankind and the environment" (paragraph 1(a)). To this end, "States should encourage and facilitate international co-operation in weather modification activities, including research" (paragraph 1(h)). As its name implies, the document calls on states to cooperate through, for example, information exchange, notification, and consultation (paragraph 1(b)). States should collect relevant information regarding their weather modification activities and share it with the WMO (paragraphs 1(c)–(d)). Finally, they are further encouraged in the document to undertake prior environmental assessment of their weather modification activities that might have transboundary impacts and to conduct those activities in a manner that prevents environmental damage in other countries or in areas beyond national jurisdiction (paragraphs 1(e)–(f)). Note that only the UNEP Governing Council, which has fifty-eight states as members, approved the Provisions.

In 2019, Switzerland, supported by ten diverse countries, introduced a draft resolution on geoengineering and its governance to UN Environment's governing body. It was modest, substantively only creating an expert committee to assess the methods and existing governance. Yet divisions emerged regarding whether scientific assessment should be left to the Intergovernmental Panel on Climate Change (IPCC). Furthermore, other states pushed changes in the resolution's preamble, adding the precautionary principle and further emphasizing a 2010 decision by the CBD's parties (see Chapter 8), each of which was unacceptable to the United States. Brazil and Saudi Arabia also opposed. Recognizing the stalemate, Switzerland withdrew its resolution (Chemnick 2019).

5.4.2 UN Education, Social and Cultural Organization

The UN Educational, Scientific and Cultural Organization (UNESCO) is a specialized UN agency with the mandate to promote international collaboration in education, science, and culture. In 2010, it hosted an expert meeting on geoengineering. The twenty participants recommended an international research program modeled on the UN-affiliated World Climate Research Programme. This would have been sponsored by UNESCO, its Intergovernmental Oceanographic Commission, the International Council for Science, and the WMO, and "could address the technological and scientific challenges of geoengineering and ensure that legitimate scientific research into this controversial issue may proceed" (United Nations Educational, Scientific and Cultural Organization 2010). A policy brief was later published, which bore the endorsements of UNEP and the Scientific

Committee on Problems of the Environment of the International Council for Science (Blackstock, Moore, and Siebert 2011).

5.4.3 World Meteorological Organization

The WMO is a UN agency that fosters international coordination and cooperation in diverse issues of weather and climate, including the atmosphere, oceans, and water resources. The WMO is referenced above in this section in the contexts of the 1980 UNEP Provisions for Co-operation between States in Weather Modification and the 2010 UNESCO expert meeting. Under the guidance of its Expert Team on Weather Modification, the WMO also worked toward a statement on geoengineering (World Meteorological Organization 2014). The draft "closely follows" the American Meteorological Society's statement on the same topic (Bruintjes 2015, 7).[5] The most recent plan within the WMO is for its Commission for Atmospheric Sciences to work toward an assessment of geoengineering, in cooperation with the World Climate Research Programme, the Intergovernmental Oceanographic Commission of UNESCO, the International Maritime Organization (IMO), and others (Bruintjes 2015, 14). Consistent with this, its current Operation Plan lists "Research to support informed decisions and policy advice on geoengineering" as an activity for 2014–18 (World Meteorological Organization 2016).

Furthermore, the WMO is responsible for the scientific assessment of the depletion and recovery of stratospheric ozone. Its 2018 report on this topic, developed in collaboration with UN Environment as well as US and European agencies, noted that possible stratospheric aerosol injection "would alter the stratospheric ozone layer" and "would delay the recovery of the Antarctic ozone hole" (World Meteorological Organization et al. 2018, 32).

5.4.4 Intergovernmental Panel on Climate Change

The UNEP and the WMO established the IPCC in 1988 to assess the state of climate change knowledge, the impacts of climate change, and possible response strategies. Its primary output has been the occasional publication of comprehensive assessment reports that collect and summarize the most recent scientific information, as well as irregular special reports. Beginning with the Third Assessment Report, published in 2001, these have devoted limited – but increasing – attention to solar geoengineering. In 2011 the IPCC convened an expert meeting on geoengineering and later published a meeting report (Edenhofer et al. 2012). The most recent Fifth

---

[5] The American Meteorological Society statement concludes, "The potential to help society cope with climate change and the risks of adverse consequences imply a need for adequate research, appropriate regulation, and transparent deliberation" (American Meteorological Society Council 2009; see also World Meteorological Organization 2014).

Assessment Report considers solar geoengineering methods in several contexts. It concludes, for example, that:

> Theory, model studies and observations suggest that some Solar Radiation Management (SRM) [solar geoengineering] methods, if practicable, could substantially offset a global temperature rise and partially offset some other impacts of global warming, but the compensation for the climate change caused by greenhouse gases would be imprecise (*high confidence*). (Boucher et al. 2013, 574; italics in original)

Although its 2018 special report on the ambitious 1.5 degrees Celsius target for global warming does not use scenarios of solar geoengineering, it dedicates a multipage box to the technologies and concludes that

> Uncertainties surrounding Solar Radiation Modification (SRM) measures constrain their potential deployment. These uncertainties include: technological immaturity; limited physical understanding about their effectiveness to limit global warming; and a weak capacity to govern, legitimise, and scale such measures ... [Stratospheric aerosol injection] is the most researched SRM method with *high agreement* that it could limit warming to below 1.5°C. (de Coninck et al. 2018, 316–7, 350; italics in original)

5.5 CUSTOMARY INTERNATIONAL LAW

Customary international law is a set of rules that are considered legally binding. They are derived from states' repeated behavior and evidence that they do so out of a sense of legal requirement. Because it is not centrally transcribed, its precise obligations are often unclear. Nevertheless, customary international law provides essential functions including preventing, resolving, and governing international disputes over environmental rights and obligations, especially in the absence of a specific multilateral agreement. In fact, two customary rules – those of the sovereign right to exploit natural resources and the obligation to prevent transboundary harm – provide the foundation of international environmental law.

Several customary rules would be salient for outdoor solar geoengineering activities: states' right to exploit natural resources, their duty to prevent transboundary harm, their responsibility for acts that are contrary to international law, and possible compensation for transboundary harm from acts that are consistent with international law (although this final one is not necessarily established as custom). A set of draft guidelines on the protection of the atmosphere, which captures some relevant customary international law, is summarized in Chapter 6, which focuses upon the international law of the atmosphere.

The International Law Commission (ILC) of the UN suggests language for some tenets of customary international law based upon state practice and the rulings of international tribunals. These will be used as guidance here, although their details do not necessarily reflect what all states' leaders, scholars, and others believe custom to be.

5.5.1 Sovereign Right to Exploit Natural Resources

If sovereignty is the foundation of international law and international relations, then states' sovereign right to exploit their own natural resources is one of two cornerstones of international environmental law. In principle, this right is the default condition among sovereign states, at least prior to the rise of international environmental law in the twentieth century (see Schrijver 2008). Developing countries that were asserting their independence in the wake of decolonization made the right explicit. It was initially recognized globally, beginning with the 1952 UN General Assembly Resolution on the Right to Exploit Freely Natural Wealth and Resources and its 1962 Declaration on Permanent Sovereignty over Natural Resources. Within international environmental law, many multilateral agreements, beginning with the 1972 Stockholm Declaration, provide for or restate it (Stockholm Declaration, Principle 21; London Convention, Preamble recital 3; CLRTAP, Preamble recital 5; UNCLOS, Article 194.2; Vienna Convention, Preamble recital 2; Rio Declaration, Preamble recital 2; UNFCCC, Preamble recital 8; CBD, Article 3). The implication is that states have a presumptive right to conduct solar geoengineering activities within their own territory, provided that they do so in a manner consistent with their other rights and obligations, particularly regarding transboundary risks and harm.

5.5.2 Prevention of Transboundary Environmental Harm

The second of the two cornerstones of international environmental law is that states have a duty to prevent transboundary environmental harm arising from activities within their jurisdiction or under their control. Although this obligation arose in the early twentieth century, parallel to the sovereign right to exploit natural resources, the two fused in the Stockholm Declaration (Principle 21) and have since generally been presented as two sides of the same coin. Many multilateral environmental agreements reiterate this duty. In 1996, the International Court of Justice (ICJ) stated that the obligation is part of customary international law (*Legality of the Threat or Use of Nuclear Weapons (Advisory Opinion)*),[6] citing a principle of the Rio Declaration, in which

---

[6] It reiterated and clarified this in its 2010 *Pulp Mills on the River Uruguay (Argentina v. Uruguay)* and 2015 *Certain Activities Carried out by Nicaragua in the Border Area and Construction of a Road in Costa Rica along the San Juan River (Costa Rica/Nicaragua)* decisions.

States have, in accordance with the Charter of the United Nations and the principles of international law, the sovereign right to exploit their own resources pursuant to their own environmental and developmental policies, and the responsibility to ensure that activities within their jurisdiction or control do not cause damage to the environment of other States or of areas beyond the limits of national jurisdiction. (Principle 2)

This duty is not an inviolable one to ensure that no transboundary environmental harm occurs. Instead, the objective is the prevention of transboundary harm, while the substance of states' obligation is to practice due diligence regarding activities, such as outdoor solar geoengineering activities, that occur within their jurisdiction or control and that pose a risk of significant transboundary harm. The due diligence standard is roughly proportional to the probability and the magnitude of the risk. In its Draft Articles on Prevention of Transboundary Harm from Hazardous Activities, the ILC concluded that the duty to prevent transboundary harm arises when an activity could pose a "risk of causing significant transboundary harm through their physical consequences ... [including] risks taking the form of a high probability of causing significant transboundary harm and a low probability of causing disastrous transboundary harm" (Articles 1, 2(a)). Each of these two formulations of risky activities could apply to large-scale outdoor solar geoengineering activities. Once this threshold has been met, states must undertake "all appropriate measures" to prevent the potential harm and to reduce its risk. These measures include requiring authorization for the activity, performing an environmental impact assessment, notifying and cooperating in good faith with potentially affected states, informing the public, and developing contingency plans for an emergency (Articles 3–4, 6–8, 13, 16). These actions' details are subject to consultations between the countries and are to be "based on an equitable balance of interests" (Article 9). Factors to consider in this balancing of interests include:

> the importance of the activity, taking into account its overall advantages of a social, economic and technical character for the State of origin in relation to the potential harm for the State likely to be affected ... [and] the economic viability of the activity in relation to the costs of prevention and to the possibility of carrying out the activity elsewhere or by other means or replacing it with an alternative activity. (Article 10(b), (e))

This balancing should thus consider both the benefits and risks of the proposed solar geoengineering or other activity for all concerned states. An activity that posed great risk to another state with little likely benefit for the state of origin would be interpreted differently than a modest, low risk one undertaken by a state whose exposure to substantial climate change risks could subsequently be lowered through the proposed solar geoengineering activity. The standards of due diligence are not static but instead vary in relation to activity's risk and change in response to scientific and technological developments.[7] If the consultations fail to produce a consensus among the states, then the state of origin is to consider the interests of potentially affected states in its decisions, such as whether to authorize the activity (Article 9.3).

Because large-scale outdoor solar geoengineering activities would pose risks of transboundary impacts, the state of jurisdiction or control would have the above-listed procedural duties. Beyond that, scholars have divergent opinions as to the potential of this customary rule to govern solar geoengineering effectively. Haomiao Du puts the obligation to assess and monitor environmental impacts at the center of her recommendations (Du 2017, 192–201). David Reichwein and coauthors argue that the risk of termination is great enough that due diligence may call for solar geoengineering not to be implemented in the first place (Reichwein et al. 2015, 169–70; see also Brent, McGee, and Maguire 2015). However, this conclusion requires sudden and sustained termination to have a high probability, which is not self-evident (Parker and Irvine 2018; Rabitz 2019). Furthermore, although the obligation to prevent transboundary harm requires states to comply with certain procedural duties and, in the absence of consensus, to take other states' interests into account, it does not prohibit activities that pose transboundary risks or even those that are known to cause transboundary harm. In fact, a passage was removed from a previous version of the ILC's draft articles that said that the state of origin shall refuse authorization of an activity that would cause unavoidable transboundary harm (Birnie et al. 2009, 180–2, especially footnote 440). Reichwein and coauthors, as well as legal scholar Daniel Bodansky, cite the difficulty in demonstrating specific risks with sufficient confidence as a barrier to the customary rule regarding transboundary harm to effectively govern solar geoengineering (Bodansky 1996, 312; Reichwein et al. 2015, 180). Elsewhere, Bodansky highlights the challenges and likely controversy in interpreting this tenet of customary international law, given that solar geoengineering itself is intended to prevent the transboundary harm of climate change (Bodansky 2011, 15).

A final, unexplored question is whether the duty to prevent transboundary harm could be interpreted as obligating states to research or even implement solar geoengineering to prevent their transboundary harm from climate change, if it is found to be effective and with acceptable adverse impacts. Because all states emit greenhouse gases, this customary rule obligates them to adopt, implement, and enforce policies that aim to abate their greenhouse gas emissions (Lefeber 2012, 333–40). Countries, especially those with greater emissions, might in principle bear an analogous duty to enact policies to consider reducing their transboundary harm of climate change through the research – and perhaps implementation – of solar geoengineering.

---

[7] So found the International Tribunal for the Law of the Sea in its 2011 *Responsibilities and Obligations of States Sponsoring Persons and Entities with Respect to Activities in the Area* Advisory Opinion.

5.5.3 Responsibility for Acts Contrary to International Law

Under customary international law, states are said to be "responsible" for internationally wrongful acts. This concept is the international counterpart to being found at fault in a civil lawsuit or guilty of a crime. The ILC has captured the customary international law of state responsibility in its Draft Articles on Responsibility of States for Internationally Wrongful Acts. A state is considered responsible when an act or an omission that can be attributed to it was contrary to an international legal rule to which the state was obligated. Attribution concerns whether the actor who undertook or authorized the action is sufficiently part of or under the direction of the state. Nonstate actors' actions are generally not attributable to the state. For example, when "the State acknowledges and adopts the conduct in question as its own," the action is attributable to the state, but not when it expresses its "mere support or endorsement" (Article 11, commentary paragraph 6). This implies that regulatory approval of solar geoengineering would not be sufficient grounds for state attribution of nonstate actors' actions.

In terms of wrongfulness, custom or a treaty to which the state is a party provide legal rules that the state could violate. Whether the state's intention is relevant depends on the specific legal rule at hand, although most address acts and impacts, not intention. In the context of solar geoengineering, a state could, for example, be responsible for undertaking outdoor stratospheric aerosol injection using a quantity of sulfur that exceeded its limits under the Oslo Protocol to CLRTAP or for failing to exercise due diligence in its regulation of a university's outdoor research activity that posed risks of significant transboundary harm. As an additional note, a state can be responsible for an internationally wrongful act in the absence of actual harm, and there can also be transboundary harm in the absence of an internationally wrongful act or omission.

If a state is responsible for a wrongful act, it should cease the activity, assure that the act will not recur, and make full reparations for any injuries (Articles 1–2, 30–1). Reparations can take the form of restitution (re-establishing the situation that existed before the wrongful act), compensation (providing something of value, usually money, to compensate for the harm), and satisfaction (usually a statement of acknowledgement and regret), in that order of priority (Articles 34–7). Restitution is often difficult for environmental damage. Thus, states are in principle liable for compensatory damages from transboundary harm, such as what could arise from large-scale outdoor solar geoengineering activities, that results from their attributable, internationally wrongful acts (see Chapter 12). Although under the Draft Articles reparations extend to injury, including material and moral, compensation is limited to "financially assessable damage" (Articles 31.2, 36.2). The Articles' commentary notes that

> Environmental damage will often extend beyond that which can be readily quantified in terms of clean-up costs or property devaluation. Damage to such

environmental values (biodiversity, amenity, etc. – sometimes referred to as "non-use values") is, as a matter of principle, no less real and compensable than damage to property, though it may be difficult to quantify. (Article 36, commentary paragraph 15)

Liability for transboundary harm of a purely environmental character remains unsettled. The ICJ recently awarded damages for environmental harm, but used a narrow methodology that included only environmental goods and services as well as restoration costs (*Certain Activities Carried out by Nicaragua in the Border Area (Costa Rica v. Nicaragua); Compensation Owed by the Republic of Nicaragua to the Republic of Costa Rica*). Regardless, in practice, states that have acted contrary to international law are often not deemed responsible and responsible states are usually not ordered to pay damages.

In the case of solar geoengineering, some writers have expressed concern that the doctrine of state responsibility would require the scientific, statistical attribution of specific environmental harms to particular solar geoengineering activities, which would be challenging (note the different use of "attribution" than above, here concerning the causal relationship between the harm and the activity) (Bodle 2010, 306–7; Reichwein et al. 2015). However, many (if not most) international environmental legal obligations are ex ante procedural duties that countries should carry out prior to a risky activity. As stated, a state that failed to do so is still responsible, independent of any ex post harm. However, demands for compensation do require that injuries be attributed to the state's wrongful acts.

Solar geoengineering should be considered in the context of the environmental harm from climate change. States might have committed wrongful acts with regard to greenhouse gas emissions, such as failing to abate pursuant to the UNFCCC, its Kyoto Protocol, its Paris Agreement (see Chapter 6), or the customary responsibility to prevent transboundary harm. In this context, where greenhouse gas emissions abatement failure is the wrongful act, solar geoengineering could have two roles pursuant to the customary law of responsibility, one for the responsible state and one for the injured state. Regarding the former, as described, responsible states must, among other things, make reparations to injured states. Solar geoengineering could contribute to fulfilling these obligations. Although restitution of environmental damage is typically difficult, if solar geoengineering could be implemented safely, then it might be able to help restore the climate, including by offsetting climatic effects that might injure affected states.

Regarding injured states, they have a right to undertake countermeasures in response to being harmed by others' acts that were contrary to international law (Articles 22, 49–54). Countermeasures are acts that would normally be contrary to international law and that are directed at the responsible state to induce compliance. Countermeasures may be taken by an injured state in retaliation for a typical breach or by any state in response to a breach of an obligation that is owed to the international community as a whole. Considering that the UNFCCC recognizes the prevention of climate change as a common concern of humanity (Preamble recital 1), the latter might be the case. In this scenario, a state might carry out or authorize solar geoengineering activities that would otherwise be contrary to international law as a countermeasure to other states' failure to abate emissions. In particular, solar geoengineering could be effective in this context due to its high leverage, speed of action, and the reversibility of its direct climatic effects. Countermeasures are subject to several limitations, among which is that "[a]n injured State may only take countermeasures against a State which is responsible" (Article 49). Yet solar geoengineering is presently understood to be global in its effects and would thus serve poorly as a directed countermeasure. On the other hand, a desperate state might argue that all countries have failed to abate their greenhouse gas emissions sufficiently and may be therefore subjected to countermeasures.

The ILC Draft Articles also describe under what circumstances acts contrary to international law should not be considered wrongful. Among these is necessity, under which a state's action is not wrongful if it is the only means for the state "to safeguard an essential interest against a grave and imminent peril" and if it does not put the essential interests of other states at risk (Article 25). The ICJ has ruled that grave and imminent environmental risks can constitute a state of necessity (*Gabčíkovo-Nagymaros Project (Hungary/Slovakia)*). For countries such as small island states that face existential risks from climate change, necessity could provide a legal excuse from wrongfulness for solar geoengineering activities that would be contrary to international law. However, an act done out of necessity may not put the essential interests of other states at risk, and a state may not invoke necessity if it has contributed to the situation that gave rise to the necessity (Article 25). All countries have emitted greenhouse gases, which are the cause of climate change and its risks and have thus each contributed to the state of necessity. At the same time, historical contributions to anthropogenic greenhouse gas emissions and actions to reduce them vary dramatically among countries. Those countries with very low historical emissions or aggressive emissions abatement policies might be able to successfully invoke the necessity defense for solar geoengineering activities that would breach international law (Markusson et al. 2014).

5.5.4 Civil Liability and Compensation for Transboundary Harm

States can experience transboundary harm, even in the absence of another state's internationally wrongful act or omission. The ILC assessed state practice and judicial decisions regarding compensation for such harm, a category that would include large-scale outdoor soar geoengineering activities. However, given the lack of consensus on this matter, these provisions were released as draft principles – not articles – and do not necessarily reflect customary international law. Nevertheless, the somewhat awkwardly named Draft Principles on the Allocation of Loss in the

Case of Transboundary Harm Arising out of Hazardous Activities includes in its scope of compensable harms "impairment of the environment" and any "reasonable response measures," that resulted from activities "which [involve] a risk of causing significant harm" (Principles 1–2). The country in whose jurisdiction or under whose control the hazardous activity was carried out is considered the state of origin.

Consistent with the customary international law of preventing transboundary harm, in the event of an activity involving a hazardous activity that is likely to cause transboundary damage, the potentially affected state must "take all feasible measures" to minimize the damage (Principle 5(d)). The resulting obligations arising from transboundary damage are founded upon access to justice and civil liability. The state of origin should ensure that its institutions for justice, such as courts and administrative bodies, are accessible to people who have been harmed in other countries (Principle 6.1). Furthermore, it "should take all necessary measures to ensure that prompt and adequate compensation is available for victims" in a nondiscriminatory manner (Principle 4.1). The ILC's Principles broadly define the eligible environmental harm to include loss of life; personal injury; property damage, including that to cultural heritage; pure environmental damage; and reasonable response and reinstatement measures by the victim (Principle 2(a)). Specifically, the state of origin should ensure that compensation is available. These arrangements could occur through a variety of mechanisms including explicit and preferably strict liability for harm on the part of the operator who carries out the hazardous activity, mandatory insurance, industry-wide and international compensation funds, as well as possibly vicarious state liability. Finally, states should make all efforts to establish compensation regimes for categories of hazardous activities. These regimes should rely, as appropriate, on industry-wide and/or state funds to supplement the resources and insurance of the operator.

6

International Law: The Climate and Atmosphere

Climate change is foremost an atmospheric phenomenon. It is caused by is elevated atmospheric concentrations of greenhouse gases, and the changes manifest most evidently in the atmosphere. Solar geoengineering would operate either directly on the atmosphere (by making it more reflective or by allowing more heat to escape) or indirectly (by making the surface below it more reflective or by reducing the amount of solar radiation that it receives). This chapter discusses how the international environmental law of the atmosphere would and could contribute to the governance of solar geoengineering.

International law does not clearly define the atmosphere and responsibility for its quality. Airspace – the volume above a state's territory extending upwards to the undefined border with outer space – is within its jurisdiction. However, the air that constitutes the atmosphere moves and mixes, including across national boundaries and through airspaces. Pollutants consequently travel, and their movements are often addressed through customary international law and various multilateral agreements. Some atmospheric pollutants, notably ozone-depleting substances and greenhouse gases, have global effects and imply global responses. Proposals for a comprehensive "Law of the Atmosphere" have surfaced occasionally, such as when the scientific understanding concerning climate change risks first emerged, but these have not encountered a warm reception (World Conference on the Changing Atmosphere 1988, paragraph 30; Meeting of Legal Policy Experts 1989; both in "Selected International Legal Materials" 1990). This chapter addresses the three leading multilateral atmospheric agreements, their protocols, and draft guidelines on the protection of the atmosphere that the International Law Commission (ILC) is presently developing.

6.1 UN FRAMEWORK CONVENTION ON CLIMATE CHANGE AND RELATED AGREEMENTS

The UN Framework Convention on Climate Change (UNFCCC) is the central international legal instrument for multilateral cooperation to limit climate change and its impacts. As a framework, it has limited commitments and sets out expectations for subsequent, more detailed protocols. All globally recognized countries are parties to the UNFCCC. Its objective is the

> stabilization of greenhouse gas concentrations in the atmosphere at a level that would prevent dangerous anthropogenic interference with the climate system. Such a level should be achieved within a time frame sufficient to allow ecosystems to adapt naturally to climate change, to ensure that food production is not threatened and to enable economic development to proceed in a sustainable manner. (Article 2)

Because the excess quantity of the leading anthropogenic greenhouse gas – carbon dioxide – is naturally withdrawn from the atmosphere at very slow rates, it is essentially a cumulative pollutant. Genuine stabilization of its concentration implies net zero emissions. This stabilized concentration should prevent "dangerous anthropogenic interference with the climate system," which is undefined in the UNFCCC itself.

The relationship between the UNFCCC's objective and solar geoengineering is unclear. The latter's deployment would not directly affect greenhouse gas concentrations, although it would have indirect effects, possibly significantly lowering them (Keith, Wagner, and Zabel 2017). Although solar geoengineering itself could be seen as "dangerous anthropogenic interference with the climate system," ex ante due to expected risks or ex post due to actual negative impacts, the UNFCCC's objective is not to prevent such interference per se but instead to stabilize greenhouse gas concentrations at a level that would do so. Because the methods appear to be rapidly effective at reducing climate change, they might be able to help address the speed criterion within the UNFCCC's objective. Moreover, solar geoengineering could be used to slow down or prevent most climate change impacts – including those on ecosystems, food production, and economic development, each of which is emphasized in the objective – while society transitions to net zero emissions. Notably, the UNFCCC does not prohibit or exclude any means to reduce climate risks.

The UNFCCC invokes several salient principles of international environmental law and tenets of customary international law (see Chapter 5). The former include the environment as a common concern of humankind, common but differentiated responsibilities, equity (both general and intergenerational), and precaution (Preamble recital 1, Articles 3.1, 3.3, 4). The latter are the sovereign right to exploit domestic natural resources and the duty to prevent transboundary harm (Preamble recital 8).

The Convention is anthropocentrically and economically oriented. The anthropocentricism is apparent both in the objective, in which two of the three criteria for the speed of greenhouse gas stabilization are for the sake of humanity, and in its first principle: "The Parties should protect the climate system for the benefit of present and future generations of humankind" (Articles 2, 3.1). That is, the UNFCCC's objective and commitments are to be pursued not with the goal of a planet or atmosphere that is less affected by human activities but instead with one that prioritizes humans' well-being primarily and ecosystems secondarily. The importance of economic activity, particularly in the developing countries, is clear throughout the UNFCCC. This is seen both in its obligatory commitments (for example, "Parties ... shall ... employ appropriate methods ... with a view to minimizing adverse effects on the economy, on public health and on the quality of the environment, of projects or measures undertaken by them to mitigate or adapt to climate change") and in its recommendatory ones (for example, "Parties should ... tak[e] into account that policies and measures to deal with climate change should be cost-effective so as to ensure global benefits at the lowest possible cost") (Articles 3.3, 4.1(f); see also Preamble recitals 10, 16, 21, 22, Articles 3.4, 3.5, 4.1(g) and (h), 4.2(a), 4.7, 4.10, 7.2(a)). Because some solar geoengineering methods appear to have low financial implementation costs, these passages could imply that solar geoengineering should be given some priority as part of a portfolio of responses to reduce climate risks.

Although the UNFCCC's commitments mainly focus on gathering and sharing information and on developing national climate plans, three themes found therein have implications for solar geoengineering. First, and most specifically, parties commit to minimizing a range of adverse effects – on the economy, public health, and the environment – in the activities that they undertake to reduce climate change risks (Article 4.1(f)). Therefore, they should practice due diligence, such as by carrying out prior impact assessments, in their solar geoengineering activities to minimize such adverse effects.

The second relevant set of parties' commitments concerns research and technology. Several of them call for parties to undertake research, to cooperate therein, and to share the results. For example,

> All Parties ... shall ... Promote and cooperate in scientific, technological, technical, socio-economic and other research, systematic observation and development of data archives related to the climate system and intended to further the understanding and to reduce or eliminate the remaining uncertainties regarding the causes, effects, magnitude and timing of climate change and the economic and social consequences of various response strategies. (Article 4.1(g); see also Articles 4.1(h), 5)

The term "various response strategies" is not defined but presumably could include solar geoengineering. Other commitments concern and support the creation and diffusion of technologies (Article 4.1(c); see also Articles 4.3, 4.5, 4.7, 4.8, 4.9).

Adapting to a changed climate is the final set of UNFCCC commitments that might be salient. Several commitments and one principle explicitly call for adaptation, through for example the development and implementation of national adaptation plans (Articles 3.3, 4.1(b), 4.1(e), (f), 4.4). Others, such as those concerning "various response strategies" and technology transfer, are implicitly adaptive. Adaptation is undefined in the UNFCCC, in its related agreements, and by the decisions of its Conferences of the Parties (COPs). The Convention's website and other documents often use a definition crafted by the Intergovernmental Panel on Climate Change (IPCC): "Adaptation refers to adjustments in ecological, social, or economic systems in response to actual or expected climatic stimuli and their effects or impacts. It refers to changes in processes, practices, and structures to moderate potential damages or to benefit from opportunities associated with climate change" (UNFCCC n.d.). Solar geoengineering could fall within this unofficial definition's latter sentence, in that it would be a change in practices to moderate potential climate damages (but see Heyward 2013).

The Kyoto Protocol of 1997 is a supplemental treaty to the UNFCCC that is in force, having been ratified by almost all UNFCCC parties.[1] It is dedicated to quantified greenhouse gas emissions abatement targets, especially for industrialized countries. It thus has little applicability to solar geoengineering, save for further commitments by industrialized country parties to implement policies to research, promote, develop, and transfer "environmentally sound technologies," and to cooperate in research to reduce uncertainties concerning "various response strategies" (Articles 2.1(a)(iv), 10(c), (d)), which could presumably include solar geoengineering (see also Articles 10(b)(i), 11.2(b)).

The second legally binding agreement under the UNFCCC is the 2015 Paris Agreement, which entered into force the following year with nearly global participation. Compared with the Kyoto Protocol, the Paris Agreement offers a fundamentally different approach to reducing climate change. It set a specific target for climate change: the parties' first aim is to keep global warming "well below 2°C," and they are "to pursue efforts to limit the temperature increase to 1.5°C" (Article 1.2(a)). This can be understood as implying the greenhouse gas concentration for which the UNFCCC's objective aims.[2] The heart of the Paris Agreement is its bottom-up emissions abatement mechanism, in which all parties are to submit their own nationally determined contributions toward the Agreement's abatement aim. In addition, the Agreement emphasizes adaptation to a greater degree than the UNFCCC and the Kyoto Protocol. In the second of its two aims, adaptation is now an explicit goal on par with abatement (Articles 1.2(b), 7). It is to be pursued in "a manner that does not threaten food production" and "with a view to contributing to sustainable development." The Paris Agreement also establishes a Technology

---

[1] The United States did not ratify, and Canada ratified and later withdrew.
[2] Such an interpretation requires knowing the climate sensitivity. If it is indeed three degrees Celsius, which is the present best estimate, then Parties should work to stabilize atmospheric greenhouse gas concentrations at about 450 parts per million of carbon dioxide equivalent to remain within two degrees of warming. The current emissions trajectory, while uncertain, implies that the carbon dioxide concentration will pass this threshold around the year 2040.

Mechanism to help "fully realiz[e] technology development and transfer in order to improve resilience to climate change and to reduce greenhouse gas emissions" (Article 10).

The Paris Agreement could contribute to the governance of solar geoengineering through each of these mechanisms. There is nothing to preclude states from including it as part of their nationally determined contribution to keeping warming to less than two degrees Celsius. If solar geoengineering were to be considered a component of adaptation, then parties to the Agreement could include it also as part of their obligatory adaptation plans, and it could constitute part of the global stocktaking. This is to occur every five years "to assess the collective progress towards achieving the purpose of this Agreement and its long-term goals," which implicitly include the 1.5 to 2 degrees targets (Articles 7.9, 7.11, 7.14, 14.1). If so, then parties' adaptive solar geoengineering activities would need to satisfy several desiderata set out by the article. They should be participatory and transparent, take into consideration vulnerable groups, be based upon the best available science, and be integrated with socioeconomic and environmental policies and actions (Article 7.5). Finally, solar geoengineering could be a technology that parties develop and transfer.

6.2 VIENNA CONVENTION FOR THE PROTECTION OF THE OZONE LAYER AND ITS MONTREAL PROTOCOL

Prior to the emergence of climate change, the destruction of stratospheric ozone by the emission of certain anthropogenic substances was the preeminent environmental problem of the global atmosphere. This depletion increases the amount of ultraviolet radiation at the Earth's surface, posing risks to humans, other species, and ecosystems. It relates to solar geoengineering because stratospheric aerosol injection could contribute to ozone depletion, or at least slow its current recovery (see Chapter 2).

The international community has responded effectively to stratospheric ozone destruction through both legal and technological means (Benedick 1998; Parson 2003). The Vienna Convention for the Protection of the Ozone Layer of 1985 is a framework convention, has limited commitments, and enjoys global participation. It is furthered by its Montreal Protocol of 1987, in which its 197 parties commit to phase out specific ozone-depleting substances. The agreements are supported by a standing secretariat and other dedicated institutions at UN Environment, regular COPs, a robust compliance mechanism, three Assessment Panels that provide scientific and technical input, and procedures for amending the agreements and updating other policies.

The Vienna Convention supports research. Its parties commit to

> [c]o-operate by means of systematic observations, research and information exchange in order to better understand and assess the effects of human activities

on the ozone layer [and] ... to initiate and co-operate in ... research and scientific assessment on: The physical and chemical processes that may affect the ozone layer; ... [c]limatic effects deriving from any modifications of the ozone layer; ... [and s]ubstances, practices, processes and activities that may affect the ozone layer, and their cumulative effects. (Articles 2.2(a), 3.1)

Solar geoengineering in general, and stratospheric sulfate aerosols specifically, are activities and/or substances that may affect the ozone layer. This text of the Vienna Convention implies that parties should research potential impacts on stratospheric ozone from solar geoengineering activities, which the World Meteorological Organization (WMO) is facilitating (World Meteorological Organization 2018). The commitment to conduct research, which the Montreal Protocol reiterates and expands, also includes a related duty to transfer technology (Articles 9, 10A).

The implications of the Vienna Convention and its Montreal Protocol for governing solar geoengineering implementation are uncertain and dependent upon the outcome of future research. If stratospheric aerosol injection (or some other form of solar geoengineering) were or were likely to cause "adverse effects resulting from modification or likely modification of the ozone layer," then the parties would be committed to adopt policies to control, limit, reduce, or prevent these activities (Article 2.2(b)).[3] The parties to the Montreal Protocol could then choose to add the ozone-depleting substance to the list of controlled substances (Article 2.10). Because they are to "take appropriate measures ... to protect human health and the environment against adverse effects resulting or likely to result from human activities which modify or are likely to modify the ozone layer," the parties' decision should balance the protective and adverse effects that occur through impacts on the ozone layer (Article 2.1). However, given that sulfate aerosols are already produced, albeit unintentionally, through industrial processes (especially coal combustion) in the lower atmosphere at a rate roughly ten times that which would be needed for stratospheric aerosol injection, the parties would in this case need to incorporate some qualifier regarding the location of prohibited emissions.[4]

6.3 CONVENTION ON LONG-RANGE TRANSBOUNDARY AIR POLLUTION AND ITS PROTOCOLS

The third multilateral atmospheric agreement that is relevant to solar geoengineering is the Convention on Long-Range Transboundary Air Pollution (CLRTAP) of 1979. Like the UNFCCC and the Vienna Convention, it is a framework convention

---

[3] Note that "significant" implies a relatively moderate threshold for the magnitude of harm, greater than "detectable" but not necessarily "serious" or "substantial."

[4] Annual global lower atmospheric pollution is roughly 100 megatons sulfur dioxide (50 megatons sulfur) and declining (Klimont, Smith, and Cofala 2013). Offsetting the climate change from doubling the preindustrial atmospheric carbon dioxide concentration would require roughly 3 to 10 megatons sulfur dioxide (1.5 to 5 megatons sulfur) (Vaughan and Lenton 2011, 764).

with operationalizing protocols. These CLRTAP agreements aim to reduce air pollution, particularly precursors to acid rain. Some forms of geoengineering, as well as the global warming that they could reduce, could be considered as pollution. Specifically, the sulfates that could be injected into the stratosphere are acid rain precursors.

CLRTAP is a regional agreement, developed under the UN Economic Commission for Europe (UNECE), addressing the problem of transboundary movement of certain pollutants. Its fifty-one parties include all industrialized countries and most emerging economies of North America, Europe, and central Asia. As in other framework conventions, CLRTAP itself has only general and qualified commitments to, for example, monitor, exchange information, consult with one another, and develop air quality management systems (Articles 3–6, 8). Among its principles, CLRTAP's parties "shall endeavour to limit and, as far as possible, gradually reduce and prevent air pollution" (Article 2). The agreement has institutional support from the UNECE, its own standing bodies, a noncompliance procedure, ongoing scientific support, and the capacity to be amended through protocols.

Here, the definition of air pollution – one which is seen elsewhere in multilateral environmental agreements – implicitly includes greenhouse gases, global warming, and any harmful substances used for atmospheric solar geoengineering:

> "Air Pollution" means the introduction by man, directly or indirectly, of substances or energy into the air resulting in deleterious effects of such a nature as to endanger human health, harm living resources and ecosystems and material property and impair or interfere with amenities and other legitimate uses of the environment. (Article 1(a); see Sands and Peel 2012, 247)

Note that, here, deleterious effects must reach a certain threshold of magnitude and must have already occurred, which is or will soon be the case with greenhouse gases and global warming. This definition highlights the recurring tension in international environmental law between climate change and solar geoengineering.

CLRTAP's consequences for the governance of atmospheric solar geoengineering resemble those of the Vienna Convention. At the very least, source parties are committed to report their emissions of acid rain precursors – which include sulfates, presently the leading candidate substance for stratospheric aerosol injection – and consult with other parties "which are actually affected by or exposed to a significant risk of long-range transboundary air pollution" (Articles 5, 8).[5] CLRTAP also commits its parties to the research and development of, among other things, technologies to reduce air pollution and "economic, social and environmental assessment[s] of alternative measures for attaining environmental objectives including the reduction of long-range transboundary air pollution" (Article 7). Given that greenhouse gases

---

[5] Note the low threshold for consultation, in which source states must consult, upon request, with states that are merely at risk or affected by long range transboundary air pollution.

and perhaps also global warming are air pollution pursuant to the CLRTAP definition, a literal reading of this provision implies a commitment to research, develop, and assess solar geoengineering that have a potential to reduce their deleterious effects. If a solar geoengineering activity were known to cause deleterious effects, then CLRTAP parties would be obligated to, among other things, endeavor to limit and to gradually reduce and prevent it, as far as possible. If these activities also reduced the deleterious effects of greenhouse gases or global warming, then the parties (perhaps operating through the Convention's Executive Body and Implementation Committee) should also take this into account as appropriate. Furthermore, if certain solar geoengineering techniques were believed to offer "the best available technology which is economically feasible and low- and non-waste" to reduce harm from greenhouse gases or global warming, then parties arguably should include them as part of their air quality management systems that are designed to combat air pollution, although it is unclear whether the parties would have to actually carry them out (Article 6).

The goals and commitments of CLRTAP are furthered by eight protocols, most of which establish emission limits for long-range air pollutants. All protocols are in force, with differing constellations of participating states.[6] Three of these – the 1985 Helsinki Protocol on the Reduction of Sulphur Emissions or their Transboundary Fluxes by at Least 30 per Cent, the 1994 Oslo Protocol on Further Reductions of Sulphur Emissions, and the 1999 Gothenburg Protocol to Abate Acidification, Eutrophication and Ground-Level Ozone – include sulfate emissions. Research on the impacts of stratospheric sulfate aerosol injection appears to be supported by the Oslo Protocol, wherein parties commit to "encourage research, development, monitoring and cooperation related to ... [t]he understanding of the wider effects of sulfur emissions on human health [and] the environment" (Article 6). At the same time, the sulfate emissions at the quantities necessary for global climate response field tests or implementation of stratospheric aerosol injection, if conducted within the territory of a single party, would exceed the limits under these Protocols.

CLRTAP has a noncompliance procedure. An Implementation Committee first reviews possible noncompliance, and the Executive Body takes decisions regarding how to respond. These bodies would be able to consider three mitigating factors when considering noncompliance with a CLRTAP Protocol due to stratospheric sulfate aerosol injection. First, these activities might reduce global warming and the harm therefrom, as described. Second, the goal of CLRTAP and its Protocols is to reduce adverse effects on humans and ecosystems from air pollution, whose definition includes greenhouse gases and perhaps global warming (CLRTAP, Articles 2–3; Oslo Protocol, Article 2.1; Gothenburg Protocol, Article 2.1). Third, the sulfate particles injected into the stratosphere would be deposited across the globe, with

---

[6] Several larger countries, including Canada, Russia, the UK, and the US, have ratified CLRTAP are not parties to one or more of the three Protocols described here.

a very small impact on the acidity of precipitation within parties' territories. Therefore, although the additional sulfate emissions from solar geoengineering research or implementation might exceed a party's limit, their effect on acid rain would be minimal.

6.4 UN INTERNATIONAL LAW COMMISSION DRAFT GUIDELINES ON THE PROTECTION OF THE ATMOSPHERE

The ILC has provisionally approved a set of twelve Draft Guidelines on the Protection of the Atmosphere. These guidelines attempt to capture existing international legal practice, some of which reflects customary international law. In their current form, they define the atmosphere as "the envelope of gases surrounding the Earth" (Guideline 1(a)). One of them states, "Activities aimed at intentional large-scale modification of the atmosphere should be conducted with prudence and caution, subject to any applicable rules of international law" (Guideline 7). The accompanying commentary notes that this includes but is not limited to "what is commonly understood as 'geo-engineering'," including solar geoengineering (paragraphs 3, 4). It also calls attention to the tension in governing these activities:

> Activities aimed at intentional large-scale modification of the atmosphere have a significant potential for preventing, diverting, moderating or ameliorating the adverse effects of disasters and hazards, including drought, hurricanes, tornadoes, and enhancing crop production and the availability of water. At the same time, it is also recognized that they may have long-range and unexpected effects on existing climatic patterns that are not confined by national boundaries. (paragraph 7)

The guidelines do not aim to authorize or prohibit such activities, or "to stifle innovation and scientific advancement" (paragraphs 8, 9). The phrase "activities aimed at intentional large-scale modification of the atmosphere" is based on the Convention on the Prohibition of Military or Any Other Hostile Use of Environmental Modification Techniques (ENMOD; see Chapter 8), and "prudence and caution" on rulings of the International Tribunal for the Law of the Sea (paragraphs 2, 9). The commentary further says that "[t]he draft guideline is cast in hortatory language, aimed at encouraging the development of rules to govern such activities, within the regimes competent in the various fields relevant to atmospheric pollution and atmospheric degradation" (paragraph 9). Beyond this, the draft ILC guidelines largely consist of restatements of existing principles and customary international law, including the obligations to exercise due diligence, to conduct environmental impact assessments, and to utilize the atmosphere equitably.

However, the draft guidelines apply these obligations to the atmosphere itself, with no explicit reference to transboundary risks to other states. The drafting process is not complete, and further changes seem likely.

7

International Law: Human Rights

In its preamble, the Paris Agreement states that "Parties should, when taking action to address climate change, respect, promote and consider their respective obligations on human rights" (Preamble recital 11). Although researching, developing, and implementing solar geoengineering would be actions to address climate change, assessing the role of international human rights law in the governance of solar geoengineering is challenging and speculative. As in the other chapters, some of this is due to uncertainty regarding states' actions, the conditions under which they are undertaken, and their environmental and social effects. Furthermore, the complexity is amplified here by the often ambiguous nature of states' commitments under international human rights law, particularly with respect to the environment and climate change. Finally, the ways in which some observers rhetorically employ human rights in environmental discourses further muddies the situation.

This chapter offers a foundation for understanding the relationships between solar geoengineering and international human rights law. It first introduces human rights law and how it relates to the environment and climate change. The existing, limited literature on human rights and solar geoengineering is briefly reviewed and critiqued. The subsequent Sections 7.3 through 7.6 discuss the relationships by viewing solar geoengineering in four ways: as scientific research, as experiments, as necessitating a set of procedures, and as having substantive impacts.

7.1 INTERNATIONAL HUMAN RIGHTS LAW

Descriptions of human rights, particularly in popular discourses, resemble the story of the blind men and the elephant. In that Indian legend, a handful of blind men touch an elephant, and their resulting descriptions differ sharply because one felt the trunk, another a tusk, and another an ear. Likewise, definitions of human rights and the resulting inventories of specific human rights vary widely. Here, I adopt a legal, rather than moral or rhetorical, view of human rights. This is because states arguably act in ways that are influenced more by international law than by morality and rhetoric. This could be because legalized rights are those to which states have committed and that some international tribunals can enforce.

In general, a right is a claim of entitlement by someone to something from someone else. That is, a right holder can claim the object of a right from a duty-bearer. Legal instruments make some rights explicit and generally vary by jurisdiction. A rights-based claim is superior to mere preferences or claims that are not based on rights because it is more likely to be enforceable. Likewise, a violation of rights is usually considered a graver injustice than a violation of preferences or of norms that are not based on rights. At the same time, rights are not absolute but instead must be balanced against other relevant, legitimate considerations, not least of which are other rights. A key function of courts is to resolve such potential conflicts among rights.

Human rights are rights that all people have by the mere fact of being human. States agree, at least in principle, that human rights are universal (all people have them), equal (people have the same ones), interdependent (they are mutually reinforcing), and indivisible (they cannot be ranked) (Vienna Declaration and Programme of Action, paragraph 5). The duty-bearers of human rights are usually states, although nonstate actors may have some responsibilities, such as through domestic law of states' implementation of international human rights agreements.

Because human rights are universal legal claims for which states bear the duties, they are a matter of international law.[1] Countries have committed to protecting human rights through, among other means, a series of international legal instruments. The most important of these legal instruments are the nonbinding Universal Declaration of Human Rights (UDHR), the International Covenant on Civil and Political Rights (ICCPR), and the International Covenant on Economic, Social and Cultural Rights (ICESCR). The latter two treaties have nearly global participation yet operate in parallel due to Cold War-fueled differences regarding the nature of human rights.[2] Each was approved in 1966 and came into force ten years later. Regional treaties such as the African Charter on Human and Peoples' Rights, the American Convention on Human Rights, and the Council of Europe's European Convention on Human Rights, as well as focused agreements concerning the rights of children, women, and the disabled, reiterate and elaborate the rights found in the three central global agreements. These regional and specialized agreements have varying degrees of state participation. Moreover, many parties to these human rights agreements approved them with various reservations, understandings, and declarations, providing exceptions and details as to how they intend to implement the agreements. Therefore, human rights are not uniform, as states' views and commitments differ. Nevertheless, human rights are not only operative; they are also powerful normative tools with which international organizations, judges, legal scholars,

---

[1] However, human rights arose nationally and regionally before being universalized.
[2] Even now, the world's two largest economies are each not a party to one of these: the US has not ratified the ICESCR, and China has not ratified the ICCPR.

and advocates create a coherent moral framework calling for progressive state action toward their visions of improved human well-being.

The boundaries distinguishing human rights from other law are not always clear. One such unclear boundary is with national constitutional rights. Many rights that are now considered human rights were developed through the English Bill of Rights, the French Declaration of the Rights of Man and the Citizen, and the US Bill of Rights. By limiting states' power over their citizens and residents, and in some cases committing states to ensure certain living conditions, constitutional rights overlap substantially with human rights. However, considering constitutional rights to be human rights as such can be misleading. One key difference is that domestic courts can directly enforce constitutional rights, whereas the judicial enforcement of human rights is often indirect, such as through international tribunals. The second unclear boundary is with other international law. For example, under customary international law, a state that is a source of a transboundary risk has the obligation to notify the potentially affected public (see Chapter 5). One could make the case that this implies a human right to such information or at least certain types of it.

Some scholars organize human rights into three rough categories. So-called first-generation human rights are the political and civil rights necessary for the functioning of liberal democracy. These were initially established during the American and French revolutions and are now found in the ICCPR. These are mostly negative rights, in that they are claims to be free from certain forms of state interference. First-generation human rights include the right to life (that is, to not be arbitrarily deprived of one's life), to speech and assembly, and to freedom of thought, conscience, and religion. States have duties to ensure that people subject to their jurisdiction can enjoy these rights, but they are not committed to enforce them elsewhere.

Second-generation human rights are economic, social, and cultural rights such as those in the ICESCR. They are mostly positive rights, in that they are claims that the state must ensure the provision of something. These include rights to the highest attainable standard of physical and mental health, to be free from hunger, to education, and to take part in cultural life. The actual provision of such positive rights is more difficult for states to guarantee because doing so is costly, while states have limited resources. Consequently, the ICESCR calls for parties to realize them progressively (Article 2.1). A decision later adopted by the ICESCR's committee detailed progressive realization as consisting of three steps (General Comment No. 12, the Right to Adequate Food (Article 11), 4–5). States have duties first to respect the human right by not directly interfering with its enjoyment, second to protect it by preventing third parties from interfering with its enjoyment, and third to fulfill it by acting toward its full realization. Part of this progressive realization is through international assistance and cooperation, implying some degree of extraterritorial applicability.

Finally, the third generation of human rights goes further, and many of these are based upon the right bearer's membership in a group. However, despite claims by some scholars and others, third-generation human rights' international legal character is generally weak and found in only nonbinding declarations and the aspirational portions of legally binding agreements.[3] As a result, these rights remain contested. An exception is the collective right to self-determination, which is contained in the first articles of both the ICCPR and the ICESCR.

7.2 THE ENVIRONMENT, CLIMATE CHANGE, AND SOLAR GEOENGINEERING

Although the environment and human rights are clearly related, the precise legal scope and nature of human rights and concomitant duties in the environmental context is unclear. Scholars make diverse claims, which are here described in six general categories. The first and arguably strongest is that people have procedural rights to environmental information, to participate in relevant decision-making, and to access legal remedies in the event of violations of these rights (UDHR, Articles 8, 19, 21; ICCPR, Article 25; Rio Declaration, Principle 10; Framework Principles on Human Rights and the Environment, Principles 7, 9, 10). The second general claim is that states should enact and enforce legislative and administrative measures to protect the enjoyment of human rights from environmental threats (Framework Principles on Human Rights and the Environment, Principle 1). The third is that states may not infringe upon human rights in their efforts to ensure a clean and healthy environment (Framework Principles on Human Rights and the Environment, Principle 16). Fourth, states may have a duty to cooperate internationally to control threats to human rights from environmental degradation (UN Charter, Articles 55–6; Framework Principles on Human Rights and the Environment, Principle 13). The fifth general category is certain groups' claims to environmental protection. For example, parties to the Conventions on the Elimination of All Forms of Discrimination against Women, the Rights of the Child, and the Rights of Persons with Disabilities have duties to protect those specific populations and their rights, including as they relate to the environment (Framework Principles on Human Rights and the Environment, Principles 14, 15). The final is a contested claim of a human right to a clean environment itself. For example, the first principles of the two seminal nonbinding multilateral agreements of international environmental law – the 1972 Stockholm Declaration and the 1992 Rio Declaration – refer to rights to live "in an environment of a quality that permits a life of dignity and well-being" and "to a healthy and productive life in harmony with nature," respectively.[4]

---

[3] Indeed, scholars dispute the mere existence of third-generation human rights (see Brownlie 1985).

[4] See also the African Charter on Human and Peoples' Rights (Article 24), the Inter-American Court of Human Rights case *The Environment and Human Rights, Advisory Opinion*, and the report accompanying the Framework Principles on Human Rights and the Environment (paragraph 14).

7.2 The Environment, Climate Change, and Solar Geoengineering

In many ways, the relationship of climate change with human rights is like that of environmental degradation more generally. For example, people may have rights to information, participation, and access to justice (UN Framework Convention on Climate Change (UNFCCC), Article 6 (a)(ii); Paris Agreement, Article 12; Report of the Special Rapporteur on the Issue of Human Rights 2016, paragraphs 50–64). Furthermore, climate change will interfere with the enjoyment of several human rights such as those to an adequate standard of living (including food and housing), to the highest attainable standard of health, to be free from hunger, and to self-determination. States' commitment to progressively protect, respect, and fulfill these points toward the importance of adaptation to a changed climate (Report of the Special Rapporteur on the Issue of Human Rights 2016, paragraphs 68–70). Their duty to not interfere with human rights in their responses to climate change is highlighted by the Paris Agreement's preambular passage, quoted at this chapter's opening (Preamble recital 11). As with environmental threats in general, there may also be special duties to groups (Report of the Special Rapporteur on the Issue of Human Rights 2016, paragraphs 81–48). A duty to cooperate internationally seems particularly important, as preventing dangerous climate change by emissions abatement necessarily requires deep and sustained international cooperation (Report of the Special Rapporteur on the Issue of Human Rights 2016, paragraphs 42–6, 73). Finally, some scholars assert that there is a right to a stable climate, although the basis for such claims is weak (Vanderheiden 2008).

Determining specific duties based upon human rights law is particularly challenging because of climate change's global causes and impacts, its delayed effects, and lingering uncertainty. The UN Human Rights Council thus concluded, "While climate change has obvious implications for the enjoyment of human rights, it is less obvious whether, and to what extent, such effects can be qualified as human rights violations in a strict legal sense" (Report of the Office of the United Nations High Commissioner for Human Rights on the Relationship between Climate Change and Human Rights, paragraph 70). The Council cited three reasons: the impossibility of connecting a given country's emissions with a particular manifestation of climate change, the difficulty in attributing a given harmful environmental phenomenon to climate change, and the fact that climate change is *expected* to cause *future* harms. Analysis is further complicated by the close relationships among the use of fossil fuels, economic development, and the capacity to fulfill human rights. Wealthier countries generally satisfy human rights better, both because the wealth directly allows the provision of requisite goods and services and because it supports institutions that can protect and enforce the rights. Economic development presently requires energy from fossil fuels. Consequently, states need to balance preventing further climate change and development, each of which would arguably advance human rights.

If climate change implies duties for states to take and to refrain from various actions, then solar geoengineering likely would as well. Yet solar geoengineering is different from anthropogenic climate change. The former has capacities both to protect as well as to interfere with human rights, while the latter will be mostly harmful. Moreover, solar geoengineering's effects will have a closer causal relationship with specific decisions made by states, largely removing one of the difficulties in applying human rights to climate change.

Many legal and other scholars have referred to the relationships between solar geoengineering, or geoengineering more generally, and human rights. With a few exceptions (Burns 2016; Adelman 2017; Svoboda, Buck, and Suarez 2019), these are brief, cursory statements. A review of the academic and grey literature indicates five rough categories of assertions from scholars regarding solar geoengineering and human rights. In the first and simplest approach, some writers merely observe that solar geoengineering and human rights are related and ask what guidance the latter could provide for decision-making regarding the former (Schäfer et al. 2015, 101). Those in the second category go further and assert that human rights could provide a normative framework to assess proposals for solar geoengineering and its governance. For example, Rider Foley and colleagues list human rights as one of several "normative anchors ... to which geoengineering discourses must remain explicitly moored" that can help us "to understand how science and technology can advance the 'public good' without creating unacceptable trade-offs" (Foley, Guston, and Sarewitz 2018, 242). In the third proposed relationship between human rights law and solar geoengineering, states and nonstate actors bear procedural duties in the case of solar geoengineering activities. Here, Neil Craik emphasizes the "emerging human right to notification of environmental risks that is held by individuals" (Craik 2015, 137). In the fourth type of claim, a few scholars have indicated that solar geoengineering could protect human rights from interference by climate change. Philosopher Toby Svoboda wrote

> Anthropogenic climate change may infringe the human rights of some, including rights to life, subsistence, and health. There may also be distinctively environmental rights to a safe environment or a stable climate, which are threatened by anthropogenic climate change. Climate engineering with SRM [solar geoengineering] could alleviate some of these threats, such as by slowing the rate of temperature increase. (Svoboda, Buck, and Suarez 2019, 399).

The final proposed relationship it that solar geoengineering could or would violate human rights, usually asserted without considering in what ways it could protect human rights through reducing climate change. For example, Sam Adelman opens his article with the claim that "manipulating the Earth's climate may provoke unforeseen, unintended and uncontrollable consequences that threaten human rights" (Adelman 2017, 119; see also Burns 2016).

7.3 HUMAN RIGHTS OF SCIENTIFIC RESEARCH

We can now consider the ways in which international human rights law could shape the governance of solar geoengineering. First, solar geoengineering is presently being explored as scientific research. Here, human rights law has implications for both the researcher and the potential beneficiaries of the knowledge. Regarding the former, some scholars claim that academic freedom is a human right.[5] They provide three bases for this claim, the first of which is the freedom of expression and opinion. This is primarily a negative right, in which people are free, among other things, to hold opinions and to seek, receive, and impart information and ideas (UDHR, Article 19; ICCPR, Article 19). The state may not interfere with these freedoms, save for legitimate limitations to respect others' rights and for reasons of national security, public order, public health, or morals. Another basis of academic freedom is the right to an education. Although this may initially seem to have little clear applicability to scientific research, countries have agreed "that education shall enable all persons to ... further the activities of the United Nations for the maintenance of peace" (ICECSR, Article 13.1; also UDHR, Article 26). To the extent that climate change or solar geoengineering could undermine peace, then improved understanding of the capabilities, limitations, co-benefits, and risks of responses to climate change would likewise further the maintenance of peace. The final basis of a purported human right of academic freedom is the freedom of scientific research and creative activity. This is provided in the ICESCR's commitment for its parties "to respect the freedom indispensable for scientific research and creative activity" and to "recognize the benefits to be derived from the encouragement and development of international contacts and co-operation in the scientific and cultural fields" (ICECSR, Article 15; also UDHR, Article 27.2). This is the strongest basis for states' duties both to refrain from interfering with scientific freedom and to cooperate internationally in scientific research in a manner consistent with the progressive realization of positive human rights. A possible freedom of scientific research is reinforced by regional human rights agreements, especially the Charter of Fundamental Rights of the European Union, in which European states are to ensure that scientific research is free of constraint and to respect academic freedom (Article 13; see also the Additional Protocol to the American Convention on Human Rights in the Area of Economic, Social and Cultural Rights, Article 14.3). To be clear, a human right of scientific research would not imply a positive right to conduct research and would not be absolute.[6] A human right to academic freedom would indicate that in the absence of a legitimate justification states may not censor,

---

[5] For a constitutional argument, see Robertson (1977).

[6] For example, the nonbinding Declaration on Science and the Use of Scientific Knowledge states that "scientists ... have a special responsibility for seeking to avert applications of science which are ethically wrong or have an adverse impact," and "scientists should commit themselves to high ethical standards, and a code of ethics based on relevant norms enshrined in international human rights instruments should be established for scientific professions" (paragraphs 21, 41).

interfere with, or hinder the generation, expression, publication, distribution, and discussion (including through international communication) of ideas concerning solar geoengineering. They should also take steps toward international cooperation in solar geoengineering research, to the extent appropriate.

Another human right of scientists is the protection of their "moral and material interests" from their research (ICECSR, Article 15.1(c)). This indicates a duty of states to progressively realize inventors' rights to benefit from their work, for which patents are the leading vehicle. Thus, states are to respect this right by refraining from infringing upon patents, to protect it by preventing infringement by third parties, and to fulfill it by implementing laws and administrative regulations that award and protect patents. Governments should also balance inventors' right to benefit with individuals' right to enjoy the benefits of scientific research, the topic of the next paragraph. For example, industrialized countries could facilitate technology transfer to the poorer ones through innovative intellectual property policies (see International Council on Human Rights Policy 2011). Therefore, states may be obligated to pursue such a balanced – and if appropriate, novel – solar geoengineering intellectual property policy that protects inventors' patents while ensuring that the inventions' benefits are widely shared (see Chapter 11).

Regarding the human rights of potential beneficiaries of scientific knowledge, international law provides for a human right to enjoy the benefits of scientific research. For example, the UDHR states, "Everyone has the right freely ... to share in scientific advancement and its benefits" (Article 27.1). Likewise, parties to the ICESCR "recognize the right of everyone ... To enjoy the benefits of scientific progress and its applications" (Article 15.1(b)). Regional human rights agreements provide for similar rights (Charter of the Organization of American States, Article 38; Additional Protocol to the American Convention on Human Rights in the Area of Economic, Social, and Cultural Rights, Article 14.1(b); UNESCO Universal Declaration on Bioethics and Human Rights, Article 15). Indeed, widely sharing the benefits of scientific research – possibly including that of solar geoengineering – can help fulfill other human rights such as those to the highest attainable standard of health, to an adequate standard of living, and to be free from hunger.

In a report, a special rapporteur to the UN Human Rights Council describes the right to enjoy the benefits of scientific progress as having four normative tenets (Report of the Special Rapporteur in the Field of Cultural Rights). First, everyone should be able to access science's benefits, without discrimination. At the least, states can fulfill this by ensuring that the data, results, publications, materials, and samples arising from scientific research are accessible, ideally without cost. This is especially important for research regarding innovations that may be necessary for a life with dignity. Second, everyone should have the opportunity to contribute to scientific research, a tenet that is roughly congruous with the freedom of scientific research. The special rapporteur further links this right to an obligation to conduct research ethically, the subject of Section 7.4. Third, people have a right to participate in science-related decision-making, particularly in the case of major decisions regarding research priorities and science policies. The special rapporteur emphasizes that this is to improve decision-making, to help protect people – particularly vulnerable populations – from possible negative impacts of scientific research and applications, and to help ensure that research priorities align with wider interests. Such procedural human rights are discussed in Section 7.5 of this chapter. Fourth and finally, states are to enable the conservation, development, and diffusion of science and technology, in accordance with the ICESCR (Article 15.2). According to the special rapporteur, this highlights industrialized countries' duty to collaborate internationally in scientific research. This is underscored by the ICESCR's commitment "to take steps, individually and through international assistance and cooperation, especially economic and technical" toward the full realization of the rights therein (Article 2.1). Meanwhile, the special rapporteur argues that developing states should prioritize technologies that would be of greatest benefit to poor and otherwise marginalized people.

When applied to solar geoengineering, this interpretation of the right to enjoy the benefits of scientific research indicates, first, that governments should ensure that results are freely available, such as through data repositories and open access publications, in a nondiscriminatory manner (see Chapter 11). Second, they should provide their residents and citizens with opportunities to participate in relevant decision-making regarding solar geoengineering, including the setting of research priorities. Third, countries, especially the industrialized ones, should facilitate international research cooperation. If modeling and empirical evidence continue to indicate that solar geoengineering could reduce climate change risks to already-vulnerable populations, then an industrialized country should prioritize its research and the international transfer of its results, while a vulnerable one should take steps toward any appropriate development, importation, and dissemination of technologies. Because public policy specific to solar geoengineering is almost entirely absent, nonstate actors may be able to play an important role in ensuring that solar geoengineering research proceeds in a manner consistent with the right to enjoy the benefits of scientific research and other human rights. For example, professional societies, funders, publishers, and scholars could develop and promulgate codes of conduct that are informed by international human rights law (see Declaration on Science and the Use of Scientific Knowledge, paragraph 41; Chapter 10).

7.4 HUMAN RIGHTS OF EXPERIMENTATION

The second way in which human rights law can inform the governance of solar geoengineering concerns the people who may experience the environmental effects of outdoor experiments. Although those planned and undertaken thus far pose negligible risks to the environment and human health, scientists may scale up and even eventually intend climatic impacts. In this case, people would be affected, and the human rights regarding the protection of research subjects could be salient.

Because modern international human rights law arose in part as a response to the atrocities of World War II, it is unsurprising that the protection of human research subjects is central. The foundational human rights treaties are supplemented by the regional Council of Europe's Convention on Human Rights and Biomedicine and by nonbinding agreements such as UNESCO's Universal Declaration on Bioethics and Human Rights. Other authoritative guidelines include the Declaration of Helsinki and the International Ethical Guidelines for Health-Related Research Involving Humans (World Medical Association 2018; Council for International Organizations of Medical Sciences and World Health Organization 2016). Although these latter documents were neither put forth as human rights documents per se nor agreed to by countries, they are conceptually informed by human rights and can provide guidance. However, these agreements and documents were intended to apply to biomedical research, which outdoor solar geoengineering research would not be. Furthermore, the people who experience climatic and environmental impacts from solar geoengineering research might or might not be research subjects as these agreements intend. For example, the ethics review committee of the cancelled Stratospheric Particle Injection for Climate Engineering (SPICE) outdoor test concluded that it did not involve research subjects (Stilgoe 2015, 1). Regardless, here I assume that these people do resemble research subjects without claiming that they qualify as such and try to convey a sense of the implications of human rights in this context.

The first of two central principles of the human rights of research subjects is the protection of their autonomy. In this, research subjects must give their prior informed consent in order to participate. The ICCPR states, "no one shall be subjected without his free consent to medical or scientific experimentation" (Article 7). The Declaration on Bioethics and Human Rights is more specific: "Scientific research should only be carried out with the prior, free, express and informed consent of the person concerned" (Article 6.2).

How to obtain consent for a large group of diverse individuals presents a challenge to implementing this requirement. Indeed, group consent – for which some of the bioethics documents provide norms – has been a disputed matter in recent bioethical controversies (Drabiak-Syed 2010) and has been discussed with respect to outdoor solar geoengineering (Morrow, Kopp, and Oppenheimer 2009; Wong 2016; Frumhoff and Stephens 2018, 5; see also Dilling and Hauser 2013). For example, the Universal Declaration on Bioethics and Human Rights states that the additional agreement by a group's legal representative may be sought, but the individual medical research subjects must still consent (Article 6.3). Yet allowing one person's objection to prevent an outdoor solar geoengineering research project, including ones that hold potential to reduce negative climate change impacts, would grant each individual a veto. One response would rely on traditional political processes of representation, in which state and substate leaders would be asked to consent on behalf of their populations (Morrow, Kopp, and Oppenheimer 2013). However, countries' governance varies in their satisfaction of democratic norms. Regardless, for a decision regarding outdoor research to be legitimate, any standards for obtaining consent should be stringent consistent with the research's risks and inversely so to its expected benefits.

At the same time, the traditional notion of consent may be inappropriate in nonmedical activities, such as solar geoengineering research, that would affect groups (Hansson 2006). Indeed, democracy is based upon the consent of the governed in constitutional terms, but not with respect to each decision (Horton et al. 2018). Otherwise, as noted, a single individual could hold out, demanding a maximal share of the social surplus and thwarting the desires of many others. The need to avoid stalemate is evident in the absence of human and constitutional rights to be free from generally undesirable conditions, including environmental ones. Furthermore, there are additional reasons that obtaining consent for solar geoengineering research would be difficult, among which are its technical complexity relative to widespread understanding of science and its possible intergenerational impacts.

The second leading principle of the human rights of experimentation is the protection of vulnerable people and groups (Universal Declaration on Bioethics and Human Rights, Article 8). At the least, any outdoors solar geoengineering experiment should be based on the best available evidence and methodology, be consistent with widely accepted scientific norms and best practices, maximize its potential benefits, and minimize its risks, especially to vulnerable people and groups. Because research can provide direct benefits to its subjects, vulnerable groups should also have equitable access to participate in solar geoengineering research. Furthermore, the researchers should be responsive to the needs and priorities of the potentially affected groups and share with them the benefits of any subsequent development or invention. If monitoring during outdoor solar geoengineering research indicates substantial harm to vulnerable groups, then the experiment may need to be altered or ended.

These applicable nonbinding agreements and documents provide other standards and norms for human subjects research that are not directly grounded in human rights but may nevertheless be helpful in developing and implementing governance for solar geoengineering research. For example, an independent body should review and arguably approve research proposals. Scientists should not conduct experiments in ways that deliberately harm people. Those who are harmed might deserve compensation, although this would be challenging in the case of solar geoengineering experiments (see Chapter 12). Finally, research should aim to fairly distribute the benefits and burdens and neither exacerbate existing inequities, create new ones, nor take advantage of vulnerable groups.

7.5 PROCEDURAL HUMAN RIGHTS

The third set of implications of human rights law for solar geoengineering governance is states' procedural duties regarding activities that pose environmental risks.

As described in Section 7.1, first-generation human rights offer the strongest basis of environmental human rights, which are largely procedural. These rights have three mutually reinforcing primary components. First, citizens and residents have a right to seek information and might have a right to obtain certain information. In the words of the ICCPR, "Everyone shall have ... freedom to seek, receive and impart information and ideas of all kinds" (ICCPR, Article 19.2; see also UDHR, Article 19). Second, they have a right to participate in public affairs (UDHR, Article 21(1); ICCPR, Article 25(a)). Third, people have a right to access legal remedies if their human rights have been violated (UDHR, Article 8; ICCPR, Article 2.3). Although these three indicate little in the way of justiciable rights when considered in isolation, they might imply more expansive rights, namely for people to have access to information that is necessary for participation in public affairs, to participate meaningfully in decisions that affect them, and to access legal remedies if they have been harmed. However, legal scholars do not agree on whether the global human rights agreements should be interpreted so broadly.

The case for these more expansive rights might be stronger in situations regarding the environment. The nonbinding Rio Declaration provides for them (Principle 10), and many European countries participate in the Aarhus Convention on Access to Information, Public Participation in Decision-Making and Access to Justice in Environmental Matters. Under the UNFCCC, parties are to "promote and facilitate" access to information and public participation (Article 6(a)), and under the Paris Agreement they are to "take measures, as appropriate, to enhance" these conditions (Article 12). The customary international law of transboundary environmental risks holds that the source state is obligated to inform the potentially affected publics and provide the same legal access to foreign potential victims as it does for its own citizens (Draft Articles on Prevention of Transboundary Harm from Hazardous Activities, Articles 13, 15). However, states' legal obligations to other states under customary international law are not synonymous with international human right law. Under the former, the rights bearer is the potentially affected state, whereas under the latter it is individual people.

Nevertheless, these procedural rights – independent of their legal status – can contribute to environmental law (Birnie, Boyle, and Redgwell 2009, 290). This includes the governance of solar geoengineering activities. States should ensure that the public that is potentially affected by proposed, ongoing, and completed outdoor solar geoengineering activities has reasonable access to the relevant information that is held by public authorities and they can participate in the associated decision-making. States should also provide means of redress for those whose rights have been violated.

7.6 SUBSTANTIVE HUMAN RIGHTS

Fourth and finally, the international law of substantive human rights may inform the governance of solar geoengineering. Here, "substantive" refers to the primarily positive, second-generation human rights such as those to the highest attainable standard of health, to an adequate standard of living, and to be free from hunger, as well as positive interpretations of the right to life and other first-generation human rights. Could solar geoengineering help protect people's substantive human rights from climate change's impacts, or might it undermine these rights? If the former, should they research, develop, and even implement solar geoengineering? If the latter, should states cooperate to prevent these activities?

Currently, such talk largely remains speculation. Models indicate that the judicious implementation of solar geoengineering could greatly reduce climate change impacts. Solar geoengineering might thus offer a means to protect substantive human rights. Indeed, industrialized countries could develop and deploy it on humanitarian grounds and claim to do so on behalf of vulnerable populations (Buck 2012). It also indicates that, under other scenarios, it would pose risks of its own to humans. Furthermore, whether and how solar geoengineering would affect conditions underlying human rights depends not only on the climate and environment but also on social responses, particularly by states. For example, those countries that would benefit from solar geoengineering could compensate those who would be harmed, at least in principle. For example, the former could fund the latter's adaptation to a changed climate.

Because some uncertainty will linger, solar geoengineering's actual effects could remain partially unknown before a large-scale activity is undertaken. Therefore, the process for determining whether and how to proceed will be important with respect to human rights. William Burns has proposed a human rights-based approach to ensure that human rights are sufficiently considered in developing and implementing solar geoengineering policy (Burns 2016). This could be operationalized in part through a human rights impact assessment, which would include determining duty holders and rights bearers, gathering relevant evidence, deliberatively identifying further particular issues, and providing analysis and recommendations. This or a similar process that systematically links policy and decisions regarding solar geoengineering with international human rights law could help ensure that human rights are protected, if not advanced, through solar geoengineering.

As a final note, some human rights treaties, such as the ICCPR, allow their parties to act contrary to their commitments in the case of an existential national emergency "to the extent strictly required by the exigencies of the situation" (Article 4.1). Although climate change does not currently threaten the existence of any country, it could in the future for some, particularly low-lying island states. Even in such a case, under the ICCPR the threatened state must officially proclaim the emergency, must act consistently with other international law, may not discriminate, and may not derogate from several of the Convention's commitments.

8

International Law: Other Agreements

Multilateral agreements address a broad spectrum of issues in which states can disagree. Many of these areas besides the atmosphere and human rights are relevant to the governance of solar geoengineering. These are reviewed in this chapter and include agreements that govern states' actions regarding the three areas beyond national jurisdiction: the high seas, Antarctica, and outer space, as well as biodiversity, the hostile use of weather modification, aviation, procedural duties with respect to transboundary environmental risks, and the institutional resolution of international disputes.

8.1 UNITED NATIONS CONVENTION ON THE LAW OF THE SEA

The world's oceans are historically the most important area beyond national jurisdiction and have been the object of much international law. In some ways, the oceans resemble the atmosphere: physically in that they consist of a large body of a fluid that mixes through currents and legally in that some portions are demarcated as within states' sovereign control. Climate change, solar geoengineering, and the world's oceans are interrelated. Climate change will warm maritime waters and raise sea level, while elevated atmospheric carbon dioxide gas concentrations will acidify their waters. In turn, the oceans moderate terrestrial effects by serving as sinks for greenhouse gases and heat. Some solar geoengineering activities could take place in, on, or over the seas, and all methods undertaken at sufficient scale would affect the oceans by reducing warming and incoming solar radiation.

Unlike the atmosphere, the oceans have a central comprehensive multilateral agreement – the 1982 UN Convention on the Law of the Sea (UNCLOS) – that governs most activities that take place there or that would affect the marine environment. It establishes a legal and institutional setting to oversee and coordinate the activities of states and – indirectly – nonstate actors in, on, and above the oceans. UNCLOS describes parties' rights, duties, and other commitments in their maritime activities, including their commitments to protect the marine environment and their rights and duties in conducting marine scientific research. As such, it

8.1 United Nations Convention on the Law of the Sea

governs solar geoengineering activities that would take place in or be likely to affect the marine environment, including the atmosphere above the oceans. UNCLOS counts most countries – but not the United States – as parties, and much of its content is considered to reflect customary international law, which applies to nonparty states as well.[1] UNCLOS is supported by regular meetings of its parties; by dedicated bodies created by UNCLOS, including the International Tribunal for the Law of the Sea; and by the International Maritime Organization (IMO), a specialized UN agency that predates UNCLOS.

Among its purposes, UNCLOS is an environmental agreement that provides that "States have the obligation to protect and preserve the marine environment" with neither qualification, exception, nor threshold of harm (Article 192). The term "marine environment" is not defined but is generally understood to include the entire space above (that is, the atmosphere and the sea surface), within, and below (that is, the seabed and subsoil) the oceans (Valencia and Akimoto 2006, 708; Frank 2007, 12). Parties' sovereign right to exploit their natural resources is explicitly subject to their commitment to protect and preserve the marine environment (Article 193). Parties are also to cooperate in developing regulations for environmental protection (Article 197).

More specifically, the agreement addresses "pollution of the marine environment," defined in a manner very similar to that of the Convention on Long-Range Transboundary Air Pollution (CLRTAP; see Chapter 6) as:

> the introduction by man, directly or indirectly, of substances or energy into the marine environment, including estuaries, which results or is likely to result in such deleterious effects as harm to living resources and marine life, hazards to human health, hindrance to marine activities, including fishing and other legitimate uses of the sea, impairment of quality for use of sea water and reduction of amenities. (Article 1.1(4))

Pollution of the marine environment can come from elsewhere, such as the land or the terrestrial atmosphere, provided that it has or is likely to have deleterious effects on the marine environment. Greenhouse gas emissions and global warming qualify under UNCLOS as pollution of the marine environment (Boyle 2012, 832–3), as could some forms of solar geoengineering, including that which is based on land. Unlike CLRTAP, the UNCLOS definition includes substances or energy that are merely likely to cause deleterious effects, not only those that have already done so. Parties' commitments concerning pollution of the marine environment include:

- "to take ... all measures consistent with this Convention that are necessary to prevent, reduce and control pollution of the marine environment from any source";

---

[1] The US objected primarily to the agreement's provisions regarding the sharing of the resources of the seabed and seems to accept the provisions discussed here as custom.

- to "ensure that activities under their jurisdiction or control are so conducted as not to cause damage by pollution to other States and their environment";
- to ensure "that pollution arising from incidents or activities under their jurisdiction or control does not spread beyond the areas where they exercise sovereign rights";
- to notify potentially affected parties and competent international organizations when they become aware of actual or imminent pollution damage;
- to cooperate in eliminating the effects of pollution and in preventing or minimizing the resulting damage;
- to "take all measures necessary to prevent, reduce and control pollution of the marine environment resulting from the use of technologies under their jurisdiction or control";
- to monitor the risks or effects of pollution and to publish the results therefrom;
- to assess and to communicate the expected effects of potential "substantial pollution of or significant and harmful changes to the marine environment"; and
- to adopt and enforce laws and regulations to reduce pollution, including from their own vessels, those that enter their territorial waters or their quasi-territorial exclusive economic zones (EEZs), land-based sources, and the atmosphere (Articles 194, 196, 198–9, 204–22).[2]

Additionally, parties commit to undertake their own measures, and to cooperate with others, to conserve living resources and mammals, to protect and preserve rare or fragile ecosystems, and to protect and preserve depleted, threatened, or endangered species (Articles 61, 117–20, 194.5).

What these and other provisions mean with respect to solar geoengineering is not obvious. Their application would depend in part upon the extent to which the activity in question would reduce or would be likely to reduce the "pollution" of global warming as well as the extent to which the activity itself would result in or would be likely to result in deleterious effects. For example, solar geoengineering could be a measure to prevent, reduce, and control pollution of greenhouse gases, while measures to limit solar geoengineering could be to prevent, reduce, and control its own pollution of the marine environment. This is another example of the tension between climate change and solar geoengineering. At one extreme, if it were certain or likely that the solar geoengineering activity would lessen the negative impacts from climate change on the marine environment while posing little risk of its own, parties with the capacity to do so might be obligated – at least in principle – to undertake the solar geoengineering activity, assuming that it would be consistent with UNCLOS and other international law. Regardless, the deploying or

---

[2] The International Convention for the Prevention of Pollution from Ships (MARPOL), a multilateral environmental agreement with wide participation, details the prevention of pollution from ships' discharge.

8.1 United Nations Convention on the Law of the Sea   117

authorizing state would need to exercise due diligence by taking all measures necessary to minimize the deleterious effects of the solar geoengineering activity itself (Articles 194, 212.2). If substantial negative impacts were expected, the party carrying out or overseeing the activity would need to assess and communicate these expected effects prior to undertaking it (Articles 204–6). At the other extreme, if the activity were unlikely to reduce climate change's impacts yet would cause or would be likely to cause large deleterious effects, then parties would be committed to take all measures necessary to prevent, reduce, and control solar geoengineering within their jurisdiction or under their control (Articles 194, 196).

The tension among the deleterious impacts of climate change, the potential for solar geoengineering to reduce them, and solar geoengineering's own environmental risks are complicated by a provision in UNCLOS: "In taking measures to prevent, reduce and control pollution of the marine environment, States shall act so as not to transfer, directly or indirectly, damage or hazards from one area to another or transform one type of pollution into another" (Article 195). Solar geoengineering could transfer hazards from one area to another, such as from the atmosphere to the ocean, and might transform one type of pollution into another, such as global warming into stratospheric sulfates that contribute to ozone depletion.[3] This article uses the imperative "shall" and makes no explicit provisions for possibly lower relative magnitude or probability of the new damage, hazard, or pollution. Legal scholar Philomene Verlaan calls it "a particularly difficult hurdle for geo-engineering projects" and asserts that "[p]roponents of geo-engineering projects must show, inter alia, why such projects do not violate Article 195" (Verlaan 2009, 458).

Solar geoengineering research could constitute marine scientific research, an undefined phrase about which the UNCLOS has numerous provisions. Most definitions considered during the drafting of UNCLOS and in legal scholarship emphasize the importance of the marine environment as the object of study but diverge on whether the research must occur in the marine environment.[4] Solar geoengineering research is complicated by the fact that there may not be sharp distinction between outdoor research and implementation. Regardless, parties and competent international organizations have a right to conduct marine scientific research, and are obliged, among other things, to

---

[3] Notably, other responses to climate change, including emissions abatement and adaptation, as well as actions to reduce pollution in general done at sufficient scale, can also transfer the location of damage or hazards and transform the type of pollution.

[4] For example, proposals discussed during negotiations included "any study or related experimental work designed to increase man's knowledge of the marine environment" (Informal Single Negotiating Text, Article 1; see United Nations Division for Ocean Affairs and the Law of the Sea, Office of Legal Affairs 2010; Verlaan 2012). In contrast, the International Law Association (American Branch) Law of the Sea Committee requires that marine scientific research be "undertaken in ocean space" (Walker 2011, 241).

promote and facilitate the development and conduct of marine scientific research ... [to] promote international cooperation in marine scientific research ... to create favourable conditions for the conduct of marine scientific research in the marine environment ... [and] to make available ... information on proposed major programmes and their objectives as well as knowledge resulting from marine scientific research. (Articles 238–9, 242–4)

Parties further commit to research pollution of the marine environment (Article 204.1). Their right to conduct marine scientific research is subject to limitations, most generally "to the rights and duties of other States" (Article 238). More specifically,

marine scientific research shall be conducted exclusively for peaceful purposes ... be conducted with appropriate scientific methods and means ... not unjustifiably interfere with other legitimate uses of the sea ... [and] be conducted in compliance with all relevant regulations ... including those for the protection and preservation of the marine environment. (Article 240; see Verlaan 2007; Hubert 2011)

As discussed in more detail later in this section, parties and sponsoring international organizations are responsible for and might be held liable for "damage caused by pollution of the marine environment arising out of marine scientific research" (Article 263).

Ships could be the sites of solar geoengineering research and implementation. They are required to fly the flag of a state that has given them permission to do so and to which the ship has a genuine link (Articles 91–2). The flag state is supposed to exercise its jurisdiction over its flagged ships, including by ensuring that its crew is familiar with regulations concerning marine pollution and by adopting laws and regulations to minimize pollution from its flagged ships (Articles 94, 211.2). This indicates that a ship engaged in solar engineering should have a state that was responsible for regulating it.

States' duties and rights regarding marine solar geoengineering would vary by where the activity would take place. UNCLOS divides the water, water surface, and atmosphere horizontally into three primary zones of jurisdiction of coastal states. A state may claim up to the first twelve nautical miles from shore as its territorial sea, which becomes part of its sovereign territory (Articles 2–3). Activities other than innocent passage are subject to the approval of the coastal state (Article 17). Solar geoengineering would not be innocent, at the very least because the foreign ship would be engaged in an "activity not having a direct bearing on passage" (Articles 18–19). Coastal states' jurisdiction in their territorial waters also includes "the exclusive right to regulate, authorize and conduct marine scientific research," and research there by foreign ships requires their express consent (Article 245). Coastal states have a right to enforce their laws and regulations in their territorial waters (Articles 27, 220, 230.2; see Franckx 2001, 47–62).

The second primary jurisdictional zone defined by UNCLOS are the coastal states' EEZs, for which they may claim up to the first 200 nautical miles from their coasts (Articles 55, 57). There, the coastal state has sovereign rights over the natural resources and activities of exploration or economic exploitation, as well as jurisdiction over installations and structures, marine scientific research, and protection of the marine environment (Article 56.1). These activities would include much solar geoengineering, particularly if the marine environment's ability to contribute to reducing incoming solar radiation were considered a natural resource. In the EEZ (which hereinafter refers to the EEZ beyond the coastal state's territorial waters), the coastal and other states must have due regard for each other's rights and duties (Article 56.2). States and, consequently, their flagged ships in an EEZ are required to comply with the laws and regulations of the coastal state (Article 58.3). A coastal state's enforcement rights and its jurisdiction over marine scientific research, such as solar geoengineering research, are in its EEZ somewhat similar to those in its territorial waters. However, the coastal state should grant its consent for marine scientific research "in normal circumstances" (Article 246; see also Articles 252–3). In turn, the researching state is committed to provide certain information to the coastal state and allow it to participate or be represented in the research project if it so wishes (Articles 248–9). Any dispute concerning activities in the EEZ "should be resolved on the basis of equity and in the light of all the relevant circumstances, taking into account the respective importance of the interests involved to the parties as well as to the international community as a whole" (Article 59). This provision implies that the severity of climate risks that the states face and the potential for the solar geoengineering activity to reduce or exacerbate risks are factors to consider.

The third and final marine zone is the high seas, which lie beyond the EEZs. This area is open to all states for peaceful purposes provided that they exercise their freedoms there "with due regard for the interests of other States" (Articles 86–8, 257). Thus, solar geoengineering and other activities on the high seas should not inappropriately interfere with, among other things, the navigation, overflight, fishing, and scientific research of other states (Article 87.1). Furthermore, states' rights in the high seas and in other states' EEZs include the right to overflight, and thus their various rights and duties in these zones would extend to atmospheric solar geoengineering activities undertaken there (Articles 58.1, 87.1(b)).[5]

One challenge is the regulation of ships on the high seas that bear no flag or that of a nonratifying state. It is reasonable to conclude that UNCLOS parties might be

---

[5] This provision, coupled with Article 56.1, seems to grant coastal states jurisdiction over atmospheric solar geoengineering in their EEZs. Reinforcing this interpretation, a group of "senior officials and analysts" concluded in their "Guidelines for Navigation and Overflight in the Exclusive Economic Zone" that "States exercising the freedoms of navigation and overflight in a coastal State's EEZ should not interfere with or endanger the rights of the coastal State to protect and manage its own resources and their environment" (Valencia and Akimoto 2006, 709).

responsible for the activities of their nationals in such circumstances. However, UNCLOS is not clear to what extent its provisions regarding the marine environment and marine scientific research extend to parties' nationals. On one hand, several of the key articles regarding preventing, reducing, and controlling pollution speak of "activities under [parties'] jurisdiction or control," whereas others refer to "their natural or juridical persons," suggesting that the former duties do not necessarily extend to the latter group.[6] On the other hand, parties can exercise some control over their nationals, and their commitments to protect and preserve the marine environment and to take all measures to minimize pollution from any source are without any qualification regarding jurisdiction or control.

In three articles, UNCLOS establishes, or at least reaffirms, liability for damage from parties' activities at sea. Some of these provisions could apply to solar geoengineering activities that cause harm, depending on the circumstances at hand and on the interpretation of the articles. For the most part, this is limited to damage from acts that are contrary to UNCLOS or other international law. The first and most generally relevant article addresses liability for harm from pollution to the marine environment. This article vaguely states that

> States are responsible for the fulfilment of their international obligations concerning the protection and preservation of the marine environment. They shall be liable in accordance with international law.
>
> States shall ensure that recourse is available in accordance with their legal systems for prompt and adequate compensation or other relief in respect of damage caused by pollution of the marine environment by natural or juridical persons under their jurisdiction. (Article 235)

The first paragraph creates no new obligations under international law and could include compensation for transboundary harm caused by activities contrary to international law, pursuant to the customary international law of state responsibility (see Chapter 5). The latter paragraph establishes only a procedural standard for pursuing compensation – not a substantive one for state liability – for harm from pollution arising from acts that are not necessarily contrary to international law.

A second article in UNCLOS establishes liability from marine scientific research (Article 263). One passage in this article seems to establish liability for harm in general, whereas another does so for "damage caused by pollution of the marine environment arising out of marine scientific research," but only pursuant to Article 235 regarding the transfer or transformation of pollution. Furthermore, the article regarding marine scientific research extends liability to international organizations.

In the third relevant UNCLOS article, parties are liable for damages and loss from their unlawful or excessive enforcement of laws and regulations with respect to pollution of the marine environment (Article 232). This could apply, for example, to

---

[6] Compare Articles 194.2, 196, 206 with Articles 139, 153, 235, 263.

parties' unlawful or excessive solar geoengineering activities, which control the pollution of greenhouse gases and global warming, and to efforts to prevent, reduce, or control solar geoengineering's deleterious impacts. Finally, it should be noted that UNCLOS leaves the scope of "damage" and "loss" undefined.

In one part of UNCLOS, parties commit to develop and transfer marine technology, especially to developing states. These obligations could apply to solar geoengineering technologies, depending on how one interprets the undefined term "marine technology" and the particular solar geoengineering technology at hand. Most generally, parties are "to promote actively the development and transfer of marine science and marine technology on fair and reasonable terms and conditions" (Article 266.1). The commitment to the "development of the marine scientific and technological capacity of States which may need and request technical assistance in this field" explicitly regards the marine environment, strengthening the case that these provisions could apply to solar geoengineering (Article 266.1). Regardless, this is to be done with due regard for the rights and duties of the holders, suppliers, and recipients of marine technology (Article 267).

Three categories of proposed solar geoengineering methods are examined more closely here due to their locations or means of operation. First, several methods would involve the placement of diffuse substances into the marine environment. This includes stratospheric aerosol injection and cirrus cloud thinning. These might (or might not) be considered dumping, which the UNCLOS defines as "any deliberate disposal of wastes or other matter from vessels, aircraft, platforms or other man-made structures at sea," but excludes "placement of matter for a purpose other than the mere disposal thereof, provided that such placement is not contrary to the aims of this Convention" (Article 1.1(5)).[7] These techniques' purpose would not be mere disposal of those substances but instead to reduce climate change and its risks. This exclusion from dumping would also require that the practice not be contrary to UNCLOS's unstated aims. Based upon its preamble and commitments, these aims implicitly include protection of the marine environment, raising the tension between preventing climate change and solar geoengineering (Preamble recital 4). For example, if a proposed solar geoengineering activity were likely to have or did have deleterious effects, it would qualify as pollution of the marine environment. If these impacts were known or expected in advance, the activity would appear contrary to UNCLOS's aims. Regardless, parties are to prevent, reduce, and control pollution of the sea by dumping (Article 194.3(a)). The parties are to regulate dumping nationally, regionally, and globally. Coastal states have the right to regulate dumping in their EEZs, where the practice requires their prior consent (Article 210). International regulation has been largely through the London Convention and London Protocol, considered in

---

[7] Note that disposal need not be into marine waters.

Section 8.2, although participation in these agreements is not as broad as that in UNCLOS.[8]

Second, some researchers believe that marine cloud brightening could be carried out by unmanned ships. Although the interchangeable words "ship" and "vessel" remain undefined in UNCLOS, they are widely understood to include unmanned ships, which would therefore be required to bear the flag of a state to which they have a genuine link. That flag state would need to issue laws and regulations to prevent, reduce, and control the ships' potential pollution of the marine environment and to conduct an inquiry if the ship were to cause serious damage to the marine environment (Articles 94.7, 211.2). Nevertheless, unmanned ships, whether for solar geoengineering or for other purposes, raise legal uncertainties (Van Hooydonk 2014). For example, UNCLOS gives certain responsibilities to a ship's master, officers, and crew. Among these, the flag state of a ship must take measures

> to ensure ... that each ship is in the charge of a master and officers who ... are fully conversant with and required to observe the applicable international regulations concerning the safety of life at sea, the prevention of collisions, the prevention, reduction and control of marine pollution, and the maintenance of communications by radio. (Article 94.4)

Unmanned ships lack a master, officers, and crew. To some extent, the shore-based vessel operator(s) could be considered as fulfilling these roles. Other questions surround who would be liable for accidents or responsible for violations of international or coastal state laws and regulations. Another concern is ensuring that unmanned ships would navigate the seas in such a manner that they would have due regard for the rights and interests of other states.

In the third specific category, some solar geoengineering methods could function through objects that are placed and left in the ocean. For example, aerosols could be injected into the stratosphere from a stationary platform through a tethered, elevated hose. Parties' rights and duties under UNCLOS regarding these are somewhat confusing in that the agreement usually – but not always – uses the undefined terms "artificial islands, installations and structures" in the general context and "installations and equipment" in the context of marine scientific research, yet in some passages omits one of these terms. Regardless, throughout the ocean, parties' obligation to take measures to minimize pollution of the marine environment explicitly extends to stationary installations and devices (Article 194.3(d)). Installations or equipment for research are required to bear the identifying markings of the state of registry or the international organization to which they belong. In the EEZs or on the continental shelves, coastal states have jurisdiction over these placed objects and due notice must be given of their construction (Articles. 56.1(b), 60, 80).

---

[8] A possible but controversial interpretation is that UNCLOS Parties that are parties to neither the London Convention nor London Protocol are nevertheless committed to some or all provisions of those dumping-specific agreements through UNCLOS (Verlaan 2012, 810–11).

Although coastal states are to grant permission under normal circumstances for other states to conduct marine scientific research in their EEZs and on their continental shelves outside their territorial waters, artificial islands, installations, and structures constitute one of four conditions under which they may clearly deny such permission (Article 246.5(c)). For marine scientific research in all locations, unused installations, structures, and equipment must be removed; artificial islands, installations, and structures may have a designated zone of safety; and all categories of placed objects may not interfere with existing sea lanes (Articles 60, 249.1(g), 260–1). On the high seas, all parties have the freedom to construct artificial islands and other installations, provided that they exercise due regard for other states' exercise of high seas freedoms (Article 87). Moreover, because objects placed in the water for solar geoengineering might be free floating, the party that oversees their placement would need to ensure that they do not migrate into the EEZ of another state.

8.2 LONDON CONVENTION AND LONDON PROTOCOL

The London Convention on the Prevention of Marine Pollution by Dumping of Wastes and Other Matter and the London Protocol thereto are a pair of multilateral treaties that focus upon reducing marine pollution from dumping. They are each in effect and institutionally supported by the IMO. As noted in Section 8.1, they might represent the global rules and standards regarding dumping to which UNCLOS refers (Article 210.6). The 1972 London Convention has eighty-seven parties, including all major industrialized maritime countries, as well as a handful of countries – including the United States – that are not parties to UNCLOS. In contrast, the 1996 London Protocol – which is intended to replace the London Convention and supersedes the conventions for those states that are parties to both – presently has fifty-one parties, a cohort that lacks major states such as the United States and Russia. The agreement's specific objectives, stated in the obligatory language of "shall," refer to the control of all sources of pollution of the marine environment, especially that from dumping (LC Article I; LP Article 2). The two agreements apply to all maritime waters, including the high seas, EEZs, and non-inland territorial waters, as well as to the parties' flagged ships and loading that occurs in their ports.

The Protocol uses somewhat stronger and more obligatory language than the Convention and, unlike the Convention, actually defines pollution (LP Article 1.10). Although this definition is mostly like that of UNCLOS, it is limited to matter, and not including energy. It could encompass elevated atmospheric and dissolved greenhouse gases and thus implicitly some solar geoengineering methods but not global warming itself. Among other things, parties to the London Protocol are committed to promote scientific research on pollution from dumping and from "other sources of marine pollution relevant to this Protocol" (LP Article 14.1). The latter presumably may include research into those forms of solar geoengineering that might constitute pollution. Furthermore, parties may not "transfer, directly or indirectly, damage or likelihood of damage from one part of the environment to another or transform one type of pollution into another" (LP Article 3.3). Although this clause resembles that in UNCLOS (Article 235), its application is limited to parties' actions in implementing the Protocol.

Most commitments in these two agreements are specific to pollution from dumping, despite their broader objectives. The definitions of dumping in the agreements are almost the same as that in UNCLOS, retaining the exception for purposes "other than the mere disposal thereof, provided that such placement is not contrary to the aims of" the agreement (LC Article III.1; LP Article 1.4). The older Convention uses a "black list" of prohibited substances that may not be dumped and a "gray list" of substances whose dumping requires a special permit, while all other substances to be dumped require a general permit (LC Article IV). Substances that are presently considered for placement into the water for solar geoengineering purposes are found on neither the "black list" nor the "gray list." However, the latter categorically includes those substances "which, though of a non-toxic nature, may become harmful due to the quantities in which they are dumped, or which are liable to seriously reduce amenities" (LC, Annex II, paragraph D). Consequently, solar geoengineering activities that would be considered dumping and of a sufficient scale would require a special permit under the terms of the London Convention. In contrast, the newer London Protocol generally prohibits dumping except for a list of substances that require a permit, while "being mindful of the Objectives and General Obligations of this Protocol" (LP Annex 1). That list includes "inert, inorganic, geological material" and "organic material of natural origin," undefined terms whose applicability to substances used in solar geoengineering could be clarified in a permitting process. Furthermore, parties to the London Protocol commit to applying "a precautionary approach to environmental protection from dumping of wastes or other matter whereby appropriate preventative measures" are taken (LP Article 3.1).

In 2013 the parties to the London Protocol approved an amendment that would govern "marine geoengineering" (see Ginzky and Frost 2014). The amendment – not yet in force – defines this as:

> a deliberate intervention in the marine environment to manipulate natural processes, including to counteract anthropogenic climate change and/or its impacts, and that has the potential to result in deleterious effects, especially where those effects may be widespread, long lasting or severe. (paragraph 5*bis*)

This definition includes solar geoengineering, at least those activities that take place at sea. Notably, it is not limited to climatic purposes.

When and if this amendment comes into effect, the parties would collectively maintain a new Annex 4 to the Protocol, listing specific marine geoengineering activities.[9] The parties would then be committed to not allowing the placement of

---

[9] Only three countries – the UK, Finland, and the Netherlands – have ratified the amendment.

matter into the sea for these activities, unless the activity's listing in the Annex allows for case-by-case authorization by the party (Article 6bis). This article narrows the amendment's applicability to solar geoengineering to proposals to make sea water more reflective through, for example, a reflective surface film. Placement of matter for marine geoengineering activities that are not listed in the Annex is implicitly permitted, provided that this placement is neither dumping nor contrary to the Protocol's aims. Presently, the Annex's list includes only ocean fertilization – a negative emissions technology – which parties should permit only if the activity is legitimate scientific research (paragraph 1). Parties would need to adopt administrative or legislative measures to ensure that pollution of the marine environment from these listed activities is prevented or minimized as far as practicable and that the activities are not contrary to the Protocol's aims. In the parties' ongoing response to marine geoengineering, they are being assisted by a dedicated working group of the Joint Group of Experts on the Scientific Aspects of Marine Environmental Protection, which advises the UN and its related bodies on the scientific aspects of pollution of the marine environment.

The case-by-case authorizations should follow both a general assessment framework as well as any specific assessment framework for the activity. The former is intended to be legally binding upon the parties that ratify the London Protocol amendment and is detailed in a new Annex 5 that the amendment would create. The party in whose jurisdiction or under whose control the proposed activity would occur is to require a detailed description of the activity, to notify potentially affected countries, and to develop a consultation plan (Annex 5, paragraph 10). The activity's proponents are to demonstrate that the activity is not mere disposal, that it would fulfill its purpose, that its "rationale, goals, methods, scale, timings and locations as well as predicted benefits and risks" are justified, and that the proponents have the financial resources to carry out the proposed activity adequately (Annex 5, paragraphs 5, 8). The responsible party is also to encourage consultation with all stakeholders, and consent – while not required – "should be sought from all countries with jurisdiction or interests in the region of potential impact" (Annex 5, paragraph 11). Furthermore, the party and any potentially affected countries should seek expert advice, including peer review of proposals (Annex 5, paragraph 12). In contrast, specific assessment frameworks for activities would be nonbinding and approved by the meetings of parties. Ultimately, a permit requires that the proposal has satisfied all assessments, impact evaluations, and consultation requirements; that it would fulfill its purpose; that the risk management and monitoring requirements have been determined; that the proposal's environmental harm would be minimized while benefits would be maximized; and that pollution would be minimized as far as practicable (Annex 5, paragraph 26).

The approved Annex offers a nonexhaustive list of reasons that some potential marine geoengineering techniques may be limited to marine scientific research in order to be approved and describes the characteristics that constitute such research and the conditions that should be imposed upon it (Annex 5, paragraphs 7–8). These required conditions are adding to scientific knowledge, an absence of direct financial gain and the influence therefrom, being subject to peer review, having an appropriate research methodology and sufficient financial resources to carry out the proposed research activity, and committing to publish in a peer-reviewed outlet and to make data publicly available. The two paragraphs concerning marine scientific research are not related explicitly to any specific obligation, but instead appear to be intended to guide the assessment of listed marine geoengineering activities that are limited to or might have alternative assessment criteria for legitimate scientific research.

This amendment to the London Protocol is notable in several regards, despite having limited applicability to solar geoengineering and not yet being in force. First, it is the only binding international legal instrument that is specifically concerned with geoengineering, and it likewise offers the first definition of geoengineering in an international legal instrument. Second, the amendment relies upon an expansive interpretation of the Protocol's scope, which is not limited to dumping but instead includes a commitment to "protect and preserve the marine environment from all sources of pollution." Third, as with marine geoengineering, criteria for marine scientific research are laid out in a binding international legal instrument for the first time (Verlaan 2012). These characteristics of and conditions for marine scientific research could help clarify that concept in the context of UNCLOS, where the phrase is undefined. Finally, the amendment adopts an explicit balancing approach to the tension between climate change and geoengineering, calling for "conditions [to be] in place to ensure that, as far as practicable, environmental disturbance and detriment would be minimized and the benefits maximized" (Annex 5, paragraph 26.5).

8.3 ANTARCTIC TREATY AND ITS MADRID PROTOCOL

The continent of Antarctica is the second of three primary areas that are typically considered beyond national jurisdiction.[10] Although solar geoengineering would be less effective there due to the angle of incoming sunlight and the already high albedo, some researchers are interested in regional solar geoengineering as a means to preserve ice sheets (Latham et al. 2014). The area south of sixty degrees latitude is governed by a few interrelated treaties that, among other things, call upon their parties both to protect the environment and to conduct scientific research. Unlike UNCLOS, they prioritize neither of these goals over the other, implying a potential tension between them.

---

[10] Antarctica is subject to territorial claims by seven states, but these are not fully recognized. Further claims are prohibited by the Antarctic Treaty.

The central Antarctic Treaty counts fifty-three parties, including the major economic powers, and has been in effect since 1961. It establishes a freedom of scientific investigation – presumably including that of solar geoengineering research – and encourages international cooperation in this area (Articles II–III). At their meetings, parties are to discuss, among other things, the facilitation of scientific research and the "preservation and conservation of living resources" (Article IX). All activities in Antarctica must be for peaceful purposes (Article I).

The 1991 Protocol on Environmental Protection of the Antarctic Treaty (the Madrid Protocol) further details states' commitments and rights concerning scientific research and environmental protection in Antarctica. For example, in its objective, parties commit to "the comprehensive protection of the Antarctic" and "designate Antarctica as a natural reserve, devoted to peace and science" (Article 2). Its first principle provides that both environmental protection and "scientific research, specifically research essential to understanding the global environment" are to be "fundamental considerations in the planning and conduct of all activities" (Article 3.1). Another principle is that "[a]ctivities shall be planned and conducted so as to accord priority to scientific research ... including research essential to understanding the global environment" (Article 3.3). The upshot of this is that solar geoengineering activities that would help protect the Antarctic environment and solar geoengineering research that would be free of adverse environmental impacts are each to be prioritized. However, it is unclear how parties should respond to proposals for solar geoengineering activities – especially research – that might have adverse impacts.

Regarding their substantive commitments, parties to the Madrid Protocol are to plan their activities in the Antarctic to limit adverse environmental impacts. This limitation includes avoiding, among other things, "adverse effects on climate or weather patterns; significant adverse effects on air or water quality; significant changes in the atmospheric, terrestrial (including aquatic), glacial or marine environments;" detrimental changes to the populations of plants and animals; putting endangered or threatened species at further risk; and degrading or putting at risk "areas of biological, scientific, historic, aesthetic or wilderness significance" (Article 3.2). These commitments all emphasize the tension between climate change and solar geoengineering, which respectively will and might have adverse effects on climate, the environment, species, and significant areas. Furthermore, parties are to plan their Antarctic activities based on sufficient information to allow a prior assessment of and informed judgments about the potential impacts on the continent's environment and its value as a site of scientific research. In line with this, parties are to cooperate as well as monitor, assess, and report the environmental impacts of their activities (Articles 3.2, 6, 8, 17). Scientific research programs are explicitly subject to prior environmental impact assessment.

Five Annexes that are in effect further the Madrid Protocol. The first details the requirements of environmental impact assessment. The second and fifth address protected animals and plants, and protected areas, respectively. If an activity such as solar geoengineering "results in the significant adverse modification of habitat," or if a research activity were planned to take place in a "Specially Protected or Managed Area," then a permit from the party's appropriate regulatory authority would be required (Annexes II, V).

8.4 OUTER SPACE TREATY AND ITS LIABILITY CONVENTION

Solar geoengineering could be done in outer space, the third and final area beyond national jurisdiction. This would be accomplished, at least in principle, by placing objects in orbit or at the L1 Lagrangian point between the Earth and the Sun.[11] Although these proposals presently appear to be prohibitively expensive, this might not always be the case. States' activities in outer space, an area that remains undefined, are governed by a set of multilateral agreements negotiated through the UN Committee on the Peaceful Uses of Outer Space. The 1967 Treaty on Principles Governing the Activities of States in the Exploration and Use of Outer Space, Including the Moon and Other Celestial Bodies (Outer Space Treaty) is foundational and counts all states with space programs as parties. In it, parties are to conduct space activities "for the benefit and in the interests of all countries ... in accordance with international law ... in the interest of maintaining international peace and security and promoting international cooperation and understanding" and "with due regard to the corresponding interests of all other States Parties" (Articles I, III, IX). Parties commit to inform the UN, the international scientific community, and the public as to their space activities and the results thereof (Article XI). All space activities must be conducted or authorized by the party's government, and the party is internationally responsible for those activities (Article VI). The Outer Space Treaty establishes a "freedom of scientific investigation in outer space," and parties should facilitate and encourage cooperation in carrying out research (Article I). These commitments would apply to solar geoengineering activities in space.

Perhaps the most interesting space law provisions regard liability for harm from space activities. The Outer Space Treaty establishes, and the Convention on International Liability for Damage Caused by Space Objects details, the provisions. Under these, the party (or parties) that launches, procures the launching, or provides the launching site has absolute liability (that is, with no need for the victim to demonstrate fault) for harm from their space objects to other parties in outer space, in the atmosphere, and on the Earth (Outer Space Treaty, Article VII; Liability

---

[11] The L1 Lagrangian point is the location between the Earth and the Sun where their gravitational forces are counterbalanced by the centripetal (center-seeking) force required for an object to orbit the Sun. An object there would be directly between the Earth and the Sun. This is approximately one percent of the distance from the Earth to the Sun, or about four times the distance from the Earth to the Moon.

Convention, Articles I(c), II). Multiple liable parties are jointly and severally liable (Liability Convention, Article V.1). This liability implicitly includes harm from accidents as well as from expected operations, and from direct contact as well as from remote effects. Space-based solar geoengineering is the only suggested technique under which the state that undertakes or authorizes it would clearly be liable through a multilateral agreement's provisions.

8.5 CONVENTION ON BIOLOGICAL DIVERSITY

The Convention on Biological Diversity (CBD) is one of the most important and far-reaching multilateral environmental agreements. Agreed upon at the 1992 UN Conference on Environment and Development in Rio de Janeiro, it now includes almost all the world's countries – except the United States – among its parties. The CBD can be described as a framework treaty whose commitments are general and often weak through qualifying language. Its objective extends beyond the conservation of biological diversity to encompass the sustainable uses of biological resources and the equitable sharing of benefits from genetic resources (Article 1). Furthermore, because many human activities, such as solar geoengineering, undertaken at sufficient scale would affect biodiversity, the CBD's broad scope and robust institutional support have led it to function as a vehicle for environmental protection in general. Indeed, its implementation has touched upon issues as diverse as economic development, trade, agriculture, tourism, and climate change.

With respect to conservation, solar geoengineering as a response to the risks of climate change could have positive, negative, or mixed impacts on biological diversity (see McCormack et al. 2016). A report from the CBD Secretariat concluded:

> Recent studies and assessments have confirmed that SRM [solar geoengineering] techniques, in theory, could slow, stop or reverse global temperature increases. Thus, if effective, they may reduce the impacts on biodiversity from warming, but there are high levels of uncertainty about the impacts of SRM techniques, which could present significant new risks to biodiversity. (Williamson and Bodle 2016, 10)

Two principles and two commitments of the CBD are relevant for solar geoengineering activities that might affect biodiversity. The Convention's singular explicit guiding principle is a restatement of states' sovereign right to exploit their own natural resources and their concomitant responsibility to prevent transboundary harm (Article 3; see Chapter 5). The CBD also invokes precaution but only by "noting" it in the document's Preamble (recital 9; see Chapter 5). In terms of commitments, the CBD parties are, "as far as possible and as appropriate," to identify activities that have or are likely to have significant adverse impacts on biodiversity and to monitor the effects thereof (Article 7). Article 14 of the CBD, the second relevant commitment, has provisions for three contexts. In terms of domestic effects, parties are, once again "as far as possible and as appropriate," to require environmental impact assessments for proposed activities that are likely to have significant adverse effects on biological diversity and to ensure that the impacts are "duly taken into account." In the case of likely significant transboundary impacts, the parties are to promote notification, exchange of information, and consultation, by encouraging multilateral arrangements as appropriate. Finally, if there is grave and imminent danger to biological diversity, CBD parties should have arrangements for emergency responses and encourage international cooperation. If these dangers are transboundary, then immediate notification and action are required.

The CBD Conferences of the Parties (COPs) have issued three nonbinding decisions concerning geoengineering in general. The first was made in 2010:

> The Conference of the Parties ... Invites Parties and other Governments ... to consider the guidance below ... Ensure ... in the absence of science based, global, transparent and effective control and regulatory mechanisms for geo-engineering, and in accordance with the precautionary approach and Article 14 of the Convention, that no climate-related geo-engineering activities that may affect biodiversity take place, until there is an adequate scientific basis on which to justify such activities and appropriate consideration of the associated risks for the environment and biodiversity and associated social, economic and cultural impacts, with the exception of small scale scientific research studies that would be conducted in a controlled setting in accordance with Article 3 of the Convention, and only if they are justified by the need to gather specific scientific data and are subject to a thorough prior assessment of the potential impacts on the environment. (Decision X/33, paragraph 8(w))

The 2012 decision by the CBD reaffirms the previous one; notes the ongoing lack of science-based, global, transparent, and effective control and regulatory mechanisms; and provides several possible definitions of climate-related geoengineering (Decision XI/20). It also points out that no geoengineering method "currently meets basic criteria for effectiveness, safety and affordability"; that "approaches may prove difficult to deploy or govern"; and that substantial gaps in understanding impacts remain. Finally, this COP decision states that customary international law, especially that regarding the prevention of transboundary harm, "may be relevant ... but would still form an incomplete basis for global regulation" (see Chapter 5). In its third decision regarding geoengineering, the CBD COP again reaffirms the key paragraph of its 2010 decision and concludes that:

> more transdisciplinary research and sharing of knowledge among appropriate institutions is needed in order to better understand the impacts of climate-related geoengineering on biodiversity and ecosystem functions and services, socio-economic, cultural and ethical issues and regulatory options. (Decision XIII/14)

Supporting this process, the CBD institutions have issued numerous geoengineering-related documents, such as reports within the Secretariat's Technical Series.

These decisions by the CBD COP are important, as they represent the only negotiated consensus among representatives of most of the world's states regarding geoengineering in general. The two substantive statements are ones of concern, calling upon all states to ensure that geoengineering activities of a certain type or scale do not take place until explicit criteria are met. For geoengineering to be covered by these decisions, the activity must be at a large enough scale that it would affect biodiversity, and an exception is made for "small scale scientific research studies that would be conducted in a controlled setting." Because "controlled setting" has not been defined, it remains unclear whether the exception is limited to indoor activities or could include low-risk outdoor ones. Each decision describes the conditions under which the request for preventing activities would be lifted, and these are highly congruous: there must be an adequate scientific basis, an assessment or consideration of risks, and (science-based) global, transparent, and effective control and regulatory mechanisms. Appropriate project justification and impact assessment procedures prior to a geoengineering activity could satisfy the first two conditions. In contrast, the need for global regulation is a challenging criterion to satisfy.

At the same time, the CBD COP decisions are limited in their effects, legal and otherwise, despite sometimes being referred to as moratoria (Chavez 2015, 17; Adelman 2017, 131; see Reynolds, Parker, and Irvine 2016). Most importantly, COPs do not have the power to create binding international law. These decisions are also hortatory and nonbinding. For example, the 2010 one merely "invites" states to "consider the guidance." Furthermore, it invokes Article 14 to clarify that the decision is limited to geoengineering activities that would have significant adverse impacts on biodiversity. Finally, negotiators' knowledge base was, at least for the 2010 decision, questionable, as one participant reported that negotiations "were characterized by a low level of understanding of geoengineering, especially from some of the parties more vehemently opposed to geoengineering research" (Parker 2014, 13; see also Sugiyama and Sugiyama 2010).

8.6 CONVENTION ON THE PROHIBITION OF MILITARY OR ANY OTHER HOSTILE USE OF ENVIRONMENTAL MODIFICATION TECHNIQUES

The Convention on the Prohibition of Military or Any Other Hostile Use of Environmental Modification Techniques (ENMOD) was completed in 1976 in order to end the use of weather modification techniques in warfare and other hostile situations. It is now in effect through the participation of its seventy-seven parties, which include almost all major industrialized states. ENMOD is salient here because its definition of "environmental modification" would encompass most, if not all, solar geoengineering proposals (Article II).

Centrally, the ENMOD parties agree "not to engage in military or any other hostile use of environmental modification techniques having widespread, long-lasting or severe effects as the means of destruction, damage or injury to any other State Party" (Article I.1). The trio "widespread, long-lasting or severe," which was later used in the amendment to the London Protocol regarding marine geoengineering, is not defined in ENMOD itself but is in an associated document. This describes them as:

> "widespread": encompassing an area on the scale of several hundred square kilometres; "long-lasting": lasting for a period of months, or approximately a season; "severe": involving serious or significant disruption or harm to human life, natural and economic resources or other assets. (Understandings regarding the Convention)

The first two criteria can be determined somewhat objectively, and large-scale solar geoengineering field research projects or implementation would most likely satisfy them. The criterion of severity, which is notably limited to life, resources, and assets, is more subjective.

Simultaneously, ENMOD recognizes and rhetorically supports peaceful environmental modification. The agreement explicitly "shall not hinder the use of environmental modification techniques for peaceful purposes" (Article III.1). Furthermore, in its Preamble, parties state "that the use of environmental modification techniques for peaceful purposes could improve the interrelationship of man and nature and contribute to the preservation and improvement of the environment for the benefit of present and future generations" (Recital 5). To that end, parties are to facilitate the exchange of information regarding such peaceful uses, and those parties "in a position to do so shall contribute ... to international economic and scientific co-operation in the preservation, improvement and peaceful utilization of the environment" (Article III.2). If solar geoengineering could counter climate change risks, then it would be such as "preservation, improvement and peaceful utilization," and parties with the capacity to do so seem committed to cooperate.

A challenge to ENMOD's implementation and enforcement is its weak institutional support. It has neither a standing secretariat nor regular meetings of its parties, although such meetings are infrequently proposed.[12] For example, the parties could clarify how ENMOD could contribute to the governance solar geoengineering.

8.7 CONVENTION ON INTERNATIONAL CIVIL AVIATION

Some solar geoengineering techniques, such as stratospheric aerosol injection and cirrus cloud thinning, could be researched or implemented using aircraft that would emit substances. In these actions, states would need to comply with international

---

[12] The ENMOD Parties have held two meetings, but most recently declined to do so in 2013.

aviation law. The central multilateral agreement, the Convention on International Civil Aviation (the Chicago Convention), which enjoys global participation, recognizes that parties have exclusive sovereignty over their airspace (Article 1). This airspace extends upward from their territory – from both land and territorial waters – to the undefined upper border with outer space (Article 2). Parties must ensure that aircraft bear a nationality mark and comply with other standards (Articles 17–21, 29–36).[13] States are to permit unscheduled flights through their airspace provided that the foreign aircraft are not military, customs, or police aircraft and that the aircraft operate with a purpose consistent with the aims of the Chicago Convention (Articles 3, 5).[14] Scheduled flights as well as pilotless, military, customs, and police aircraft require prior authorization (Articles 3(c), 6, 8). Flights over the high seas are required also to follow the Convention's rules (Article 12). Parties may establish their own regulations for aircraft operating in their territory, as long as these rules are enforced without distinction to the aircraft's country of registration (Article 11). However, parties are expected the keep the rules relatively uniform and consistent with guidelines established by the International Civil Aviation Organization, a UN body established by the Chicago Convention (Articles 12, 37, 43–79).

Substances intentionally injected into the atmosphere for solar geoengineering could be considered aircraft emissions. The International Civil Aviation Organization has guidelines regarding specific pollutants from emissions, but these do not address materials such as sulfates that are presently considered for solar geoengineering (International Civil Aviation Organization 2008). Therefore, atmospheric solar geoengineering – even in the airspace of countries other than that of the aircraft's nation of registry – appears to be compliant with the letter of international aviation law. However, the state in whose airspace the solar geoengineering activity occurred could claim that such flights are inconsistent with the aims of the Chicago Convention or the international law of territorial sovereignty.

8.8 CONVENTION ON ENVIRONMENTAL IMPACT ASSESSMENT IN A TRANSBOUNDARY CONTEXT

The Convention on Environmental Impact Assessment in a Transboundary Context (the Espoo Convention) is the first of two procedurally-oriented multilateral environmental agreements developed through the UN Economic Commission for Europe (UNECE). Finalized in 1991, its current forty-five parties include the European Union, several former Soviet countries, and Canada. If a proposal, such

---

[13] Under the related Convention on Offences and Certain Other Acts Committed on Board Aircraft, the state of registration can exercise jurisdiction over aircraft (Article 3).

[14] The Chicago Convention's aims are unstated beyond the Preamble's exhortations regarding "friendship and understanding among the nations and peoples of the world," the desire "to avoid friction and to promote ... cooperation between nations and peoples," and the goal to develop civil aviation "in a safe and orderly manner" (Recitals 1–3).

as one for solar geoengineering activity, is subject to a decision of a national competent authority and would likely have adverse transboundary impacts, then the authorizing party is subject to the Espoo Convention's commitments (Article 1).

The only commitments that apply to all significant adverse transboundary impacts require parties to "take all appropriate and effective measures to prevent, reduce and control significant adverse transboundary environmental impact from proposed activities" (Article 2.1). The Espoo Convention parties are also to give special consideration to establishing and expanding research programs for better understanding environmental impacts, a commitment that could encourage solar geoengineering research programs (Article 9). The other commitments mostly apply to only the specific proposed activities listed in an Appendix, none of which could be reasonably construed to include any currently proposed solar geoengineering technique. The Espoo Convention also calls on its parties, "to the extent possible," to conduct environmental assessments on their relevant policies, plans, and programs, implicitly only those that concern activities covered by the Convention (Article 2.7).

A proposed activity that is not listed but is deemed likely to have significant adverse transboundary impacts can still be subject to the obligations of the Espoo Convention. For this to be the case, a concerned party (presumably the state of origin of the unlisted proposed activity) is obligated to enter discussions with another party (presumably a potentially affected state), and if they both so agree, they are to treat the proposed activity as if it were listed (Article 2.5). When and if parties agree to this, then the party of origin is subject to several procedural duties. These include, prior to a decision to authorize, conducting an environmental impact assessment that provides particular information, such as a description of the proposed activity, details of its likely impacts, possible steps to mitigate adverse impacts, potential alternative activities, and uncertainties (Articles 2.2–2.3, 4, Appendix II). Furthermore, the party of origin is obligated to notify and consult with other affected parties, and the public in the likely affected areas are to have the opportunity to participate (Articles 2.4, 2.6, 3, 5). The party of origin should take the outcome of the assessment, the public comments, and the international consultations into account in making the final decision regarding whether and how to proceed with the proposed activity (Article 6).

The Espoo Convention is supplemented by its Protocol on Strategic Environmental Assessment, which entered into force in 2010 and presently has thirty-two European parties, including the European Union. They commit to undertake strategic environmental assessments of certain listed categories of official draft plans and programs that are likely to have significant environmental or health effects (Article 4.1). This obligation might extend to plans and programs for solar geoengineering. Notably, the definitions of likely effects to be considered include, among other things, those on climate (Article 2.7). Parties to the Protocol also commit to ensuring the public availability of relevant information, and to providing opportunities for public participation and consultation. If a plan or program would likely have significant transboundary environmental and/or health effects, the party of origin is to notify and consult with the potentially affected party, including providing an opportunity for public comment in the affected state.

8.9 CONVENTION ON ACCESS TO INFORMATION, PUBLIC PARTICIPATION IN DECISION-MAKING AND ACCESS TO JUSTICE IN ENVIRONMENTAL MATTERS

The 1998 Convention on Access to Information, Public Participation in Decision-Making and Access to Justice in Environmental Matters (the Aarhus Convention) is another procedurally oriented UNECE multilateral environmental agreement. In this case, its forty-seven parties pursue the three objectives given in the Convention's title and generally guarantee these as rights (Article 1). In contrast to most multilateral environmental agreements, the effects need not be transboundary to trigger obligations. Furthermore, parties are to offer these rights to the public of the state of origin and of other parties in a nondiscriminatory manner (Article 3.9). In these ways, the Aarhus Convention resembles a human rights agreement.

Concerning the first objective, "environmental information" is information that concerns the state of the environment and its elements as well as factors – including public policies, plans, and programs, and the analyses upon which they are based – that affect or are likely to affect the environment (Article 2.3). Information regarding solar geoengineering activities such as outdoor tests and implementation clearly falls within this definition. Parties' public authorities are obligated to make relevant environmental information available to the public in accordance with requirements, such as timeliness, and with limitations, such as preventing adverse effects on intellectual property rights (Article 4). They are also to proactively establish and maintain systems for the collection and dissemination of environmental information (Article 5). Requests for information need not demonstrate a specific interest, such as actual or potential harm.

Regarding the second objective, the public that might be affected as well as environmental nongovernmental organizations have the right to participate in decision-making regarding whether to permit proposed activities that could affect the environment. This right could include any activities specifically listed in the Convention's Annex, any activity for which domestic impact assessment legislation provides public participation, or unlisted activities that might have a significant effect on the environment (Article 6; Annex I). Many outdoor solar geoengineering activities would fall into the second or third category. Once this obligation is triggered, parties must ensure that the affected public and environmental groups are informed and given access to relevant information and the opportunity to participate, including by providing comments. Public authorities are to take participation and comments into account in their decision-making.

Some of these provisions extend, with qualifications, to plans, programs, and policies (Articles 7–8).

The third objective of the Aarhus Convention, access to justice, addresses deficiencies in how parties could carry out the first two objectives. The Convention sets minimum standards for redress for members of the public who have been denied environmental information or who wish to challenge a prior decision concerning environmental matters in which they have a sufficient interest (Article 9). Parties are to ensure that courts or other independent tribunals can enforce the rights granted in the Aarhus Convention.

The Aarhus Convention is furthered by its Kiev Protocol on Pollutant Release and Transfer Registers, which is in effect through the participation of thirty-two states as well as the European Union. Its parties are to ensure the public availability of information regarding transfer and release of pollutants (Article 1). In this context, a pollutant is any substance that may be harmful to human health or the environment due to its introduction into the environment, and includes both accidental and deliberate release (Article 2). Thus, substances intended for use in solar geoengineering might qualify as pollutants, depending upon their expected impacts. The Protocol's parties are to establish and maintain a publicly accessible register of such information, which is to include information from nonstate actors acquired via mandated reporting (Articles 3–11).

8.10 RESOLUTION OF INTERNATIONAL DISPUTES

States sometimes disagree and might do so over solar geoengineering. Most often, they resolve their disputes through nonlegal and political means. Dispute resolution occasionally moves to more legalized approaches, such as negotiation, mediation, and arbitration. Some multilateral agreements have specialized mechanisms, including compulsory arbitration or special tribunals, to try to resolve disputes within their scope.

Four international legal fora have broad ranges of issues that they can address, potentially including solar geoengineering (see also Chapter 4). First, the UN General Assembly has universal participation and can consider almost any matter, but its resolutions are nonbinding. Second, the UN Security Council can issue resolutions that are considered legally binding, even on states that object (see Armeni and Redgwell 2015c, 12; Chalecki and Ferrari 2018). The Security Council could thus serve as a forum of last resort to address problematic uni- or minilateral implementation of solar geoengineering. As described in Chapter 4, its mandate is limited to the "maintenance of international peace and security," which could reasonably include disputes concerning solar geoengineering (Charter of the UN, Article 24.1.) The Security Council has also gradually widened its scope, including by considering how climate change relates to security (United Nations Security Council S/PRST/2011/15, S/RES/2349, S/RES/ 2408). At the same time, five of the world's most powerful countries have veto power, and any enforcement, such as sanctions or military action, must be performed by willing states. Third, the International Court of Justice (ICJ) can resolve international disputes. However, for its rulings in contentious issues to be legally binding, each state must consent to the court's jurisdiction prior to the trial. By majority vote of the UN General Assembly, a legal question can be referred to the ICJ for an advisory opinion. Such an opinion could speak to the obligations of states that have not consented to its jurisdiction for such matters, but it would not be binding on them. Finally, states can agree to submit a dispute to the Permanent Court of Arbitration, although it lacks standing judges and is thus better considered as an institution for facilitating arbitration.

8.11 SUMMARY AND CONCLUSION

States make explicit and implicit mutual promises to manage many of the same problems that individuals in a society face. These international promises have coalesced into something that resembles law. Some of these rules from custom, treaty, and principles as well as international institutions contribute to the governance of solar geoengineering. In this domain, international law provides some evaluative norms and principles, institutional sites of potential future decision-making, and rules of conduct that are primarily – but not entirely – procedural, such as notification, assessment, and consultation. Its substantive implications are mixed, in part due to the tension between climate change and solar geoengineering. Because international environmental law is anthropocentric in its normative orientation and supports scientific research and technology development, and because solar geoengineering presently appears able to reduce the greater risks of climate change, international law is on the whole tilts favorably toward solar geoengineering research. Despite its gaps – such as liability and compensation for harm – it provides both substantial governance of solar geoengineering and a foundation upon which future norms, rules, procedures, and institutions can be built.

9

US Law

Chapters 5 to 8 described how, because the large-scale field testing and implementation of solar geoengineering would have transboundary impacts, international law will be salient in the long run. However, before then, the techniques would presumably be researched at smaller scales, and national and subnational would be law applicable. These legal regimes are more specific, adaptable, and consistently enforced than their international counterpart. National and subnational law is an essential component of the existing governance framework within which solar geoengineering is developing.

To offer an example of how national law would govern outdoor solar geoengineering activities, this chapter reviews the applicable US law. It uses this country because it is the world's leading site and funder of scientific research and home to two research groups that are presently planning outdoor solar geoengineering experiments (see Chapter 2). Furthermore, US environmental law has set an example for other countries and arguably remains the most well-developed, although not necessarily the most effective, in the world.[1] This chapter first considers three major pieces of federal environmental legislation: the Clean Air Act, the National Environmental Policy Act (NEPA), and the Endangered Species Act. Liability for harm, which is found in the common law of torts, is only briefly introduced due to its variance among the states. Thereafter, I describe laws relevant to weather modification and marine pollution, as well as a federal geoengineering bill.

This chapter addresses, to the extent appropriate and possible, the extraterritorial application of US law. Although there is a default presumption against extraterritorial application, solar geoengineering activities within the territorial US could have transboundary impacts, and they could take place outside its territory, such as in the maritime exclusive economic zones (EEZs), the high seas, Antarctica, and outer space. US courts have identified two relevant characteristics that point toward the applicability of federal environmental law for certain US activities in areas beyond

---

[1] At the same time, I do not intend to further the United States' over-representation in scholarly writing and encourage others to research how diverse countries might regulate solar geoengineering (see Armeni and Redgwell 2015a; Armeni and Redgwell 2015b).

national jurisdiction. First, a criterion is the presence of a domestic effect, that is, within the United States.[2] Large-scale solar geoengineering activities conducted in an area beyond national jurisdiction would likely have a domestic effect, although such questions have not been before the courts with respect to environmental laws. Furthermore, a federal appeals court assessing NEPA's application in Antarctica argues that "where the US has some real measure of legislative control over the region at issue, the presumption against extraterritorial application is much weaker" (*Environmental Defense Fund, Inc. v. Massey*). Legal scholar Randall Abate ultimately concludes that in areas beyond national jurisdiction, "the presumption [against extraterritorial application] applies with little or no force absent any other foreign policy concerns that a court may find" (Abate 2006, 104).

I do not consider three aspects of US law here. First, the American legal situation is made more complex through federalism, in which the states retain legal competences in numerous areas, including some environmental ones.[3] For brevity, this chapter does not consider state law. Nevertheless, it is notable that the first proposed legislation to govern geoengineering specifically was introduced to the Rhode Island House of Representatives in three consecutive years, and a special House commission has been established to consider the matter.[4]

Second, some laws are insufficiently interesting for explication. On one side, whether the federal government would regulate stratospheric aerosol injection as a form of air pollution, for example, is clearly relevant and interesting. On the other are numerous aspects of solar geoengineering that would be mundane. The National Science Foundation is implicitly authorized to fund solar geoengineering research. Pilots who might fly aerosol-injecting aircraft are required to be licensed. Marine cloud brightening ships would need to have vessel identification numbers. I focus on the aspects of solar geoengineering that cause it to be distinct from other, more day-to-day outdoor and scientific undertakings.

Third, some legislation considered here, including NEPA and the Clean Air Act, contain exceptions for emergency situations. Scenarios of solar geoengineering have sometimes included emergencies, in which it is quickly implemented in response to climate change impacts (see Chapter 4). However, sudden climate change would unlikely be an emergency under federal law due to the different timescales at hand. Even at its most rapid, climate change would still take years to unfold, whereas typical US regulatory and review processes require only several months to a couple

---

[2] For example, see *United States v. Aluminum Co. of America (Alcoa)*; *Steele v. Bulova Watch Co., Inc.*; *Hartford Fire Insurance Co. v. California*.

[3] In this chapter, "state" means one of the fifty constituent states of the United States, whereas elsewhere it is synonymous with country or nation.

[4] Rhode Island H 7655 (2014); H 5480 (2015); H 7578 (2016); H 5607 (2017); H 6011 (2017); H 6011 Substitute A (2017).

of years (Alley et al. 2003). Consequently, this chapter does not review these and other provisions for emergencies (but see Lin 2018).[5]

9.1 CLEAN AIR ACT

The Clean Air Act is among the most important US environmental laws, regulating air pollutants from stationary and mobile sources and from new and existing ones. Its purposes are to maintain ambient air quality for public health and welfare as well as to address specific environmental problems such as acid rain and visibility impairment. The Environmental Protection Agency (EPA) carries out the Act federally, while the states have substantial responsibility for developing and implementing specific regulations to meet the standards. Notably, in light of Congress's ongoing inability to pass climate change legislation, the Barack Obama administration used its executive authority to regulate greenhouse gas emissions under the Clean Air Act.

The materials used in stratospheric aerosol injection or cirrus cloud thinning could satisfy the Clean Air Act's broad definition of "air pollutant," which is "any air pollution agent or combination of such agents ... which is emitted into or otherwise enters the ambient air" (§7602(g)). In turn, the phrase "air pollution agent" is not defined. Furthermore, the Clean Air Act does not require consideration of a substance's altitude in determining whether it is an "air pollutant," at least according to a majority ruling of the Supreme Court. There, in a pivotal climate change case, the Court's majority rejected a dissenting judge's assertion that the stratosphere is beyond the "ambient air" and instead noted that the Act "uses the phrase 'the ambient air' without distinguishing between atmospheric layers" (*Massachusetts v. Environmental Protection Agency*, footnote 26). However, the EPA regulates pollutants only once its Administrator has decided that they "cause or contribute to air pollution which may reasonably be anticipated to endanger public health or welfare" (§7408(a)(1)(A)). Thus, although air pollution's altitude does not matter, its impacts on the public do matter, and these impacts must occur on the ground. Regardless, if the EPA Administrator rules that an air pollutant endangers public health or welfare, then the agency will typically develop national ambient air quality standards for the pollutant. Each state must subsequently create and institute a state implementation plan for areas that fail to meet the air quality standard and have it approved by the EPA (§7410(a)).

The EPA presently regulates six substances as such "criteria air pollutants." Aerosol injection could, in principle, increase the atmospheric loads of two of these substances, sulfur dioxide and particulate matter. Sulfates are presently the most considered material for stratospheric aerosol injection, and aerosols are a form

---

[5] However, an extreme weather event that is partially caused by climate change could engender technological responses, such as suppressing a hurricane, in ways that resemble solar geoengineering.

of particulate matter. Scientists might inject relatively small amounts of sulfates or other aerosols at lower altitudes for early research, which would have little impact on atmospheric concentrations, public health, and welfare. Larger amounts could be later injected at higher altitudes, where the substances would disperse over wider areas, diluting their ground-level concentrations. It thus seems that outdoor solar geoengineering activities are unlikely to have significant effects on public health and welfare through such air pollutants. Nevertheless, whether through the extant regulation of sulfur dioxide and particulate matter or through a new criteria pollutant, aerosol injection could be subject to state implementation plans, at least in principle. Among other things, these require major stationary sources – those that emit at least ten tons per year – to obtain and abide by a permit (§§7412(a)(1), 7661). These provisions substantially bound the Clean Air Act's capacity to govern aerosol injection research and implementation.

There are four ways in which the EPA could further regulate aerosol injection under the Clean Air Act but could not yet do so due to either administrative decisions or the letter of the law. First, as noted, the EPA administrator could rule that aerosol injections through substances other than sulfur dioxide and through physical manifestations other than particulate matter endanger public health and are criteria air pollutants. However, the EPA generally considers only the potential pollutant's chemical nature, not the means or intention of its release. Thus, this means of EPA action would likely either have regulatory impacts wider than aerosol injection or require a novel criterion for air pollutants.

Second, the Clean Air Act authorizes the EPA to regulate mobile sources of pollution, including nonroad engines and vehicles (§7547). Aircraft and marine vessels could be used as sources of aerosol injection. However, the relevant regulations currently limit neither sulfur dioxide nor particulate matter (except for the latter pollutant from marine compression-ignition engines, which are found in larger ships). Furthermore, these apply to the emissions from only the vehicles' engines and fuel systems, whereas aerosols would most likely be injected via dedicated hardware. The EPA's regulatory authority over US mobile sources of pollution notably extends to their operation outside of US territory. This could therefore include aircraft and marine vessels that are registered in the United States and conduct solar geoengineering activities outside of it.

Third, the EPA is to regulate the emission of acid rain precursors from stationary facilities that combust fossil fuels, whose operators must purchase a tradable permit (§7651). Stratospheric injection of sulfate aerosols for solar geoengineering testing or implementation would contribute to acid rain. Yet their injection would not be subject to EPA regulation based upon this characteristic, as aerosols or their precursors are unlikely to be emitted from stationary fossil fuel facilities.

Fourth, the EPA has the authority under the Clean Air Act to regulate ozone depleting substances (§7671). This is the domestic implementation of the Vienna Convention and its Montreal Protocol (see Chapter 6). Some substances, including sulfates, that are being considered for aerosol injection could deplete stratospheric ozone. However, none of these are presently among those that the EPA lists for this category of regulation.

Notably, the Clean Air Act allows not only for federal enforcement of its regulatory provisions but also for potentially affected US residents to bring civil actions (§§7413, 7604). Although solar geoengineering seems unlikely to harm ground level air quality in the context of the Clean Air Act, this could present an opportunity for opponents to use the courts to thwart solar geoengineering activities.

The Clean Air Act's implementation can consider transboundary impacts of US sources (§7415). This requires either that an international agency reports transboundary risks from US sources of air pollution or that the US Secretary of State requests the EPA Administrator to act. If the transboundary risks are to a country that offers reciprocal consideration of its own risks to the United States, then the EPA can require that the US state that is the source of the pollution modifies its implementation plan accordingly.

Finally, the Clean Air Act provides for the promotion and coordination of research regarding the prevention and control of air pollution (§7403). This includes the monitoring, analysis, modeling, inventory, health effects, ecosystem effects, prevention, and control of air pollutants. The EPA conducts some research in-house, such as through its National Risk Management Research Laboratory, and supports activities elsewhere through grants, conferences, and other activities. This promotion and coordination could support research of solar geoengineering's potential effects.

9.2 NATIONAL ENVIRONMENTAL POLICY ACT

The NEPA is central to US environmental governance due to its broad scope. It requires that the federal government consider the environmental impacts of any major project or program that it undertakes, funds, or approves that might "significantly affect the quality of the human environment" (§4332(c)). As a first step, the involved federal agency must either categorically exclude the proposed project or program as lacking significant environmental impact based upon past assessments, prepare a preliminary environmental assessment, or directly conduct a full environmental impact statement (40 CFR 1501–8). In turn, a preliminary environmental assessment should find either that the action will have no significant impact or that the agency will need to prepare a full environmental impact statement. The environmental impact statement should describe expected environmental impacts, which adverse environmental effects are unavoidable, alternatives to the proposal, and irreversible commitments of resources. Before the environmental impact statement is finalized, a draft version must be made public for comments, to which the agency should respond.

Most outdoor solar geoengineering activities of even a modest scale would likely be at least partially federally funded and require an environmental impact statement. Furthermore, as argued in Chapter 4, national governments including that of the United States would insist that efforts to modify the climate are within their control, not those of nonstate actors. The US government can therefore be expected to be increasingly involved if and when outdoor solar geoengineering activities increase in scale substantially.

Like the Clean Air Act, NEPA could offer a platform for opponents of a federally supported solar geoengineering program or project to intervene. In the past, lawsuits regarding whether an agency had complied with the Act in its consideration of environmental impacts have delayed and, in some cases, prevented projects from going forward. Opponents of solar geoengineering could seek to do likewise.

NEPA requires assessment of possible environmental impacts that might occur outside of US territory and that might arise due to US activities in areas beyond national jurisdiction. A guidance document on transboundary impacts in federal environmental impact assessments cites the lack of territorial limitation in the Act, case law, and customary international law and concludes that "agencies must include analysis of reasonably foreseeable transboundary effects of proposed actions in their analysis of proposed actions in the United States" (Council on Environmental Quality, Guidance on NEPA Analyses for Transboundary Impacts). Furthermore, the United States has granted standing in court for potentially affected foreign persons to challenge impact assessments. Regarding activities in areas beyond national jurisdiction, an official from the State Department stated in Congressional testimony during NEPA's original consideration that the law could and should apply to the seas, outer space, and Antarctica (Herter 1971). Since then, courts have held such with regard to activities in the EEZs, the high seas, and Antarctica (*Environmental Defense Fund, Inc. v. Massey*; *Natural Resources Defense Council v. US Department of Navy*; see Abate 2006; Nash 2009). US space activities are also subject to environmental impact assessments, and these consider global environmental impacts (Viikari 2008, 274).

9.3 ENDANGERED SPECIES ACT

The Endangered Species Act governs US activities that could have negative effects on protected species and their habitats. Federal agencies are to consult with one of two wildlife agencies (the National Marine Fisheries Service for marine species or the Fish and Wildlife Service for those on land and in freshwater) to ensure that the activities that they plan to authorize, fund, or undertake would neither jeopardize endangered or threatened species' existence nor have destructive or adverse impacts on their critical habitat (§1536). If the wildlife agency finds that the proposed action would indeed jeopardize the species or adversely modify its habitat, then it issues an opinion that includes alternatives. In addition, under the Endangered Species Act, legal persons may not "take" – that is, harass, harm, pursue, hunt, shoot, wound, kill, trap, capture, or collect, or attempt to do so – an endangered species (§1532). Such prohibited takings also include habitat modifications that would result in injuring or killing protected species (50 CFR 17.3). Federal agencies as well as persons can request, and might receive, a determination that their actions would be a permissible incidental taking. The Act can be enforced through both federal action (including criminal charges) and citizen lawsuits (§1540). The Endangered Species Act not only restrictively regulates the actions of public and nonstate actors but also calls for them to take proactive steps. The federal government must, in most cases, prepare a recovery plan for endangered or threatened species (§1533 (f)). Moreover, a person who seeks a determination of an incidental taking is required to include a habitat conservation plan with steps to minimize and mitigate the impacts (§1539(a)(2)).

In the solar geoengineering context, the Endangered Species Act's most likely role appears to be that of a potential regulatory barrier to proposed large-scale outdoor activities. If a planned experiment posed substantial risks to threatened or endangered species or their habitats, then a federal agency authorizing, funding, or undertaking it would need to explore alternatives, and a nonstate actor might be prohibited from performing the solar geoengineering activity. Furthermore, citizen suits could offer a means for opponents to impede the proposal. However, the challenger would need to demonstrate a causal relationship between the planned solar geoengineering activity and threats to the species, which might be difficult due to remaining uncertainty.

At the same time, the fact that solar geoengineering is being researched with the hopes of reducing dangerous climate change and might consequentially help protect endangered species raises the question of whether such activities could be part of an endangered species recovery plan or a habitat conservation plan. Ultimately, doing so would face challenges of uncertain causation, sufficient precision of effects on the species, and the ability of the federal agency or legal person that submits the plan to perform the solar geoengineering activity.

The Endangered Species Act applies beyond US territory. The threatened and endangered species that it governs are not limited to those with US habitat. The Act also authorizes and allocates funds for international cooperation in the conservation of endangered species and helps implement the Conventions on International Trade in Endangered Species of Wild Fauna and Flora and on Nature Protection and Wildlife Preservation in the Western Hemisphere (§1537A). Federal agencies' consultation requirement and the prohibition on the taking of endangered species by persons subject to US jurisdiction apply to actions in both American territory and the high seas (§1538(a)(1); 50 CFR 402.02). That is, they implicitly do not apply to actions in other countries, Antarctica, and – for what it is worth – outer space. Although a federal appeals court ruled that federal agencies should, in fact, consult regarding their actions' impacts in foreign countries, the Supreme Court overturned this based upon the plaintiff's standing (*Defenders of Wildlife v. Lujan*; *Lujan v. Defenders of Wildlife*). Ultimately, any solar geoengineering activity with federal involvement that would pose significant risks to endangered species or their critical habitat in US territory or the high seas would require interagency consultation, and any such activities by persons under US jurisdiction that would result in a taking of threatened or endangered species in the same areas would be prohibited in the absence of a finding of an incidental taking.

Finally, other US laws that protect wildlife could be applicable to solar geoengineering, at least in principle. However, the two most important ones have narrow definitions of the prohibited actions. The Migratory Bird Treaty Act applies only to actions or attempts to "pursue, hunt, shoot, wound, kill, trap, capture or collect," and the Marine Mammal Protection Act only to those "to harass, hunt, capture, or kill" (50 CFR 10.12; 16 USC 1362).

9.4 WEATHER MODIFICATION LAWS

Outdoor solar engineering activities of sufficient scale would satisfy the definition of weather modification under US law.[6] In particular, the Weather Modification Reporting Act defines it as an activity "with the intention of producing artificial changes in the composition, behavior, or dynamics of the atmosphere" (§330). Subsequent administrative rulemaking strengthened the implicit inclusion of solar geoengineering by including "Modifying the solar radiation exchange of the earth or clouds, through the release of gases, dusts, liquids, or aerosols into the atmosphere" and "Seeding or dispersing of any substance into clouds or fog, to alter drop size distribution" (15 CFR 908.3). Any legal person who undertakes or attempts to engage in weather modification within the United States must report it to the Secretary of Commerce, who has delegated this responsibility to the National Oceanic and Atmospheric Administration (15 CFR 908). A person who attempts to modify the weather and fails to report it is subject to a fine. Furthermore, the Secretary is to maintain records and publish summaries of weather modification activities. Finally, pursuant to administrative regulations, if the National Oceanic and Atmospheric Administration concludes that a report of a weather modification activity "may significantly depart from the practices or procedures generally employed in similar circumstances to avoid danger to persons, property, or the environment," then it is to notify both the operator of the activity and its home state, as well as to offer recommendations, as appropriate (15 CFR 908.12). All these provisions would apply to outdoor solar geoengineering projects that would modify incoming solar radiation, even locally.

---

[6] For more depth on the similarities between weather modification and solar geoengineering and how they relate to governance, see Hauser (2013).

The 1976 National Weather Modification Policy Act amended the Weather Modification Reporting Act to authorize the Secretary of Commerce to develop a national weather modification policy. It has not done so to date.

A handful of courts have ruled on liability for harm from weather modification. Although none were federal courts, they drew on common law principles that would have wide applicability. Of six cases in which plaintiffs sought injunction or ex post damages from those engaged in weather modification, only two were granted temporary restraining orders, and of these, only one established causation. According to Gregory Jones, "The primary reason plaintiffs have failed in court is that it is difficult, if not impossible to prove causation" (Jones 1991, 1169).

9.5 LIABILITY FOR HARM

A person who harms another can be held legally liable. In the United States, this law of torts is largely governed by common – or judge-made – law and varies somewhat by state. Furthermore, numerous states have legislation that governs weather modification, including liability for harm that it causes (Jones 1991). Some of these could apply to solar geoengineering activities (see Hester 2018). Because of this variance, this section introduces only the key points (see also Chapter 12).

Most liability rules require that the injurer acts (or fails to act) in a way that causes a harmful result. The bases of liability of harm from outdoor solar geoengineering activities could include bodily harm, private nuisance (that is, interference with enjoyment of private property), or public nuisance (that is, interference with enjoyment of a commonly held resource or right). For these, the victim would need to demonstrate harm.

To be held liable for harm, the injurer usually must also breach a duty of care. The most common of these standards is negligence, which is the level of care that a reasonable person would exercise in similar circumstances. This often includes a duty to notify potential victims. The negligence standard applies to most accidents and would include the more mundane aspects of solar geoengineering. For example, if large equipment damaged a house during transportation, the homeowner could bring a civil suit. The judge would ask whether the driver or transportation company was negligent in his or its behavior that caused the harm.

Accidents from abnormally dangerous activities do not require negligence or the breach of some other duty of care for the injurer to be liable. Instead, she is strictly liable. An abnormally dangerous activity is one that "creates a foreseeable and highly significant risk of physical harm even when reasonable care is exercised by all actors; and ... is not one of common usage" (American Law Institute 2009, 229). Courts have also considered the activity's location and whether its public benefits are outweighed by its risks. Solar geoengineering appears to satisfy these criteria of an abnormally dangerous activity, at least presently, in its climatic and other environmental impacts. Yet what is and is not an activity of common usage changes over time. In the future, solar geoengineering could be sufficiently routine and/or its risks of physical harm might not be highly significant when reasonable care is exercised that it would no longer be considered abnormally dangerous, and victims would need to demonstrate negligence.

Regardless of the duty of care, a liability suit would need to demonstrate causation to be successful. This would be difficult, as noted above in the cases for harm from weather modification. In the US common law of torts, causation usually has two necessary conditions. The first of these, factual causation, is relatively less daunting in that it requires only that the harm would not have occurred but for the injurer's actions. Yet even this could be challenging, given weather's variability, which will be more complex as the climate changes due to anthropogenic greenhouse gases. The second, proximate causation, requires that the harm be sufficiently related to the injurer's actions, and its criteria are less clear.[7]

An injurer accused of causing harm from solar geoengineering could offer defenses that would excuse her from all or some potential liability. Among these are contributory and comparative negligence. In these cases, the question would be whether the victim had taken reasonable care to prevent or minimize expected harm. For example, if it is well known that solar geoengineering would reduce precipitation in a given region, a farmer who fails to appropriately change which crops he plants and his irrigation practices might be deemed contributorily or comparatively negligent. Another possible defense is necessity, although this typically applies to intentional harm done in emergency situations.

If found liable, remedies usually include financial damages for harm and sometimes punitive awards. In addition, a court can grant injunctive relief and order the injurer to cease an ongoing harmful activity.

9.6 MARINE LAWS

Marine solar geoengineering activities could be undertaken by US persons, on US-flagged vessels, or in American territorial waters or EEZs. The most important law governing such activities is the Marine Protection, Research and Sanctuaries Act – sometimes called the Ocean Dumping Act – which regulates the disposal of materials at sea and implements the London Convention (see Chapter 13). It broadly defines both dumping ("a disposition of material") and material ("matter of any kind"), each with some exceptions (§1402). Stratospheric aerosol injection and cirrus cloud thinning in the marine environment could conceivably qualify as "dumping," as they would be dispositions of a material that would subsequently enter marine waters. If so, such dumping would require a permit, which the EPA is to grant if it "will not unreasonably degrade or endanger human health, welfare, or

---

[7] Proximate causation can also be understood as determining the scope of harms for which an injurer may be liable.

amenities, or the marine environment, ecological systems, or economic potentialities" (§1412(a)). The specific criteria for evaluating a permit request include the need for the dumping activity and its expected effects on human health and welfare, fisheries, wildlife, and marine ecosystems. The information received pursuant to the request is public (§1414(f)). Violators of the Ocean Dumping Act can be subjected to substantial fines and imprisonment (§1415) .

Vessels' introduction of materials into the marine environment for solar geoengineering could also be considered pollution, discussed in the context of the Clean Air Act's regulation of ships' emissions. The Act to Prevent Pollution from Ships also regulates emissions from maritime vessels and implements the Convention for the Prevention of Pollution from Ships (MARPOL) and Annex IV on Marine Pollution of the Antarctic's Treaty Madrid Protocol. The Act applies to US flagged ships; to ships that enter US navigable waters, including to US ports; to vessels that bear the flag of another party to MARPOL's Annex VI that are in either a US EEZ, an Annex VI emission control area, or federally designated area of concern regarding emissions; and to platforms (§1902). Such vessels are required to possess a certificate from a MARPOL party that demonstrates their compliance with that agreement and its annexes. They are subject to inspection while at US ports and terminals to determine whether they comply. A ship either that does not or whose condition is contrary to the certificate may be detained (§1904).

9.7 GEOENGINEERING RESEARCH EVALUATION ACT OF 2017

In December 2017, Representatives Jerry McNerney and Eddie Bernice Johnson, Democrats of California and Texas, respectively, introduced the first national bill in any country concerning solar geoengineering. The Geoengineering Research Evaluation Act was modest. Under it, the government would have contracted with the US National Academies to develop and issue two reports. The first would have described a possible federal solar geoengineering research agenda. The second would have proposed governance mechanisms for the solar geoengineering research. The latter "should seek to maximize the benefits of research while minimizing risks" (Section 3(b)(1)). The proposed legislation would have also directed the Office of Science and Technology Policy, an office within the White House, to issue a plan to implement the research agenda and governance mechanisms. The bill was not voted upon and is now moot, as in 2018 the National Academies began a project to produce a report on these two topics.

9.8 SUMMARY AND CONCLUSION

After moving out of the laboratory and before having potential transboundary impacts, solar geoengineering research would most likely be of modest scales. These outdoor experiments might present some risks to humans and other species.

Existing national legislation and common law, such as that in the United States described in this chapter, would govern solar geoengineering. In some cases, relatively minor changes to administrative regulations or legislation would further enable national law to do so. This situation would be similar in most other countries with robust environmental governance regimes.

10

Nonstate Governance

Most discussions of solar geoengineering and its governance – including this book thus far – understandably focus on states as actors.[1] For example, countries might subsidize, authorize, implement, and otherwise govern solar geoengineering activities. Yet most decision-making occurs outside the public sector. This chapter examines nonstate actors' potential roles in governing solar geoengineering, while Chapter 11 takes on nonstate actors – including commercial ones – developing, patenting, and deploying relevant technologies.

States are not the only source of governance, and not all governance is legal in character. Nonstate governance can sometimes fill roles that state law cannot or does so poorly. Indeed, the extant governance that is specific to solar geoengineering is largely nonstate. As early as 2008, David Victor advocated "concentrate[ing], today, on laying the groundwork for future negotiations over norms rather than attempting to codify immature norms now" (Victor 2008, 332). The following year, the landmark Royal Society report on geoengineering suggested some principles for governance and called for a code of conduct for research that would be internationally approved and initially voluntary (Shepherd et al. 2009, 61). Since then, various nonstate actors have developed principles and a code of conduct, some of which appear to have already influenced solar geoengineering researchers.

This chapter considers the extent to which nonstate actors do, could, and should contribute to the governance of solar geoengineering. It first offers a conceptual basis in Section 10.1. Sections 10.2 and 10.3 review principles and codes of conduct, respectively, that have been put forth. Section 10.4 provides a brief analysis of the potential of nonstate governance for solar geoengineering.

10.1 CONCEPTS OF NONSTATE GOVERNANCE

Many people often equate regulation with binding, legal rules that one or more states develop, implement, and enforce through sanctions of the threat thereof.

---

[1] This chapter draws from Reynolds (2017).

10.1 Concepts of Nonstate Governance

Indeed, until now, this book has largely been limited to international law, national law, and administrative regulations. This widespread implicit equivalence is understandable but is not necessarily the case. In this book, governance is understood to be the goal-oriented, sustained, focused, and explicit use of authority to influence behavior. This definition allows for some breadth. The governing actors and targets can be states, businesses, nongovernmental and intergovernmental organizations, communities, and individuals. Governance can target another actor or the self and can operate through diverse modalities such as legal rules; social norms; market incentives; design (that is, technology and architecture); and the production and (re)distribution of goods, services, money, information, and other resources (see Lessig 2009; Brownsword 2008). Enforcement mechanisms of nonstate governance include social and reputational sanctions or praise; rewards; subsidies; granting of property; offers or revocation of permission, license, and membership; and threats of escalation and offers of de-escalation. In contrast, enforcement through appropriation of property (including through fines and taxes), imprisonment, and – in the most extreme cases – execution of a human (or other legal person) are solely the state's purview.

Nonstate governance is that which is developed, implemented, and/or enforced by nonstate actors. These actors include businesses, nongovernmental organizations such as advocacy organizations and professional societies, and individuals. It varies by the relations among the governing actor, the target, and the state. One category of these relationships is self-regulation. This is when the governing actor and target are congruent, such as when a firm develops, monitors, and enforces a code of conduct.[2] Another is private regulation, in which a nonstate third party acts as the governing actor. Examples of this include auditing and certification. To the extent that the state monitors nonstate governance, one can speak of meta-regulation. In fact, sometimes a state will legislatively or administratively delegate some governing authority to a nonstate body. The distinctions among these terms are not clear, and actual manifestations are diverse. For example, suppose that the members of an industry establish a body to govern them, and only compliant firms may advertise this fact. Although the industry as a whole is self-regulating, each individual firm is governed by the nonstate body, a manifestation of private regulation.

The instruments of nonstate governance also vary. Norms are general values that are widespread but not codified. Principles are explicit values, usually few in number, that the governing actor asserts that the targets should follow. These are the subject of Section 10.2. A code of conduct is when a governing actor offers

---

[2] Although I use "governance" broadly and "regulation" to refer to binding rules that are developed and enforced by authoritative institutions – especially state ones – that can punish violators, I speak of self-, private, and meta-regulation because these terms are much more widely used than their "governance" counterparts.

a longer, more detailed set of nonbinding explicit rules. Such a code for geoengineering research is discussed in Section 10.3. There are other means of nonstate governance. For example, an environmental advocacy organization could organize a boycott, a nonstate funder could attach conditions to grants, and a community could establish and advertise a neighborhood watch.

Nonstate governance has some advantages compared with state governance. Nonstate bodies can be more dynamic and responsive than public agencies, allowing them to adapt to rapidly changing conditions. In contrast, state governance is conservative, can get locked-in, and can be counterproductive due to the dynamics of large institutions. Nonstate governing actors may have stronger incentives to be effective, innovative, and efficient with respect to administrative and compliance costs. In some cases, public governing actors – which are usually politically constrained – may not have sufficient incentives to act. For example, if the domain is highly contested, politicians might face censure if they were to legislatively govern. Activities and targets of governance that cross national boundaries can often be better governed through nonstate mechanisms because states are geographically bounded and sometimes unable to effectively cooperate in transboundary governance. Self-regulation can sometimes incorporate expert knowledge that is superior to that of government bodies and can have better established, less adversarial relationships with the targets, which can facilitate compliance. Furthermore, governance's administrative costs are more likely to be borne by the targets in cases of self- and meta-regulation, which can be more efficient and equitable than state governance. Nonstate governance can also serve as a sort of governance laboratory in which proposals are tested, possibly before later becoming legalized. Finally, self-regulation has inherent value in allowing people and institutions the freedom to resolve problems on their own, without state coercion or the threat thereof.

Nonstate governance is – perhaps unsurprisingly, given its advantages – often seen in scientific and technological domains, particularly in controversial ones. For example, the leading international guidelines for medical research on human subjects are promulgated by the World Medical Association, a confederation of national professional medical associations (World Medical Association 2018). Likewise, the principles for biosafety in biotechnology research were issued by some of the early scientists who developed genetic modification (Talbot 1980). Notably, the primary US federal funder of biomedical research subsequently incorporated much of these principles into its own regulations. More recently, several businesses that are commercializing synthetic biology issued a code of conduct (International Association Synthetic Biology 2009).

This is not to suggest that nonstate governance will always arise when it would be effective and otherwise advantageous. Other conditions must be satisfied (Green 2014; see Zelli, Möller, and Asselt 2017, 687). For example, there must be sufficient expertise to supply nonstate governance, and the targets typically must recognize its benefits. Moreover, one of more norm entrepreneurs are often required to catalyze the governance's development.

There are drawbacks as well. Nonstate governing actors may be more vulnerable to influence from and capture by the targets and may be insufficiently accountable and transparent. Furthermore, nonstate governance can be more difficult to enforce and may be perceived as lacking legitimacy.

Governance targets comply with nonstate and other nonbinding governance for diverse reasons. For one thing, they might be concerned that they would lose customers, clients, or supporters if they developed a reputation for noncompliance with widely shared norms of responsibility. A second reason is that nonstate governance can exist under the explicit or implicit threat of more restrictive state regulation. That is, the state is not absent but instead "casts a shadow" over the nonstate governance. In fact, targets may create self-regulation or another form of nonstate governance to prevent state regulation and to set the initial terms for debating any possible future state governance. Third, targets may comply due to internal community or peer pressure. In particular, when governing actors, politicians, and voters seem unlikely to differentiate targets' identities. If so, then the targets may realize that if one of them fails to comply, then their shared reputation will suffer and they will all be punished. That is, they constitute a community of shared fate. It can thus be in each target's interest to encourage and facilitate others to comply. These reasons for compliance are sometimes evident when the governed activity is (or has the potential to be) controversial. In such situations, the targets might be aware that they require an implicit social license to operate and that being seen as an irresponsible actor could jeopardize that (Gunningham, Kagan, and Thornton 2004). In fact, this can give them incentives to go beyond compliance with explicit rules.

The US nuclear power industry offers an example. After the 1979 accident at Three Mile Island, industry leaders feared aggressive state regulation or even nationalization of plant operations. They thus quickly formed an industry association, the Institute of Nuclear Power Operations, to promote safety and reliability. Its chairman observed that "an event at a nuclear power plant anywhere in our country ... could and would affect each nuclear power plant ... Each licensee is a hostage of every other licensee ... [W]e truly were all in this together" (Rees 1994, 2). In other words, the industry recognised itself as a community of shared fate. The Institute of Nuclear Power Operations developed and implemented a form of industry self-regulation that was, by many measures, successful. Joseph Rees describes it as communitarian governance, a system that has "well-defined industrial morality that is backed by enough communal pressure to institutionalize responsibility among its members" (Rees 1994, 6).[3]

---

[3] To be precise, Rees called it communitarian *regulation*.

10.2 PRINCIPLES

The Royal Society report not only suggests an international nonbinding code of conduct but also offers some initial principles for geoengineering research: "Research activity should be as open, coherent, and as internationally coordinated as possible and trans-boundary experiments should be subject to some form of international governance, preferably based on existing international structures" (Shepherd et al. 2009, xiv). Some of these themes, especially that of transparency, recur in the sets of principles that various groups have put forth since then, which are reviewed here. Notably, for the most part, these overlap in their substance (Chhetri et al. 2018, 11).

Soon after the report, a group of British academics drafted the first full set of principles for geoengineering research and implementation (Rayner et al. 2013). Initially as a submission to a UK House of Commons committee that was investigating geoengineering, the "Oxford principles" were endorsed generally by both the committee and the UK government (Rayner et al. 2010; House of Commons (UK), Science and Technology Committee, The Regulation of Geoengineering, 35; Great Britain Department of Energy and Climate Change, Government Response to the House of Commons Science and Technology Committee, 5–7). Their authors intend them to be "high-level and abstract" and "as laying down the basic parameters for decision-making" due to the proposed geoengineering methods' diversity, their stages of development, and social contexts (Rayner et al. 2013, 504).[4] They were subsequently refined and still serve as a reference point in the geoengineering governance discourse. This section therefore reviews them in some depth.

> **Principle 1: Geoengineering to be regulated as a public good.** While the involvement of the private sector in the delivery of a geoengineering technique should not be prohibited, and may indeed be encouraged to ensure that deployment of a suitable technique can be effected in a timely and efficient manner, regulation of such techniques should be undertaken in the public interest by appropriate bodies at the state and/or international levels. (Rayner et al. 2013, 502)

The first half of this first principle expresses concerns regarding commercial actors in geoengineering and particularly the awarding of patents. The text that accompanies the principles in a scholarly journal claims that "the distribution of intellectual property rights can result in, or exacerbate existing, injustices. There should therefore be a presumption against exclusive control of geoengineering technology by private individuals or corporations" (Rayner et al. 2013, 505; see Chapter 11). In response, the UK House of Commons countered that restrictions on patents could deter investments in research and development (House of

---

[4] Note that the Oxford group, like most of those here, included negative emissions technologies within geoengineering.

10.2 Principles

Commons (UK), The Regulation of Geoengineering, 31). The principle's second half calls for geoengineering governance to promote the wider public interest. The justification for this – that all people have an interest in the climate – echoes the UN Framework Convention on Climate Change's opening declaration that "change in the Earth's climate and its adverse effects are a common concern of humankind" (Preamble recital 1; see Chapter 5). The authors acknowledge that further defining "the benefit of all" is difficult and speculate whether this principle would require that all parties would benefit, or only that some parties would benefit and could compensate those who do not (that is, Pareto and Kaldor-Hicks criteria, respectively).

> *Principle 2: Public participation in geoengineering decision-making.* Wherever possible, those conducting geoengineering research should be required to notify, consult, and ideally obtain the prior informed consent of, those affected by the research activities. The identity of affected parties will be dependent on the specific technique which is being researched . . . [A] technique which involves changing the albedo of the planet by injecting aerosols into the stratosphere will likely require global agreement. (Rayner et al. 2013, 502–3)

This second principle emphasizes public participation in decision-making in order to establish and maintain legitimacy of the geoengineering endeavor. The Oxford group notes that "affected parties" could include those who might experience either only physical impacts or also cultural and moral ones. The authors stress that disadvantaged groups should receive specific attention. This principle reflects the growing norms of public participation in environmental decision-making, as seen in the Rio Declaration and the Aarhus Convention on Access to Information, Public Participation in Decision-Making and Access to Justice in Environmental Matters as well as the possible right to participate in science-related decision-making as part of the human right to benefit from scientific research (see Chapters 5, 7, and 8). However, this Oxford principle goes further than typical public participation in calling for "ideally" obtaining prior informed consent of those affected (see Horton et al. 2018). Although "consent" and "likely requir[ing] global agreement" imply authority to deny consent and to reject agreements, respectively, the principles' authors acknowledge that such veto power is questionable and that what is considered proper public participation and consent vary among cultures and countries.

> *Principle 3: Disclosure of geoengineering research and open publication of results.* There should be complete disclosure of research plans and open publication of results in order to facilitate better understanding of the risks and to reassure the public as to the integrity of the process. It is essential that the results of all research including negative results, be made publicly available. (Rayner et al. 2013, 503)

This principle of transparency appeals to geoengineering researchers to disclose their results. Like the previous one, this is consistent with the principle of access to information in environmental matters in the Rio Declaration (Principle 10) and the Aarhus Convention (Articles 4–5). Here, the principle emphasizes instrumental reasons for transparency: better understanding could facilitate more effective public participation, and public reassurance could help prevent the growth of potentially unfounded concerns. The accompanying text adds the intrinsic value of informing others as well as the importance of publishing both positive and negative results. In apparent response to the comments of the UK House of Commons committee, the scholars agree that disclosure is unqualified, with no exception for national security. However, the UK government later argued that there should be some exceptions to transparency, including "the need to respect commercial rights to certain data, security considerations and the need to protect personal confidentiality" (Great Britain Department of Energy and Climate Change, Government Response 6; see Chapter 11).

> ***Principle 4: Independent assessment of impacts.*** An assessment of the impacts of geoengineering research should be conducted by a body independent of those undertaking the research; where techniques are likely to have transboundary impact, such assessment should be carried out through the appropriate regional and/or international bodies. Assessments should address both the environmental and socio-economic impacts of research, including mitigating the risks of lock-in to particular technologies or vested interests. (Rayner et al. 2013, 503)

Here, independent assessment of geoengineering's impacts is justified as more likely to be impartial and unbiased. The principle's accompanying text offers a few challenging issues, such as how to assure genuine independence, which kinds of possible impacts to include, and what quantity of resources to invest in independent assessment. Regarding impacts, the authors argue that, in general, both environmental and social impacts should be considered. The House of Commons committee connects such impact assessment to the question of compensation or liability for harm. In turn, the Oxford group suggested that a compensation mechanism would further accountability, as part of the final principle (House of Commons (UK), The Regulation of Geoengineering, 32–3; Rayner et al. 2013, 508; see Chapter 12).

> ***Principle 5: Governance before deployment.*** Any decisions with respect to deployment should only be taken with robust governance structures already in place, using existing rules and institutions wherever possible. (Rayner et al. 2013, 503)

The motivation behind this final Oxford principle is to ensure that some sort of governance is in place before the implementation of geoengineering. The authors acknowledge that the distinction between field research and deployment might not be clear and that the former at sufficient scale could effectively amount to the latter. Perhaps appropriately, the principle and its text remain vague. Although states' involvement in governance is implied, the Oxford group addresses neither the institutional breadth of participation in governance nor whether implementation would require the active approval of governing institutions. However, the scholars assert that any governance institutions should be accountable, including through an appeals process and compensation for those who have been harmed.

Finally, the Oxford group offers mechanisms for how its principles could be operationalized as "part of a flexible architecture" and suggest that they could "shape a culture of responsibility among researchers," inform bottom-up self-regulation, and contribute to top-down legal governance (Rayner et al. 2013, 508). More specifically, the scholars suggest that geoengineering research proposals should state how the project would addresses the principles, and an independent body should review this.

The second set of proposed principles were developed in 2010, when approximately 165 experts met at the Asilomar Conference Center in California to further develop governance of geoengineering research. "Drawing particularly" from the Oxford principles, the meeting's scientific organizing committee recommends five somewhat similar principles while offering greater detail in its report (Asilomar Scientific Organizing Committee 2010, 8). The Asilomar committee's first, "Promoting collective benefit," resembles the first Oxford principle while the report text more directly addresses the possible trade-offs between near-term, localized impacts and longer-term, wider effects. The second principle of the Asilomar report is its most novel, that of establishing responsibility and liability. The former aspect concerns primarily oversight. For the latter, the committee asserts that "Government establishment of a system for liability and compensation for definable and inadvertent harms caused by large-scale research activities is also likely to be needed," and recommends strict liability for the sake of public perceptions of potential risk (Asilomar Scientific Organizing Committee 2010, 20, 4). The third, "Open and cooperative research," is similar in its contours to the Oxford group's call for disclosure of research and open publication of results. The Asilomar report emphasizes the benefits of interdisciplinary research, assessing the risks of *not* conducting research, and disclosing funding sources. The Asilomar committee's fourth principle is "Iterative evaluation and assessment." In terms of the latter aspect, the report considers that although existing institutions such as the Intergovernmental Panel on Climate Change (IPCC) may currently suffice, a standing international body dedicated to assessing geoengineering research might be advisable. The last principle is "Public involvement and consent." However, that potentially problematic final word is not elaborated upon in the report, which instead refers to the more widely used phrase "public participation and consultation" (Asilomar Scientific Organizing Committee 2010, 9).

As an additional note, although the Asilomar conference was not intended to consider issues of implementation, the report does refer to it. On the one hand, the report says that geoengineering activities with transboundary effects might require new governance mechanisms and institutions. On the other, it concludes that concerns regarding "the large-scale implications of fully undertaking geoengineering and how to weigh the potential risks and benefits ... are legitimate and will require a serious response, even though their consideration by the appropriate governance structures may complicate decision-making regarding research activities" (Asilomar Scientific Organizing Committee 2010, 23).

The third set of principles for geoengineering was developed soon after the Asilomar conference by an ad hoc committee of the Bipartisan Policy Center, a Washington think tank (Bipartisan Policy Task Force on Climate Remediation Research 2011). Its primary objective was recommendations for how a US geoengineering research program should begin, be organized, and engage internationally. Nevertheless, a few of its principles concern governing geoengineering and have substantive overlap with those from Oxford and Asilomar. The first of these principles states that research should consider all people and ecosystems. It also emphasizes that geoengineering and its governance should aim to reduce the negative impacts of both climate change and geoengineering, goals that sometimes compete. Furthermore, the report recommends that US geoengineering research be based upon external advice, be informed by public engagement, and maintain transparency. Its remaining principles offer additional perspectives that are specific to national research programs. For example, the committee suggests that programs should be sufficiently adaptive to respond to new knowledge and be coordinated with other countries' efforts.

The fourth set of principles for geoengineering was not called as such. Instead, after the Royal Society published its report, it initiated a process to develop guidelines for solar geoengineering research. It partnered with the Environmental Defense Fund and The World Academy of Sciences (TWAS) to form the Solar Radiation Management Governance Initiative (SRMGI). The resulting report, however, does not offer guidelines per se but does identify the relevant issues and recommends approaches to governing solar geoengineering. Its list of "cross-cutting governance considerations" resemble the principles of the Oxford and Asilomar groups: transparency; public engagement; legitimacy; monitoring, compliance and verification; and compensation and liability (Solar Radiation Management Governance Initiative 2011, 39–44). Regarding the final point, the SRMGI report emphasizes the difficulty of establishing liability due to uncertain attribution.

The penultimate set of principles is suggested in a report by the Heinrich Böll Foundation, which is affiliated with the Germany Green party. Five of its eight principles are largely the same as the Oxford principles, although the latter's positive emphasis of "Geoengineering to be regulated as a public good" is presented in the Heinrich Böll Foundation report as a negative inverse, "Geoengineering must not be driven by economic profit-making" (Kössler 2012, 52–4).[5] Two of the others – international cooperation in research and the primacy of emissions abatement – are

---

[5] The translations are mine.

10.2 Principles

often seen in other sets of principles and recommendations for geoengineering governance. Only one of the Heinrich Böll Foundation's principles stands out, that "An internationally binding moratorium on deployment and experiments along the lines of the Convention on Biodiversity (CBD) is necessary." Ultimately, what's most remarkable is the extent to which a report that elsewhere concludes that "geoengineering would be the extreme form of unsustainable climate policy" concurs with the other sets of principles (Kössler 2012, 56).

Finally, in response to Oxford principles, a pair of philosophers suggests their own set of ten "Tollgate Principles" for stratospheric aerosol injection (Gardiner and Fragnière 2018). Stephen Gardiner and Augustin Fragnière are critical of the Oxford principles' instrumentality, procedural emphasis, and ambiguity and consequently ground theirs deontologically. Their principles do have a certain appeal deriving from their basis in justice, respect, and legitimacy. However, reflection on their substance reveals that the principles, at best, largely shift the difficult questions without resolving them. Nine of their ten principles speak of "ethics" in some way, such as "ethically necessary" or "ethically relevant norms." As a specific example, Gardiner and Fragnière's final principle is that "Geoengineering policy should respect well-founded ecological norms, including norms of environmental ethics and governance," citing "respect for nature" as among these norms (Gardiner and Fragnière 2018, 166). But on whose ethics and respect for nature should the governance of solar geoengineering rely? The world is diverse and ethical systems and views vary, yet the authors do not suggest how to reconcile them. At worst, Gardiner and Fragnière's principles would, if adopted, paralyze decision-making regarding solar geoengineering. For example, another principle is,

> Geoengineering decision-making . . . should be done by bodies acting on behalf of (e.g. representing) the global, intergenerational and ecological public, with appropriate authority and in accordance with suitably strong ethical norms (e.g. justice, political legitimacy) . . .
> Institutions would be needed that are ethically authorized to carry out, and capable of managing, such a task in light of and in accordance with appropriate norms of global, intergenerational and ecological ethics, including those pertaining to political legitimacy, justice, the human relationship to nature and especially the perspectives of the most vulnerable groups. (Gardiner and Fragnière 2018, 155)

They state neither what bodies can represent "the global, intergenerational and ecological public" nor who would authorize them. In fact, if these global, legitimate, and authorized decision-making bodies existed, then emissions abatement would not be so suboptimal and solar geoengineering might be unnecessary. Furthermore, Gardiner and Fragnière do not say how geoengineering decision-making could proceed in the absence of their unprecedented proposed bodies. Their high, arguably impossible thresholds seem consistent with their strident criticisms elsewhere of geoengineering (for example, Fragnière and Gardiner 2016).

10.3 PROPOSED CODE OF CONDUCT

As noted, the Royal Society report recommends that it and other scientific bodies develop an international nonbinding code of conduct for geoengineering research (Shepherd et al. 2009, xiv, 51–2). Specifically, it states that such a code should include which forms and what scales of geoengineering research should be governed (including a scale at which no additional governance is necessary), monitoring and reporting standards, and evaluation criteria and methods.

A voluntary code of conduct for geoengineering research has been developed, albeit through a process that includes neither the Royal Society nor other scientific organizations. Legal scholars Anna-Maria Hubert and David Reichwein, then at the Institute for Advanced Sustainability Studies, drafted an initial code in 2015 (Hubert and Reichwein 2015), and two years later the former revised it through the Geoengineering Research Governance Project (Hubert 2017). The Project was a partnership of the University of Calgary, the Institute for Advanced Sustainability Studies, and the University of Oxford. It was supported by, among others, three authors of the Oxford principles, one of whom was also a member of the Royal Society report's committee.

The code of conduct is based on international law and presented as a set of articles with commentary. It "aims to promote the responsible conduct of geoengineering research," especially for outdoor experiments (Article 2). This is to minimize harms, to promote responsible research, and to enhance legitimacy (Hubert 2017, 4). In the code, geoengineering research should be promoted to better understand its potential efficacy, benefits, and risks (Article 2). Although based on international law, the code is directed toward a full spectrum of actors: states, intergovernmental organizations, and nonstate actors. It builds upon the 2010 decision of the CBD Conference of Parties, here centrally requesting that

> In view of the lack of science-based, global, transparent and effective control and regulatory mechanisms for geoengineering, and in accordance with the precautionary approach and requirements for environmental impact assessment, no geoengineering activities should take place, until there is an adequate scientific basis on which to justify such activities and appropriate consideration of environmental and other effects.
>
> An exception … can be made for responsible geoengineering research conducted in accordance with all applicable laws and regulations and on the basis of the guidance in this Code of Conduct. (Article 4)

The code's subsequent substantive articles address cooperation, public participation, access to information, prior assessment, project authorization, and monitoring of experiments in ways that are consistent with relevant international law (see Chapters 5 to 8). This could help ensure that future large-scale outdoor geoengineering activities are consistent with international law.

Although the code of conduct reflects the various proposed principles – especially the Oxford principles – it does not address some of the more contested issues to which they refer, such as intellectual property, consent, and liability and compensation. This appears to be due, in part, to the code's grounding in international law, where these issues are very much unresolved. Furthermore, the ultimate utility of guiding geoengineering research that would not pose risks of significant transboundary harm with international law is unclear.

10.4 ANALYSIS

Nonstate governance has contributed, can contribute, and should contribute to the governance of solar geoengineering. Extant nonstate governance is evident in the progressive development and the consensus among the Royal Society report, the multiple sets of principles, and the proposed code of conduct. Moreover, both the UK House of Commons committee and the UK government generally endorsed the Oxford principles, implying an important role for nonstate governance in the absence of substantial state action.

The influence of this nonstate governance has already been felt. When the UK decided to publicly fund geoengineering research after the Royal Society report, two of its Research Councils hosted a "sandpit" – a sort of brainstorming session – to develop submissions. In response to the proposal for a project that would include a field test of solar geoengineering equipment, one of the Oxford principles' authors – Tim Kruger, who was also a reviewer at the sandpit – put the principles to "practical use" (Kruger 2018, 196). He suggested that the project should be subject to a "stage gate," in which certain criteria would need to be satisfied for the field test to proceed. The ultimately funded Stratospheric Particle Injection for Climate Engineering (SPICE) project would, per its collaboration agreement, "publish its results in accordance with these [Oxford] principles" (SPICE n.d.; see Chapter 2). Later, both it and its primary funder went on to "support" the principles, and its team agreed to "(a) put all results into the public domain in a timely manner and (b) not to exploit (that is, profit from or patent) results" (Engineering and Physical Sciences Research Council 2012). At the stage gate, the field test was cancelled due to an unsatisfactory stakeholder engagement process and a related patent for which one of the sandpit mentors and one of the SPICE researchers had applied (Kruger 2018). This indicates that nonstate principles can have substantial "compliance pull" and impact, despite their general and nonbinding nature, particularly when the targets of governance see the principles as legitimate.

Nonstate governance can continue to contribute to the governance of solar geoengineering because the circumstances are conducive to its effectiveness. Solar geoengineering and its research are technically complex and rapidly changing, and nonstate governance can draw upon experts' knowledge and be more dynamic than state governance. Furthermore, the field's practitioners, activities, materials, and knowledge readily cross boundaries. National governance thus might not be an ideal fit for managing such a transboundary set of practices. At best, it could be inconsistent and complicated. At worst, state regulation could unnecessarily stifle activities that might otherwise be able to reduce climate change and could drive research – including that which is otherwise responsible – to less institutionally robust jurisdictions. Meanwhile, governance targets have strong incentives to comply with nonstate governance of solar geoengineering. Like that of US nuclear power industry, the researchers' work is controversial, and they fear burdensome state regulation. The researchers recognize that they need a social license to operate and, due to undifferentiated reputations, that they are a community of shared fate. Moreover, communitarian governance would be enhanced by the fact that the researchers are generally reluctant supporters – not promotional boosters – of solar geoengineering and show "unusual self-reflexivity" (Anshelm and Hansson 2014, 135).

This was evident in the reaction of the geoengineering research community when an entrepreneur undertook the largest field test of ocean fertilization. This potential but controversial negative emissions technology (NET) resembles solar geoengineering in that it is a proposed form of geoengineering that would pose uncertain transboundary risks. He operated outside of geoengineering's norms, including the Oxford principles, by seeking to profit, not engaging the public, not sharing his research protocol and results, not having his work independently assessed, and contravening existing governance mechanisms. Many geoengineering researchers, including those active in solar geoengineering, publicly condemned him. One said, "this will make legitimate, transparent fertilisation experiments more difficult," and another that "This is extremely unhelpful for those of us wanting to do some serious work on iron fertilisation" (Marshall 2012; see Fountain 2012). Geoengineering researchers are aware that transgressing norms would result in condemnation from within the community as well as increasing the chances of external criticism and burdensome state regulation.

Finally, nonstate governance should continue to contribute to the governance of solar geoengineering because some form of governance is necessary, yet states are presently not acting (Chhetri et al. 2018, 11, 26–7). Solar geoengineering and the need for its governance have been discussed seriously in prominent scholarly and policy forums for more than a decade. In that time, governments have responded with only a handful of reports, a couple of national positions, and no substantial steps toward state or international governance (see Chapter 4). Although this is arguably insufficient, at least in the long term, it is also understandable and predictable. The politicians who would be necessary to move state governance forward have little incentive to do so. Voters are largely unaware of solar geoengineering and might perceive its introduction into climate politics as fringe if not questionable. Until major environmental advocacy organizations clearly and assertively call for solar geoengineering research, the topic will likely remain off state actors' agendas. Even when (and if) state legislators and administrators do act, they will do so only slowly. Substantial international progress – which might ultimately be needed – would be even slower. Furthermore, when national and international governing actors take these steps, they would be able to do so more effectively and rapidly if they have a foundation of norms, principles, rules, institutions, and procedures on which to build. In the meantime, nonstate governance has an important gap to fill. I suggest nonstate governance (at least initially) of intellectual property in Chapter 11 and of small-scale research in Chapter 13.

A few closing caveats are in order. First, nonstate governance does not necessarily imply self-regulation. External actors can develop and enforce governance through various forms of private regulation. For example, it may be in the interests of research institutions, professional societies, nonstate funders, and academic publishers for solar geoengineering research to proceed responsibly. They could play important roles in governance, especially in the absence of state and international governance. Second, nonstate governance does not rely merely on the targets' benevolence and goodwill. Instead, responsibility might be in their own self-interests and those of the related nonstate actors listed above. Well-crafted nonstate governance can align these self-interests with the wider interests of the public. Third, participants in and observers of nonstate governance should remain vigilant. It can be vulnerable to regulatory capture, lack accountability and transparency, be perceived as insufficiently legitimate, and be underenforced. Any institutionalized nonstate governance of solar geoengineering should monitor for such pitfalls and respond accordingly.

10.5 SUMMARY AND CONCLUSION

Nonstate governance is too often undervalued. It can sometimes fill roles that state law cannot fill or does so poorly. Nonstate governance has contributed to the governance of solar geoengineering, particularly through the development and apparent compliance with principles such as the Oxford principles. Notably, for the most part, these various sets of principles substantively agree. Furthermore, it can contribute because solar geoengineering's characteristics are favorable to nonstate governance. Finally, it arguably should do so because some governance, especially as research moves outdoors, will be warranted, yet states are not acting.

11

Nonstate Actors and Intellectual Property

Chapter 10 opened with the observation that most decision-making occurs in the private sector and proceeded to consider nonstate actors' governance of solar geoengineering. Some observers have argued that nonstate actors could deploy solar geoengineering. Although national leaders will dominate choices regarding any implementation, nonstate actors – including commercial ones – will play important roles in solar geoengineering activities.

This chapter explores solar geoengineering research, development, and implementation by nonstate actors. It begins with possible nonstate deployment, which has been a widespread concern. It then moves on to commercial entities' roles as providers and innovators of goods and services for solar geoengineering activities, as well as their potential to influence states' decision-making. The leading way that innovation by commercial actors is governed is through intellectual property, which dominates the second half of this chapter. I review the current landscape of intellectual property related to solar geoengineering and proposals for such intellectual property policy. The chapter closes with a recommended mechanism to govern such intellectual property. This is this book's first prescriptive content, which continues in Chapters 12 and 13.

11.1 IMPLEMENTATION BY NONSTATE ACTORS

In the scholarly and popular discourses, the most frequently considered role of nonstate actors in solar geoengineering is that of a rogue implementer. Given the facts that the direct financial costs of deployment could be as low as a few billion dollars per year and that some wealthy individuals have expressed concern and acted regarding climate change, this might seem feasible. David Victor first suggested this, in a colorful manner: "A lone Greenfinger, self-appointed protector of the planet and working with a small fraction of the [Microsoft founder Bill] Gates bank account, could force a lot of geoengineering on his own. Bond films of the future might struggle with the dilemma of unilateral planetary engineering" (Victor 2008, 324).[1] Since then,

---

[1] "Greenfinger" is a play on Goldfinger, the villain in the eponymous James Bond film.

this possibility has often been invoked to emphasize the problematic nature of solar geoengineering as well as the difficulty in governing it (Shepherd et al. 2009, 40; Bodansky 2013, 547–8; Hamilton 2013, 61, 153; Pasztor, Scharf, and Schmidt 2017, 2310). The specter of a rogue nonstate actor changing the world's climate is indeed powerful and seemed more probable when a businessman of questionable reputation undertook an ocean fertilization field experiment in 2012 (Craik, Blackstock, and Hubert 2013; see Chapter 10).

However, global implementation of solar geoengineering by nonstate actors is quite unlikely (see Parson 2014; Rabitz 2016). For one thing, while the financial costs are low in terms of climate change economics, they would nevertheless be great in terms of individuals' wealth. If the annual direct financial costs are twenty-five billion dollars, then ultrarich people such as Gates would deplete their assets after only a few years. It might also be difficult for a nonstate actor, even a wealthy one, to acquire the necessary technology, hardware, and expertise. Furthermore, implementation would not be in a nonstate actor's economic self-interest. After all, because it is a nonexcludable global public good, there would be no meaningful way to charge for the solar geoengineering services (see Chapter 4).

Second, given the importance of climate to essential national interests such as food production, countries' leaders will consider its alteration to be their sole domain in which to act, not one for nonstate actors. A nonstate implementer would need to act from the territory or a flagged ship of a country, whose leadership would probably have little tolerance for unauthorized solar geoengineering. Rogue implementation could not be kept a secret, as it would involve large-scale machinery and materials, and those countries with space programs could detect it through satellites and other means of observation.

Third, even if a state were to ignore or tacitly approve the solar geoengineering deployment, it would come under international pressure to end the activity. In this scenario, nonstate implementation would come to resemble problematic uni- or minilateral deployment. Indeed, in the implementation context, nonstate actors' actions would be linked to states in multiple ways that could include the activities' territorial location, regulatory approval, and international political support. I argue that this is unlikely because few actors would have the interest and the financial, technical, and political capabilities to deploy solar geoengineering in a sustained way and because the international system is increasingly dominated by a logic of multilateralism. However, it is not out of the question (see Chapter 4). If the nonstate actor were to undertake it from an entirely nonstate space, such as an unflagged marine vessel or platform on the high seas, then powerful states would likely act to end it once the activity was detected.

This is not to claim that wealthy individuals or other nonstate actors could have no roles in solar geoengineering implementation that might be problematic. I suggest three scenarios here. First, the objective of a wealthy nonstate actor who tries to implement might not be to reduce climate change per se but instead to catalyze international action through a sort of civil disobedience. Alternatively, the actor's goal might be, by implementing at a low level, to demonstrate a proof of principle or to establish a new set of "facts on the ground," with the hope that some state would subsequently continue the solar geoengineering. In these cases, the nonstate actor could act in concert with one or more states, such as by funding deployment in a country that is highly vulnerable to climate change. For example, Victor elsewhere suggests that a nonstate actor could provide "intellectual and financial help" for unilateral state deployment (Victor 2019, 43). This, coupled with the likely linkages among actors given in the previous paragraph, blurs the distinction between nonstate and state deployment.

A second possible nonstate implementation scenario is one in which a large number of nonstate actors distributed among numerous countries would each undertake a small amount of solar geoengineering, which would have cumulative climatic effects (Davies 2013, 189–90). In the case of stratospheric aerosol injection, this could be accomplished through many high-altitude balloons that could deliver an aerosol precursor (Reynolds and Wagner 2018). It would avoid both the international political pressures and international legal consequences, because any resulting transboundary impacts could not be attributed to a single state's actions. Although this might be feasible in principle, the scenario faces substantial barriers in practice. At the very least, it presents its own collective action problem: who would pay for these efforts, and why would they, when they could free ride on others' efforts? On the other hand, perhaps there are globally enough individuals who would be willing to make such highly-leveraged contributions to ending climate change.

A final feasible scenario of nonstate solar geoengineering deployment lies between the extremes of one extremely wealthy individual and millions of people. In this, a small group of nonstate actors could implement solar geoengineering for their own interests. The economic interests that appear to be most directly threatened by climate change impacts, such as the agriculture and coastal real estate sectors, involve very large numbers of actors and are thus more like the scenario described in the previous paragraph. However, aggressive emissions abatement could threaten the power and fossil fuel industries, which are relatively concentrated. Abatement policies that might limit global warming to two degrees Celsius could cause these sectors to lose roughly one trillion dollars in asset valuation (International Energy Agency and International Renewable Energy Agency 2017). This risk might motivate these companies to implement solar geoengineering. However, such action would face several challenges. First, countries would need to adopt such aggressive climate polices, or at least credibly threaten to do so, even though they have not after more than a quarter century of international discussions. Second, the companies would need to believe that their solar geoengineering deployment would prevent these policies, which it might or might not accomplish. Third, the commercial actors might refrain from threatening or undertaking solar geoengineering out of an expectation that it would trigger a strong political backlash. Fourth, the firms would need regulatory authorization by the states in which they would operate. As discussed, such approval might be perceived by other states as equivalent to the implementation itself. Finally, even though the power and fossil fuel industries are relatively concentrated, they still consist of hundreds of companies, which would need to agree how to distribute the costs of solar geoengineering among themselves, posing a collective action problem.

Nonstate actors can have roles in solar geoengineering other than implementation. For example, they could fund research that is undertaken at traditional research institutions. Indeed, much solar geoengineering research in the United States is currently funded privately due to an absence of federal support (Necheles, Burns, and Keith 2018). If public funds are later forthcoming, then the privately funded portion would likely shrink. Although nonstate funding of research at traditional research institutions is usually philanthropic, funders sometimes ask for a stake in resulting intellectual property, which can blur the distinction between philanthropist and investor. However, there is no evidence that the nonstate funders of solar geoengineering research have taken such steps, and there is some evidence to the contrary.

11.2 COMMERCIAL ACTORS

I argued in Section 11.1 that national leaders will insist that decision-making regarding altering the climate is their responsibility. However, this does not preclude substantial roles for nonstate actors – such as commercial ones – in solar geoengineering activities, including in implementation and large-scale outdoor field experiments. Public actors make decisions in many domains while commercial actors implement them. This is often done through procurement, in which a government body requests bids for a project and then enters into contracts for services with commercial actors.

If solar geoengineering activities increase in scale and move outdoors, state and quasi-public institutions, such as traditional research universities, will generally make the decisions regarding whether and how to conduct the activities while contracting with commercial entities for various necessary goods and services. For example, the proposed Stratospheric Controlled Perturbation Experiment (SCoPEx) outdoor experiment will be conducted by Harvard University researchers, who will need approval from the university and possibly public regulators and are working with a commercial launch services provider on a contractual basis.

Goods and services for solar geoengineering provided by commercial actors through procurement could grow into a moderately sized industry. Current estimates of the annual direct financial costs of implementation are tens of billions of dollars. The additional expenses of modeling, monitoring, securing, and system redundancy would increase this. To put this in perspective, more than 200 individual firms have annual revenues exceeding 50 billion dollars (Fortune, 2018). If a substantial portion of these expenditures were procured, then commercial entities would have opportunities to profit. The prospect of profits would in turn give commercial actors incentives to innovate in order to be competitive in obtaining contracts (Davies 2013). Jack Stilgoe correctly asserts that "Few could imagine a scenario in which global SRM [solar geoengineering] could become a marketable commodity. There would be industrial interests in the components and infrastructure, but this situation would be more mundane than a climate-for-sale" (Stilgoe 2015, 139; but see Buck 2012, 262).

Some scholars and other observers have expressed concerns about the involvement of nonstate actors, especially commercial ones, in solar geoengineering. Besides possible implementation by nonstate actors, a common claim is that entities with vested interests in solar geoengineering would seek to influence decision-making in a way that would be contrary to wider welfare. This could contribute to a "slippery slope" or lock-in (see Chapter 2). For example, policy analyst Robert Olson rhetorically asks, "Could patents on proprietary technologies allow the private sector to gain too much influence over [geoengineering] research and development decisions? Could the drive for shareholder profits override the public interest and lead to inappropriate deployments?" (Olson 2011, 17; see also Davies 2013; Long and Scott 2013; Morgan, Nordhaus, and Gottlieb 2013, 43)

In general, there is abundant theoretical and empirical evidence that nonstate actors influence public policy through what can be called rent-seeking behavior (Mueller 2003; Campos and Giovannoni 2007). In the case of solar geoengineering, interests – including commercial actors that provide goods and services – might desire that more solar geoengineering takes place, that they conduct a larger share of this, that governance be less stringent, that particular forms of solar geoengineering are prioritized, and even that less emissions abatement and adaptation be undertaken. They could lobby for such policies through information campaigns to influence public opinion; through advocacy groups to express, amplify, and focus support for these policies; through donations to political campaigns; and through more overtly corrupt practices. If a relatively small sector of society were to have much at stake in solar geoengineering decision-making while most of society had little, then it could take on the form of client politics. In such a structure, special interest groups can sometimes successfully lobby for policies that benefit them at the expense of the general public (Wilson 1980). There is a strong argument that governance should aim to prevent rent-seeking and undue influence by commercial and other actors that have vested interests in solar geoengineering (see Chapter 4).

At the same time, the extent to which commercial interests might unduly influence solar geoengineering decision-making could be shaped by other, more important factors. Numerous diverse actors, both nonstate and state, would have substantial interests in whether and how solar geoengineering is developed and possibly deployed. For example, agriculture is highly sensitive to climatic conditions and is a three-trillion-dollar industry (World Bank n.d.). It and other climate-sensitive interests might seek to shape solar geoengineering decision-making, and their efforts could dwarf those of the providers of solar geoengineering goods and services. This wider influence would be a problem if the outcome were to differ substantially from the interests and desires of the wider public. Yet such negotiation among divergent interests is how democracies function, at least in the pluralist vision of them. This is not to imply that attempts by commercial providers of solar engineering goods and services to influence decision-making will be negligible or should be dismissed but instead to emphasize that these efforts would be part of a larger – and possibly much larger – set of efforts, which might or might not be problematic.

Another apprehension regarding the prospect of commercial actors involved in solar geoengineering is that they would act with insufficient transparency. For example, the authors of the Oxford principles assert that commercial interests might be selective in which results they make public. This "could create a culture of secrecy and may lead to the concealment of negative results [which] has been observed in the pharmaceutical industry" (Rayner et al. 2010; see also Keith and Dykema 2018). This could lead to inaccurate assessment of solar geoengineering, undermine public support, and inhibit subsequent innovation. Ultimately, commercial actors' involvement would raise legitimate concerns regarding transparency of solar geoengineering activities, and governance should address this to the extent appropriate.

11.3 INTELLECTUAL PROPERTY

Commercial actors are leading innovators of emerging technologies. One of the main ways in which states govern commercial actors' innovation is through intellectual property rights, particularly for patents. Public bodies help establish and enforce intellectual property to overcome knowledge's public good nature and the resulting collective action problem. In this, knowledge producers usually cannot restrict access to their outputs to only those who have paid for it. They consequently have little incentives to generate valuable knowledge, which will be produced at a socially suboptimal quantity. States therefore grant them patents (and copyrights, which are unimportant here), which are temporary, legally enforceable rights to determine who else, if anyone, may use their intellectual works. The prospect of the resulting monopoly (or more precisely, supracompetitive) profits offers an incentive for innovative commercial actors to generate more useful knowledge.

Solar geoengineering raises several potential problems with respect to intellectual property policy. Some of these are common to many emerging technologies. When policy-makers define patents' scope, they should balance incentivizing innovation with avoiding a hindering of later-stage work that builds upon the earlier efforts. This is particularly challenging at the earliest stages of research when future inventions and applications remain highly uncertain. Excessive early patenting can create a so-called patent thicket that unduly restricts subsequent inventions (Merges and

Nelson 1990; Scotchmer 1991). Early patenting could also facilitate technological lock-in, in which early design decisions cause the later adoption of superior techniques to be difficult (Cairns 2014, 652). It is also unclear how much commercial actors should profit from publicly subsidized research. Furthermore, some scholars express a general unease with nonstate actors holding exclusive rights to such powerful technologies (Rayner et al. 2013, 505). Yet if states were to grant few or narrow patents, then inventors would increasingly rely on trade secrets, which would likely further hamper later research and be even less transparent. Finally, some activists and others have argued that solar geoengineering is inherently immoral. As such, they could make the case that it is contrary to the public order and should not be considered patentable subject matter (Hamilton 2013, 178; Armeni and Redgwell 2015c, 14).[2]

Nevertheless, there is presently very little patenting activity that is clearly related to solar geoengineering. In 2016, Jorge Contreras, Joshua Sarnoff, and I searched and found thirty-three international patents and patent applications that were directly relevant to solar geoengineering (Reynolds, Contreras, and Sarnoff 2017; see also Oldham et al. 2014; Chavez 2015). Most of these have little potential to be relevant. Seventeen of them were related to space- and surface-based techniques, which are generally regarded as prohibitively expensive, of limited capacity, or otherwise infeasible (see Chapter 2). Of the sixteen related to aerosol injection or marine cloud brightening, only four were issued, valid patents (all for aerosol injection) and two were pending applications (one for each technique). The remainder found in our study were either abandoned applications or issued but expired patents. Since our research, an additional patent for marine cloud brightening to protect coral reefs was applied for and issued in Australia, as an eight-year innovation patent. Furthermore, some of those issued patents and pending applications related to aerosol injection were dubious by, for example, operating in the lower atmosphere. Finally, some of applicants claim that they do not intend to acquire revenue from the patents, and instead say that they applied for the patents in order to prevent fossil fuel companies from doing so or as part of an implicitly philanthropic endeavor.[3] Such "defensive patenting" is discussed in the Section 11.4.

---

[2] Parthasarathy et al. (2010, 9) imply this.
[3] Hugh Hunt is among the applicants to patent GB2476518, issued in 2013. He said, "We're not expecting any revenue from any of the IP [intellectual property], but if there is any then it will go into a trust fund for supporting climate-change related charities etc. If Big Oil held the patents I doubt they'd do such a thing" (Science Media Center 2012). The Australian project "Marine Cloud Brightening for the Great Barrier Reef," whose primary institution holds a relevant patent, asserts on its website that "Patenting protects IP from inappropriate exploitation but makes the information widely available" (Marine Cloud Brightening for the Great Barrier Reef n.d. "Intellectual Property"). An application for a patent on a conduit to deliver material to the stratosphere is associated with Intellectual Ventures, one of the largest nonpracticing, patent-holding entities. Its website claims, "We do not expect or intend that our climate technology inventions will make money. Rather, they are part of our work applying the brain power and creativity of our inventors to the serious problems confronting society" (Intellectual Ventures 2009).

Although the governance of intellectual property is primarily through national law, international law performs two primary functions that operate in some tension with each other. The first of these is harmonization. This is presently accomplished largely through the Agreement on Trade-Related Aspects of Intellectual Property Rights, which establishes minimum standards for member states of the World Trade Organization in their enforcement of intellectual property. The second is improving developing countries' access to essential technologies. Although this has been most visible in the case of medicines, provisions for technology transfer are prominent in multilateral environmental agreements, including those that address climate change (see Chapter 6). Technology transfer includes a variety of processes that facilitate the international movement of knowledge and materials, largely from industrialized to developing countries. If solar geoengineering becomes important to reducing climate change, then the transfer of related technologies from the global North to the South would likely be salient to governance (Armeni and Redgwell 2015c, 14).

11.4 PROPOSALS FOR INTELLECTUAL PROPERTY POLICY

Several observers, including myself, believe that widespread adoption of intellectual property policies specific to solar geoengineering would be beneficial (see also Chhetri et al. 2018, 32). This could potentially help avoid the negative consequences of excessive early patenting, prevent technological lock-in, enhance transparency, and ensure that research and development align with the broader public interest while maintaining commercial actors' incentives to innovate. Notably, there is presently an opportunity to develop such a policy, as evidenced by the low level of patenting activity and the fact that almost all research is currently conducted at traditional research institutions. This opportunity may diminish if patenting activity increases or if commercial entities become more active in this domain.

Before considering possible policies, some relevant characteristics of solar geoengineering and intellectual property must be highlighted. First, its research, development, and possible implementation will be transnational. This is due to both the physical effects of large-scale outdoor activities as well as the collaboration and migration of scientists and entrepreneurs. Second, intellectual property is largely a matter of national law, as noted. Third, those who are responsible for intellectual property policy are unlikely to implement rules that are specific to solar geoengineering. Not only does law move slowly in general, but public policy-makers appear reluctant to engage with solar geoengineering. Finally, there is no clear distinction between technologies that are and are not related to solar geoengineering. Most technologies used for solar geoengineering, including some of those developed specifically for it, will have other purposes, and some developed for other applications will later find use in solar geoengineering.

What could intellectual property policies specific to solar geoengineering be? In the first of four proposals considered here, some writers have called for limitations on the patentability of solar geoengineering inventions or even for their full exclusion. For example, the article that accompanied the Oxford principles says:

> There should therefore be a presumption against exclusive control of geoengineering technology by private individuals or corporations. This does not mean that there can be no intellectual property in geoengineering, but that there might be a need for restrictions to ensure fair access to the benefits of geoengineering research. In some cases, this might result in a refusal to patent (as happened with the Human Genome Project) but we need not expect this to obtain universally. (Rayner et al. 2013, 505)

This passage is mostly ambiguous as to whether the restrictions would come from the researchers themselves, national policy, or elsewhere. If this were to be bottom-up, in which inventors voluntarily forego patents, then in the absence of prior art, nothing would stop others from applying for the patents.[4] If this were to be top-down, in which government regulators would restrict patents on solar geoengineering inventions, any such public policy would require legislative or administrative action by most countries, or at least the industrialized ones, or by an international treaty with similarly broad participation (as the Oxford authors seem to recognize; Parthasarathy et al. 2010, 9–12; Morgan et al. 2013, 43; Ghosh 2018, 13). These processes are unlikely to occur within a sufficiently short time. Furthermore, both means would also require determining whether a given patent application is sufficiently related to solar geoengineering to justify nonpatentability. As noted, this distinction is not clear. Also, some inventors would respond by concealing their inventions' potential solar geoengineering applications or by utilizing trade secrets, each of which would reduce transparency and impede future innovation.

Another possible intellectual policy for solar geoengineering would be for states to assert their powers of compulsory licensing or march-in rights to ensure that relevant technologies are widely accessible. Under the former, the government can compel patent holders to license their inventions, sometimes to specific licensees. Under the latter, public bodies may use the invention. In each, the state usually offers reasonable compensation for the patent holder, although there are variants of march-in rights – such as Crown rights in the British Commonwealth – in which this is not required. Exercise of these powers requires a justification, such as public health concerns. However, governments have been reluctant to use them, likely because doing so would cause patents to generally appear less secure, lowering their value

---

[4] As part of its open access principle, Harvard's Solar Geoengineering Research Program aims to "Discourage patents and any form of IP [intellectual property] protection," implying a researcher-led process (Harvard's Solar Geoengineering Research Program n.d.). Elsewhere its leaders restrict its concern to "technologies that are core to the deployment or monitoring of solar geoengineering" (Keith and Dykema 2018). In practice, they have done this through defensive publishing, discussed later in this section.

11.4 Proposals for Intellectual Property Policy

and reducing incentives to innovate. Furthermore, such a policy would push inventors toward greater use of trade secrets. In the case of solar geoengineering, these drawbacks of relying upon compulsory licensing and march-in rights would be substantial. At the same time, these powers do serve useful functions that would be particularly important for solar geoengineering technologies. The possibility that governments might utilize the powers encourages patent holders to ensure that their inventions are well exploited, particularly when they concern issues of public health, safety, or broad well-being. They also act as a backstop against nonstate actors exclusively controlling the rights to important technologies.

A third possible policy for intellectual property would be a patent pool for inventions related to solar geoengineering (Chavez 2015, 32–5). A pool is a contractual arrangement among the holders of interrelated patents that permit them to use each other's inventions. The pool members then share licensing revenue in some agreed-upon manner. Furthermore, those outside the pool can often obtain licenses to all of the patents in the pool through a single process. Pools are usually voluntary but can be compulsory through state intervention. They can facilitate further innovation and utilization of the constituent technologies, especially when the patents are complementary. However, complementarity is not presently the case with solar geoengineering. In fact, the opposite may apply. Which solar geoengineering techniques, if any, will ultimately prove to be useful remains uncertain. Establishing a patent pool for, say, marine cloud brightening using self-propelled rotor ships, might unduly encourage further development of that technique to the neglect of others. In other words, patent pools can encourage technological lock-in. Furthermore, a pool that includes inventions that are substitutes of each other – such as marine cloud brightening and stratospheric aerosol injection – would raise antitrust concerns among regulators. Finally, as with other state policies for intellectual property, a compulsory patent pool would necessitate the difficult distinction between technologies that are and are not related to solar geoengineering. Ultimately, patent pools might later be beneficial for one or more specific solar geoengineering techniques, but currently they appear to be both counterproductive and difficult to implement.

A fourth policy approach would be to rely upon defensive patenting and defensive publication. The former is when a researcher obtains a patent with the intention not of enforcing it against other scientists but of preventing others – such as commercial entities – from getting it. This was introduced in Section 11.3, noting that some solar geoengineering researchers claim to be defensively patenting their inventions so that fossil fuel companies do not. The latter is when a researcher publishes results to, among other things, establish sufficient prior art so that others could no longer obtain a patent. There are a few cases of solar geoengineering researcher engaging in defensive patenting (Science Media Center 2012; Marine Cloud Brightening for the Great Barrier Reef n.d. "Intellectual Property") and defensive publication (Keith 2010; David Keith, personal communication 2016; Keith and Dykema 2018). These actions can improve the solar geoengineering intellectual property landscape while neither relying upon state action nor requiring a determination of what is and is not a solar geoengineering invention. However, they have shortcomings. Defensive patenting can open researchers to accusations of personal financial interests, as they would become patent holders.[5] Also, the extent to which a defensive publication establishes prior art cannot be known with certainty ex ante but instead only after others test it by applying for similar patents.

11.5 A RESEARCH COMMONS

Jorge Contreras, Joshua Sarnoff, and I propose a four-part approach to intellectual property policy for solar geoengineering (Reynolds, Contreras, and Sarnoff 2017; Contreras, and Sarnoff 2018). Taking advantage of the current noncommercial orientation of solar geoengineering research and the absence of patenting activity, we emphasize bottom-up organization among nonstate actors and minimize relying on state actors. We include research data into the proposal because it can, in some ways, be treated as a type of intellectual property, can give rise to some of the related problematic phenomena, and can offer an incentive to participate in the bottom-up organization.

The first component of our proposal is for researchers and their host institutions to ensure than their data is publicly and freely available. This could initially be satisfied by placing data online and publishing articles with open access. In the longer run, this would require some degree of central organization to develop uniform formats, to manage these standards, and to coordinate data repositories. This process should involve a wide array of state and nonstate actors, and could be overseen by an international body such as the UN-affiliated Working Group on Coupled Modeling of the World Climate Research Programme.

The second component of our approach is an intellectual property pledge community. In this, solar geoengineering researchers and their institutions would commit either not to assert their patents related to solar geoengineering against others under certain conditions or to grant licenses to them under favorable terms. The pledge community, we hope, could facilitate access to the technologies needed for solar geoengineering research, development, and possible deployment while avoiding the need for state intervention and antitrust concerns associated with legalized patent pools. It could also be linked to the depositories of research data. Specifically, the pledge would have these tenets:

---

[5] In fact, the patent that the researcher claimed to seek for defensive purposes, in footnote 3, is the one that caused the Stratospheric Particle Injection for Climate Engineering (SPICE) outdoor test of solar geoengineering to be cancelled. One of Hunt's co-applicants for the patent was among the reviewers of the proposal, and Hunt was a researcher in the project. See Chapters 2. and 10.

11.5 A Research Commons

- The pledge would apply to all the pledgor's patents for technologies necessary to research, develop, or implement solar geoengineering, as well as to any other patents resulting from the pledgor's solar geoengineering activities.
- With respect to other pledgors' legitimate solar geoengineering research and development activities, the pledgor would commit either to not assert covered patents or to license them nonexclusively at reasonable royalty rates.
- The pledgor would make any future sales or transfers of covered patents conditional upon acceptance of the pledge by the recipient.
- The pledgor would produce and make available solar geoengineering data in a manner consistent with international standards and contribute such data to the shared body of research data. If other legitimate solar geoengineering researchers could not, for some reason, access the data from the commons, the pledgor would similarly commit to share data related to solar geoengineering with them.
- The pledgor would submit results of solar geoengineering research to peer reviewed scientific journals, preferably with open access.
- The pledgor would commit to not retain or assert valuable technical information or know-how regarding solar geoengineering as trade secrets.
- The pledgor would cooperate with international efforts to monitor patents related to solar geoengineering, discussed below.

Because the intellectual property pledge community would be a researcher-driven, bottom-up initiative, it could begin relatively soon and grow thereafter. Researchers might join to access patented inventions and research data as well as to communicate their motivations to the public and policy-makers. Research institutions – which could require that their employed researchers adopt the pledge – may have similar reasons to participate, but would move more slowly than individuals. If a pledge community were to take root, then it could expand to include research funders, scientific publishers, nongovernmental organizations, and even governments and intergovernmental organizations, each of whom could help strengthen the norms and practices embodied in the pledge community. Eventually, the pledge community could adopt a more legalized form, such as a contractual one.

The solar geoengineering intellectual property pledge community would call for significant institutional support. Given the reluctance of state actors to engage with solar geoengineering and its governance, non- and quasi-governmental scientific bodies such as the Royal Society, the American Geophysical and European Geosciences Unions, and the International Council for Science, as well as nonstate research funders, could help establish and maintain the community.

Relying upon unilateral (at least initially) promises might strike some as a questionable centerpiece of an intellectual property policy. Yet pledges have been used with some success in a range of fields (Contreras 2015). Furthermore, given conditions described in this chapter, this has the advantages of avoiding patents' restrictive effects on subsequent research and developments, limiting the undue commercialization of solar geoengineering research and development, not relying on slow and heterogeneous state action, not raising antitrust concerns, and permitting the prospect of profits to strengthen incentives for innovation. Also, in this proposal, the *uses* of patented inventions – not the inventions themselves – would be considered as either solar geoengineering or not, a determination that is easier.

The third component of our proposed intellectual property policy would be the monitoring of patents for inventions that are related to solar geoengineering (see Parthasarathy et al. 2010, 12–13). At the very least, patent offices should exercise scrutiny and caution when examining such applications. More is arguably warranted. Internationally coordinated monitoring could, at an early stage, identify the growth of broad, uncreative, or trivial patents as well as trade secrets. To guide this, an international solar geoengineering patent monitoring panel could be established and include members of patent offices supplemented by experts drawn from academia and industry, both within and beyond the solar geoengineering research community. This panel could also coordinate efforts to collect and share relevant prior art to aid patent offices in their examination of solar geoengineering patent applications. Furthermore, it could make policy recommendations if patenting activity or trade secrecy appeared to pose a threat to solar geoengineering technologies' responsible research and development.

The final constituent of our proposal would be for governments to clarify their intentions in assuring that patented technologies related to solar geoengineering are utilized. As noted, governments' powers of compulsory licensing and march-in rights serve as backstops that help prevent potentially beneficial innovations from remaining unused. This might be due to, among other things, excessive licensing demands by patent holders. Although these are important mechanisms, the uncertainty regarding under which circumstances national authorities would exercise these powers diminishes patents' value and consequently reduces incentives for innovation. We thus suggest state-led, multi-stakeholder meetings to publicly detail the criteria under which states would use their compulsory licensing and march-in powers in this domain. Because solar geoengineering activities and related patenting will likely be international, cooperation among multiple patent offices would make this process more beneficial.

Two final additional notes are in order. First, the social benefits of solar geoengineering may be substantially greater than the possible profits from the commercial development of its related products and services (see Chapter 4). This implies that these technologies would, if left to the market, likely be suboptimally developed. States should therefore consider other ways of promoting solar geoengineering research and development. This could include not only traditional research funding but also alternative mechanisms such as prizes for innovation (Parthasarathy et al.

2010, 13). Second, if the pledge community grows, it could become an attractor. That is, with sufficiently large numbers of participants and patents, researchers, entrepreneurs, and other nonstate actors would benefit by pledging due to the greater access to intellectual property that it would offer. Not only would this help the pledge community grow, it could offer leverage to facilitate the adoption of best practices for research and development, particularly in the absence of state action (see Chapters 10 and 13).

11.6 SUMMARY AND CONCLUSION

Although some observers are concerned that a nonstate actor could deploy solar geoengineering, this appears unlikely. Nevertheless, nonstate actors will likely play roles in its research, development, and possible implementation. States will probably retain control over decision making regarding large-scale outdoor field tests and implementation, as well as procure various goods and services from commercial actors. This could be problematic, and governance should vigilantly manage their involvement.

Commercial and other nonstate actors will participate in solar geoengineering through innovation, which states govern through intellectual property policies. One that is specific to solar geoengineering appears to be justified. Other scholars' proposals have various shortcomings. Along with some colleagues, I propose a research commons for intellectual property that is related to solar geoengineering. Importantly, it does not require state action and could instead arise bottom-up among researchers.

12

International Compensation and Liability

Solar geoengineering holds the potential for both relative benefit and harm. Prior to an outdoor activity, some actors, including states, might be concerned that they will be harmed. They could ask for ex ante assurances of compensation, possibly as a precondition for supporting the activity or at least not opposing it. Without such assurances, solar geoengineering might lack the political base to be considered legitimate. Afterwards, actors might claim that they had been harmed and could demand ex post compensation. Legal rules might indicate that those who conducted or approved the activity would be liable to pay damages to compensate for the harm.

Compensation for harm has been a consistent topic in the solar geoengineering governance discourse.[1] The authors of the first academic article on solar geoengineering said, "If a large segment of the world thinks the benefits of a proposed climate modification scheme outweigh the risks, they should be willing to compensate those (possibly even a few of themselves) who lose their favored climate" (Kellogg and Schneider 1974, 1170). Likewise, both the first analysis of the international environmental law of geoengineering and the important Royal Society report briefly brought up the issue (Bodansky 1996, 310, 19; Shepherd et al. 2009, 41–2, 51). Three of the sets of principles for geoengineering, discussed in Chapter 11, suggest compensation or liability for harm (Rayner et al. 2013, 508; Asilomar Scientific Organizing Committee 2010, 18; Solar Radiation Management Governance Initiative 2011, 44). More recently, a handful of publications in law, ethics, and economics have focused on the issue.

This chapter analyzes the main aspects of compensation for transboundary harm from solar geoengineering, reviews existing international law, and offers both a proposal for compensation for harm from outdoor research as well as some initial thoughts for deployment. Like much of the rest of this book, it focuses on international governance. One reason for this is that countries' domestic liability laws vary

---

[1] See Banerjee (2011); Bunzl (2011); Heyward (2014); Reynolds (2014b); Svoboda and Irvine (2014); Wong, Douglas, and Savulescu (2014); Goeschl and Pfrommer (2015); Horton, Parker, and Keith (2015); Reynolds (2015b); Saxler, Siegfried, and Proelss (2015); Svoboda (2017); Brent (2018); and Pfrommer (2018).

substantially. Discussing them in sufficient detail is beyond the scope of this text (for the United States, see Chapter 9).

In this chapter, terms that have broader meanings are used in specific ways. Here, "harm" is an undesired impact to the body, a community, property, or environment of one party that is due to the actions of another. Even when harm may have only a probability of manifesting – that is, it remains a "risk" – I simply use "victim" and "injurer" for these two parties without specifying "potential." An injurer or other actor, such as an insurer or the government, may "compensate" a victim with something of value in lieu of re-establishing the victim's conditions before the harm, which is often prohibitively difficult. When an injurer is legally obligated to compensate a victim, she bears "liability" for "damages," which means the amount of financial compensation awarded. An exception to this is the use of "damage" to mean what I call "harm" where it occurs in a legal document, such as "property damage" or the Convention on International Liability for Damage Caused by Space Objects.

12.1 ANALYSIS

As noted, compensation for harm from solar geoengineering is often discussed but faces numerous substantial political, institutional, and theoretical challenges. Some of these are similar or analogous to compensation for climate change, which also has been a persistent topic within the wider discourse yet has seen negligible progress in practice (see Faure and Peeters 2011; Lord et al. 2011). This section discusses the harm to be compensated, the identities of injurers and victims, possible mechanisms of compensation, and the reasons for compensation. In each subsection, I draw from and contrast with the case of compensation for climate change harm, as appropriate.

12.1.1 Harm to Be Compensated

Anthropogenic climate change will, and solar geoengineering could, cause substantial transboundary environmental and economic harm. If victims are to be compensated, for what harms should they be? Part of this is regards which forms of harm should be eligible. Property that has been directly harmed is a relatively clear case. Other forms of harm are less so. People might have lost future opportunities, including income. They might have been aware of the risk, invested in precautions that reduced their harm, and now seek compensation for the precautions' costs. Negative impacts to health, life, ecosystems, and communities can be genuine harm yet difficult to evaluate. Nevertheless, these are not new issues in environmental liability and existing law can provide some guidance. Under the customary law of state responsibility, compensation is limited to financially assessable harm to persons and property, including lost future income and reasonable measures taken to remedy or mitigate the harm (ILC Draft Articles on Responsibility of States for

Internationally Wrongful Acts, Article 36 and commentary paragraph 15; see Chapter 5). The International Law Commission (ILC) Draft Principles on the Allocation of Loss in the Case of Transboundary Harm Arising out of Hazardous Activities, which might or might not reflect customary international law, explicitly includes environmental harm. The existing three international legal regimes for liability for transboundary harm define compensable harm similarly, albeit with minor differences. It seems that the question of which forms of harm could be resolved.

The question of whether a given harm was caused by climate change or solar geoengineering is more difficult. Legally, an act usually requires both factual and proximate causation to be considered as causing a subsequent harm. The former asks whether the harm would have occurred but for the initial action. In other words, could an environmental event, such as severe weather, be attributed with confidence to climate change or solar geoengineering? There has been some progress in the case of climate change, although it remains challenging and probabilistic (Stott et al. 2016). In the case of solar geoengineering, Joshua Horton and colleagues as well as Pfrommer and colleagues point toward the use of fraction attributable risk methods, in which an event's probabilities are compared in climate scenarios with and without an anthropogenic alteration of the Earth's energy balance, such as through greenhouse gases or stratospheric aerosols (Horton et al. 2015, 261–2; Pfrommer et al. 2019; but see Svoboda and Irvine 2014, 162). Ultimately, attribution would be a substantial challenge for compensation for harm from solar geoengineering.

Proximate causation requires that the act be sufficiently closely related to the harm. For solar geoengineering, harm from equipment accidents would be the most proximate. Those from environmental effects, both climatic and nonclimatic ones (such as ozone depletion), would be less but arguably adequately proximate. Socially mediated risks, such as those of sudden and sustained termination and of international conflict, would be less and probably insufficiently so. The distinction is not clear but could be clarified through explicit provisions, court rulings, and practice.

12.1.2 Identities of Injurers and Victims

The second set of questions regarding compensation for harm concerns who the injurers and victims are (see Svoboda and Irvine 2014). For each group, asking these questions in solar geoengineering is more challenging than doing so for climate change. In both contexts, distinguishing aspects of authority, temporal sequence, and (blame) worthiness is useful. I consider these three aspects of identity in turn, addressing both injurers and victims.

In terms of authority, most actions are undertaken by individuals or other legal persons, that is, nonstate actors. However, in matters of transboundary harm – which is particularly important in cases such as greenhouse gas emissions or outdoor solar geoengineering – these nonstate actors are represented by their states of nationality, of residence, or that are linked by other means of establishing jurisdictional responsibility. This is clearer with respect to injurers, as states have a duty under international law to require authorization for activities under their control that pose a risk of significant transboundary harm. States can also require nonstate actors within their jurisdictional authority to pay compensation, including to reimburse the state itself if it had to compensate another state due to responsibility for transboundary environmental harm (see Chapter 5). In this way, states are vicariously liable for the risky actions of nonstate actors under their control, have a duty of care defense, and have a resulting incentive to properly govern them.[2] Although states have a mixed record of regulating their numerous greenhouse gas emitters, they are more likely to exercise effective control over outdoor solar geoengineering activities (see Chapter 11). This is because of, among other things, the fewer actors to be governed, the greater possible transboundary risk per nonstate actor, the actors' intentionality, and other countries' likely, politically mediated perceptions of the risks. At the same time, this arrangement is imperfect, in part because it creates principal-agent problems.[3] For example, a state could approve (or directly undertake) risky activities that are unpopular among its citizens. With respect to authority and victims of transboundary harm, in order to receive compensation, their state must seek compensation on their behalf. Yet states generally do not have legal obligations to do so, and again principal-agent problems are clearly possible. Nevertheless, states' leaders face some, albeit imperfect, political incentives to represent their citizens and residents in the international arena, especially in democracies.

Temporal sequence is a particularly challenging aspect of identifying injurers and victims of harm. These include well-known barriers in the case of anthropogenic climate change. For example, because climate change and its harm occur substantially later than the emissions that cause them, present victims need to point to past injurers who might no longer exist or who might not have been aware of emissions' impacts. Likewise, future victims have not yet been harmed and might not even exist yet. The case of solar geoengineering is more difficult, largely because solar geoengineering is being researched in response to another risky anthropogenic phenomenon, climate change. From which baseline to assess harm is not clear, which I will illustrate through examples. A simple first example is a country that had been negligibly affected by climate change and then harmed by solar geoengineering. This seems to be a clear case of harm. Second, now imagine one that had benefitted from climate change and was then returned to its original condition of well-being by solar geoengineering. Whether it was harmed depends on whether the baseline is before or after climate change. In turn, this might depend on the length of time between the onset of climate change and the use of solar geoengineering. If this period was long, then the case for using the post-climate change baseline would be

---

[2] An actor can often be held vicariously liable for harms arising from others' behavior if the former was in a position to exercise control over the latter.

[3] A principal-agent problem arises when one actor (the agent) has a duty to act in the interests of another (the principal) but might have incentives to not do so.

stronger, and vice versa. As a third example, suppose that climate change had harmed a state and the subsequent solar geoengineering reduced but did not eliminate this harm. Suppose further that models indicate that additional solar geoengineering could fully return the country to its pre-climate change condition, but the implementer refuses to carry this out. Although an intuitive answer might be that this refusal is not a harm, note that states are responsible under international law for their omissions as well as their acts. The victim state at hand could argue that it was harmed by the omission of additional solar geoengineering.

The third aspect of injurers' and victims' identities is their worthiness of blame or compensation. In the case of climate change, there are several reasons that emitters might not be blameworthy. For example, the contribution of each to elevated atmospheric greenhouse gases is relatively small and has almost insignificant effects on the timing and intensity of climate change. Furthermore, much of the emissions occurred before scientists were sufficiently certain about causes, probability, and impacts of climate change to establish culpability. Even then, states have duties to act in their populations' best interests, and in many cases economic development and activities that emitted greenhouse gases were justified from that country's perspective.

For injurers' blameworthiness for harm from solar geoengineering, we can turn to liability law. There, an injurer is often not liable for harm if she had to perform the risky activity to protect herself, the wider community, or a specific other actor. The defenses for these are called private necessity, public necessity, and Good Samaritan, respectively.[4] A state (or other actor) may have believed, perhaps based on the best available evidence, that solar geoengineering was necessary to prevent severe or even existential harm from climate change to itself, to multiple other states, or to a particular other state. In these cases, the injurer could argue that others, through their large greenhouse gas emissions, caused its behavior to become necessary. By extension, if solar geoengineering were to benefit some states and harm others, the case could be made that the beneficiaries – not the implementer – should be responsible for any compensation (Heyward 2014). Furthermore, an injurer is generally not liable for harm if she took reasonable care. In solar geoengineering, the injurer may have complied with prevailing standards and expected widespread benefits based upon the best available knowledge.

Victims' worthiness for compensation can be shaped by their behavior through several mechanisms. First, the victim might have contributed to the solar geoengineering. This could be by explicitly consenting to it or by politically, financially, or materially supporting it. Framed legally, the victim might have "assumed the risk." Second, the victim might have interfered with the solar geoengineering in some way that contributed to the harm. For example, a country that had previously

---

[4] Note that an injurer can claim to be a Good Samaritan with respect only to harm caused by trying to aid the victim, not to others.

protested against solar geoengineering might subsequently have destroyed the implementing facilities. The resulting sudden cessation harmed the country. In this case, the argument that it should be compensated for this harm would be a weak one. Third, the victim might have failed to take reasonable precautions in advance of solar geoengineering's predicted effects. Again using legal terms, it could have been contributorily negligent. Fourth, the victim might have contributed to the apparent need for solar geoengineering by emitting large amounts of greenhouse gases. Finally, the victim state might be relatively wealthy and seen by others as not needing compensation.

12.1.3 Mechanisms for Securing Compensation

The third set of questions concern how compensation could be secured. This has two aspects, the first of which is legal. Domestically, a victim can generally request that a court find an injurer liable and, if she refuses to pay damages, forcefully appropriate her property. In contrast, due to the consensual nature of international law, transboundary liability requires the consent of the injurer state. This presents a barrier to international liability. Nevertheless, it could be in an injurer state's interests to pay compensation for harm from solar geoengineering, independent of whether it would be liable under international law. Indeed, countries that wish to undertake solar geoengineering might offer some form of compensation to broaden political support for their activities.

The other aspect of securing compensation is financial. The harm from both climate change and solar geoengineering could be immense.[5] Those who might be willing or (somehow) required to compensate for this might lack the assets to do so. This could be the case even if the solar geoengineering were to produce net benefits and those who benefitted were willing to compensate those who were harmed. The benefits would be widely dispersed and largely nonpecuniary. To compensate victims, benefitting states would need to raise sufficient revenue to pay the damages, which could present a collective action problem.

12.1.4 Reasons for Compensation

The final set of issues is why victims should be compensated for harm from solar geoengineering. The arguments that no one should suffer harm due to others' actions and that injurers should compensate for the harm that they have caused are intuitive and defendable, especially if the harmful actions were unjust. However, these initial arguments do not always provide clear answers. What if the injurer justifiably believed that her activities would help others and acted with due care, but

---

[5] As a rough estimate, suppose that solar geoengineering could largely negate anthropogenic climate change but continuously results in harm to some countries at a magnitude one-tenth of the expected impacts of climate change. That could still be hundreds of billions of dollars annually.

accidentally caused harm? What if the victim's behavior contributed to the harm, either by making an accident more likely or severe or by making the injurer's risky activity necessary in the first place? What if the injurer's actions created net benefits, although they harmed a few, and holding her liable would have caused her to not act? What if the injurer is poor and the victim is wealthy?

Various frameworks of justice offer some guidance as to whether an injurer should compensate a victim. I consider four simple versions of them here.[6] Under retributive justice, those who have done wrong should be punished through legitimate means that are proportional to the wrong (Perry 2005). Here, an injurer who acted wrongly, such as by going against the international community's explicit wishes or by being reckless, should be punished, for example by paying damages that are even more than the harm. Compensatory justice asserts that those who have been harmed – whether by anthropogenic climate change, solar geoengineering, or some other activity – through no fault of their own should be made whole through compensation, either by the injurer or someone else, such as the state or an insurer (Dobbs, Hayden, and Bublick 2015, 22). Corrective justice focuses on the relationship between parties, which can be undermined by harm, especially if the injurer's reckless behavior caused it (Coleman 1995). In this framework, it would be important that the injurer pay the victim to restore their relationship. Finally, distributive justice encompasses diverse norms regarding how resources should be distributed among actors and thus often considers the actors' relative wealth. Unlike the first three frameworks, distributive justice might or might not be advanced through liability for harm, depending on the relative starting positions of the injurer and victim (Dobbs et al. 2015, 19). All four justice frameworks share an ex post approach, looking backward at the cause of, the magnitude of, and the response to harm. They are sometimes compatible with one another but other times come into tension and conflict.

An alternative normative framework for assessing and proposing liability for harm is to maximize well-being, or welfare. In this, these legal rules should seek to maximize the total benefits of actions minus the costs of accidents, precautionary measures, and administration (see Shavell 2007a). In contrast to the justice frameworks, welfare maximization uses an ex ante perspective, looking forward at the incentives and expected behaviors that various rules can be expected to cause. For example, liability for harm generally causes an injurer to exercise care and reduce her level of activity. The injurer would maximize her welfare by balancing her benefits from the risky action with the costs of precautionary measures and possible liability. Meanwhile, the victim's welfare is constant if the compensation is great enough that he would be ambivalent as to whether he had been harmed and compensated versus not being harmed in the first place (that is, "perfectly

---

[6] These considerations are admittedly thin due to space constraints. Numerous books have been written on specific justice frameworks, and more could be authored with respect to solar geoengineering (see Preston 2017).

compensated"). This analysis can be extended to account for administrative costs, insurance, and imperfect information, compensation, and judgments.

These two sets of normative frameworks – justice-oriented ones and the welfarist one – are not necessarily mutually exclusive. In many cases, liability and other rules can satisfy multiple objectives. However, there is often a trade-off between equity and effectiveness that should be acknowledged.

12.2 INTERNATIONAL LAW

The influential Royal Society report on geoengineering states, "At this point it is not clear whether liability for damage caused by geoengineering beyond national jurisdiction is best resolved through new or existing mechanisms" (Shepherd et al. 2009, 41–2). These existing mechanisms are limited. International law generally focuses ex ante on preventing problems and ex post on offering platforms to resolve disputes, not on liability or other processes that involve potential blame. One reason for this is that many countries have harmed many others; attempting to enumerate the harms, attribute them, and demand damages would increase political tension and divert limited resources from more productive endeavors. Environmental issues are not an exception to this. Another reason is that states are very reluctant to commit to a supranational authority that could compel payment of damages. Nevertheless, the international law regarding liability for environmental harm might, in some situations, be applicable to solar geoengineering. More generally, it can shed light on how states could and could not be expected to cooperate.

International legal provisions for transboundary environmental harm are given in principles, customary international law, and specialized treaties (see Chapters 5 and 8). First, the polluter pays principle calls for the injurer to be responsible for the costs of preventing, reducing, and remediating environmental harm. However, this is operationalized only weakly in international law and usually calls for domestically requiring actors to pay for pollution, not for state liability. Second, the ILC has drafted principles for liability – called "allocation of loss" – for transboundary harm from hazardous activities that do not violate international law. These might reflect customary international law but are of uncertain legal status. Regardless, states should ensure that both domestic and foreign victims of environmental harm have access to judicial remedies and should domestically impose strict liability on actors who undertake activities that pose significant transboundary risks. Third, under the customary law of responsibility for internationally wrongful acts, a state that has harmed another due to an act or omission that was contrary to international law is obligated to, among other things, pay compensation for the harm, particularly if the harm cannot be restituted. The inclusion of omissions as grounds for liability gives states incentives to effectively govern nonstate actors. Placed in more traditional legal language, a state is liable – at least in principle – for transboundary harm from the risky activities, such as solar geoengineering, that it negligently undertakes, approves, or fails to control. The standard of care for negligence is compliance with its international legal obligations.

In addition to principles and customary international law, a handful of specialized treaty regimes provide for compensation for harm arising from specific sets of activities: the maritime transport of oil (CLC 1969; CLC 1992; FUND92), nuclear power (Paris Nuclear Liability Convention; Vienna Nuclear Liability Convention; Joint Protocol; Convention on Supplementary Compensation; see Reynolds 2014a), aircraft (Rome Convention), genetically modified organisms (Nagoya-Kuala Lumpur Supplementary Protocol), and space activities (Outer Space Treaty; Space Liability Convention). In general, these hold the operator of the activity liable for damages, and parties must or may require nonstate operators to carry sufficient insurance or another financial security instrument. Specifically, the regimes for the maritime transport of oil and nuclear power offer useful examples of how liability for activities that pose large transboundary risks can be managed. Again grossly generalized, they utilize three tiers of responsibility. In the first tier, strict but limited liability for harm is channeled to the operator, who is required to be sufficiently insured up to the first tier's limit. If damages exceed this limit or if damages cannot be obtained from the operator or its insurer, then an industry-wide fund (in the case of maritime oil transport) or the source state (in the case of nuclear power) provides a second tier of compensation up to some limit. If damages exceed the second tier's limit, then a multistate-supported international fund provides the remaining compensation.

In contrast, space activities are among proposed solar geoengineering methods, although these presently appear prohibitively expensive. The Outer Space Treaty establishes, and the Space Liability Convention details, that the party (or parties) that launches, procures the launching of, or provides the territory for the launching of a space object is strictly liable for harm by that object to other parties on the Earth's surface and to aircraft in flight (Outer Space Treaty, Article VII; Space Liability Convention, Articles I(c), II; see Chapter 8). Therefore, space-based solar geoengineering is the only such technique for which the state that undertakes or authorizes it would clearly be liable for transboundary harm, in this case even if done consistently with international law. The only available defense is if the harm was due to the victim's gross negligence or hostile intent (Space Liability Convention, Article VI). Multiple liable parties are jointly and severally liable (Space Liability Convention, Article V). Harm that is eligible for damages is defined narrowly to include only loss of life, injury, impacts on health, and property damage (Space Liability Convention, Article I(a)). The amount of compensation should be sufficient to return the victim to the condition before the accident, "in accordance with international law and the principles of justice and equity" (Space Liability Convention, Article XII). This liability implicitly includes harm from accidents as well as expected operations and from direct contact as well as remote effects. If the injurer and victim states cannot reach a diplomatic settlement, then a Claims

Commission can be established (Space Liability Convention, Articles XIV–XX). Its decision is binding only if the involved parties have agreed so.

Liability for harm from space objects has been invoked only once, when a Soviet nuclear-powered satellite fell onto a largely uninhabited part of Canada in 1978 (Viikari 2008, 72, 294). The latter requested, based upon the Outer Space Treaty, the Space Liability Convention, and general principles of international law, that the former pay six million Canadian dollars for the clean-up of the scattered radioactive debris and environmental remediation. The Soviet Union eventually paid and Canada accepted three million Canadian dollars, even though Canada claimed that its costs were at least fourteen million. Notably, the Soviet Union's offer was ex gratia, acknowledging neither its legal obligations nor the space treaties. This reinforces the observation that states are reluctant to commit to liability for harm.

As an additional note, the UN Convention on the Law of the Sea (UNCLOS) briefly addresses liability for environmental harm (see Chapter 8). UNCLOS does not seem to substantially expand the scope of liability, instead providing only that "States are responsible for the fulfilment of their international obligations concerning the protection and preservation of the marine environment. They shall be liable in accordance with international law" (Article 235). That article goes on to provide that its parties have two additional obligations. First, they must ensure access – presumably including by foreigners – to their judicial systems for compensation for environmental harm that was caused by legal persons under their control. Second, they should cooperate in the implementation and further development of international law regarding responsibility and liability for environmental harm.

Finally, it is worth considering what states have declined to include within the scope of international liability for environmental harm. The ILC had to abandon efforts toward establishing strict state liability for transboundary environmental harm that originated from activities in a state's territory or under its control because "few governments, in whatever context, have shown any enthusiasm" (Birnie, Boyle, and Redgwell 2009, 223). Even its more modest suggestion that states ensure access to justice for civil liability is limited to draft principles, not articles, and there, strict liability for hazardous activities is merely recommended. Most pertinent to solar geoengineering is that negotiators rejected – at times quite forcefully – any notion of liability in the Paris Agreement (UNFCCC COP Decision 1/CP.21, paragraph 51). Furthermore, international law is littered with numerous multilateral agreements that would establish liability for environmental harm but failed to attract sufficient participation to come into effect. These include treaties concerning transboundary movements of hazardous waste (Basel Protocol), accidental damage to the Antarctic environment (Annex VI to the Madrid Protocol), industrial accidents that have impacts on shared waters (Protocol on Civil Liability and Compensation for Damage Caused by the Transboundary Effects of Industrial Accidents on Transboundary Waters; International Convention on Liability and Compensation for Damage in Connection with the Carriage of Hazardous and Noxious Substances by Sea), and environmentally risky activities in general (Convention on Civil Liability for Damage Resulting from Activities Dangerous to the Environment). International legal scholar Alan Boyle concludes that "the reluctance of states to ratify those same treaties may indicate a less than wholehearted commitment to the idea of shifting the focus away from state responsibility for transboundary harm in favour of civil liability and individual access to justice" (Boyle 2005, 26).

Taken together, existing international law offers a basis for liability for harm from solar geoengineering activities. Specifically, a state could be liable for compensation for transboundary harm that occurred due to its negligence, including by failing to comply with its international legal obligations. At the same time, states' limited acceptance of international law and their practice indicate that they are unenthusiastic about liability – both state and civil – for transboundary harm, consenting to it under only limited circumstances. This implies that a possible liability regime specific to solar geoengineering would face a challenging future.

12.3 A PROPOSAL FOR OUTDOOR RESEARCH

Some observers have asserted that compensation for transboundary harm should be part of international solar geoengineering governance, perhaps linked to authorizing deployment.[7] This section and Section 12.4 offer initial, admittedly somewhat idealistic proposals for compensation for outdoor research and for deployment, respectively. Because states are notably reluctant to commit to international compensation or liability, these might be only starting points for future discussions. Nevertheless, it is not out of the question that states that wished to undertake large-scale solar geoengineering activities that posed transboundary risks might suggest compensation for harm to reduce international opposition.

As discussed in Chapter 1, normatively I adopt a welfarist approach in this book, and do so for these sections as well. Welfarism's usefulness is particularly evident in the case of compensation and liability. Here, rules can increase welfare by offering incentives for injurers and victims to adopt optimal levels of activity and care. At the same time, they can cause undesirable second-order effects and have unclear and even contradictory distinctions between the deontologically just and unjust. For example, retributive, compensatory, and corrective justice – briefly introduced in Section 12.1 – each generally favor an injurer's liability for harm to others. Suppose that a country that is willing to undertake solar geoengineering that is expected to reduce climate change and the associated risks for all countries does not proceed because of possible liability for an accident or unexpected result. Or suppose that a poor state that is vulnerable to climate change deploys solar geoengineering and then must pay damages to a wealthy one that argues that it has lost its pleasantly warmed climate. Would these scenarios be just?

[7] This section is based on Reynolds (2015b).

12.3 A Proposal for Outdoor Research

Therefore, the objectives of the proposals here are primarily to maximize welfare and secondarily to compensate victims while remaining as politically feasible as possible. Bearing in mind that states' consent is required in order for them to be bound by international legal agreements, political feasibility implies that the countries that would be necessary for an international regime would benefit by participating. At the same time, I acknowledge the present low appetite for new multilateral environmental agreements and especially for commitments to compensate.

Harm from solar geoengineering would arise primarily from large-scale outdoor activities such as field tests and deployment, which are distinct from each other in several ways. Large-scale field research could be performed by state, quasi-state, or nonstate actors; is less likely to follow widespread international consultations; would take place under relatively low knowledge conditions; and would presumably be intended to generate shared knowledge. In contrast, implementation would likely be undertaken or authorized by a state, group of states, or an intergovernmental organization; be preceded by international consultations and perhaps some consensus; take place with relatively more knowledge; and be intended to alter the world's climate. Compensation or liability for harm from outdoor experiments is simpler because the possible scenarios are more constrained than those of implementation, which would depend upon – among other things – the international political power of implementation's supporters and opponents. This section considers large-scale outdoor solar geoengineering experiments – here called simply "research" – whereas Section 12.4 discusses deployment.

There are four relevant aspects, which together imply that liability for transboundary harm could be both beneficial as well as counterproductive. First, research currently appears to present a public good with a large positive value, despite possibly posing transboundary risks. It is a public good in that the knowledge that it would generate would be nonexcludable, assuming that it would be shared through scientific publications and would not be subject to extensive intellectual property claims (Foray 2004, 113–29; Saxler et al. 2015, 146; see Chapter 11). It would be beneficial in that the knowledge would reduce uncertainty regarding options to reduce harmful climate change (Moreno-Cruz and Keith 2013). Although it is possible that outdoor research could produce net harm, rough estimates of the potential values of solar geoengineering and of research, as well as estimates of the potential environmental harm, indicate that it would most likely not (Reynolds 2015b, 187, 190–1). Holding the researcher (who would lack sufficient assets regardless) or the state liable for harm would discourage the provision of this beneficial public good. Second, a market for liability insurance would be difficult and unlikely to develop, and a risk-sharing pool among researchers would be insufficient and challenging to maintain. Third, leaders who believe that their state had been harmed by research would likely demand compensation. Fourth and finally, compared to nonstate actors, states would better serve vicariously as both injurers and victims. A regime for compensation or liability for transboundary harm would require some basis in international law, where countries are the primary actors. On the injurer's side of the issue, states' financial assets are much greater than those of nonstate actors, who could go bankrupt before paying damages. Governments already have obligations to regulate and authorize projects that pose risks of significant transboundary harm and could also control moral hazard, adverse selection, and free riding among researchers.[8] On the victim's side, a country's leadership is more likely to be aware of harm, which might be widely dispersed; to be aware of its possible cause; and to have sufficient incentive to pursue damages.

I propose an international agreement to establish a compensation fund for harm from large-scale outdoor solar geoengineering research. Its parties would support the fund, with their contributions determined by some combination of their expected benefit from solar geoengineering, their net greenhouse gas emissions to date, and their ability to pay. Parties could claim damages from the fund for harm, but in doing so, they would relinquish any other international legal claim to reparations. The scope of compensable harm could resemble that found in the ILC's Draft Principles on the Allocation of Loss in the Case of Transboundary Harm: bodily injury, loss of life, property damage, economic losses, impairment of the environment, and actual reasonable preventative and response measures (Principle 2(a)). Using the best available methods, a board of international experts would review claims and estimate the probability that the activity caused the harm and what portion thereof. This would allow awarded damages to be proportionally reduced, but only above a certain threshold in order to discourage frivolous claims. The victim's failure to take reasonable precautions could reduce or eliminate damages. Meanwhile, the injurer state (or states) would be obligated to reimburse the compensation fund up to some limited amount – perhaps on the order of several hundred million dollars – only if it conducted or authorized the outdoor experiment in a way that was contrary to international law and to prevailing research standards. That is, it would be indirectly liable with a duty of care standard. The injurer state would have some other defenses, such as if its violation of international law or other standard was due to events outside of its control. A state that was liable (in the sense of obligated to reimburse the compensation fund) could pursue domestic legal action against the researchers if they seemed to be at fault. If this was the case, the state could be considered vicariously liable.

This proposal would both incentivize maximizing social welfare and compensate victims. Furthermore, it could be in all relevant states' interests to participate in the agreement. Researching countries could be attracted by the agreement's capacity to reduce, clarify, and share their potential liability. If states that appeared vulnerable to harm from research participated, it could also give researching countries political cover for a possibly controversial practice. Potentially harmed states' incentive to

---

[8] Here, "moral hazard" has its traditional meaning in the insurance context, not that of solar geoengineering displacing greenhouse gas emissions abatement. See Chapters 3 and 4.

participate is rather clear: they would have an internationally agreed-upon means to seek compensation.

12.4 INITIAL THOUGHTS ON IMPLEMENTATION

Compensation or liability for harm from solar geoengineering deployment is more complicated and speculative than that for research. As described, the possible political landscape remains uncertain in multiple dimensions, such as which states, international organizations, and other influential actors would participate in, support, remain neutral toward, and oppose the implementation; the distribution of expected and actual impacts of climate change and solar geoengineering; whether the implementation was consistent with international law and other widely shared norms; the perceived legitimacy of the decision-making process; and the confidence in the resulting impacts' attribution. Because of these uncertainties, this section is limited to offering rough contours about how we could think about compensation for harm from solar geoengineering implementation.

To establish boundaries, let us first consider two extreme scenarios. In that of global support for deployment, there would be no liability for harm under international law due to the victims' consent to the risky action. The world's countries might have agreed to compensate victims, perhaps to broaden political support for the implementation. Some sort of international fund to share this burden seems logical. The remaining challenges would include attribution, which harms to compensate, what baseline to use, and determining the precise burden sharing among countries. Yet if support was indeed global, then at least the latter three of these (that is, not attribution) seem surmountable.

In the other extreme scenario of a single state deploying or authorizing solar geoengineering contrary to global opposition, any injured states could demand damages.[9] The implementing state would be obligated, at least in principle, to pay if it had acted contrary to international law, which might or might not be the case. Independent of such possible violations, the injurer might concede to paying damages, especially if it faced international sanctions and political isolation, which appear likely. However, obtaining compensation would ultimately depend upon the injurer's consent due to the absence of centralized enforcement of international law. Moreover, if the country was willing to implement solar geoengineering in the face of global opposition, then it might also be willing to reject the calls for compensation. It is for this reason that, to be effective, any proposed compensation mechanism would need to include as participants all states that might deploy solar geoengineering. To achieve this, those countries would need to expect to benefit by participating.

---

[9] See Chapter 4 for the case that problematic unilateral implementation seems unlikely.

Most possible implementation scenarios – and arguably the more probable ones – lie between these extremes. In these, various constellations of states, intergovernmental organizations, and other actors would engage in, support, oppose, and expect to benefit from and be harmed by solar geoengineering. These sets of states and other actors, along with their historical contributions to climate change and their financial abilities to pay, would overlap in complex ways.

In these diverse scenarios, liability might not be an appropriate mechanism to both incentivize optimal levels of activity and precaution among both injurers and victims as well as provide compensation for victims. For liability to be effective – or better said, efficient – several characteristics need to be met. The injurers must be identifiable, few in number, and have sufficient financial assets, insurance, or other security instrument to pay damages. The victims need to be aware of their injuries, injured enough to warrant seeking damages, and face transaction costs to reaching an ex ante contractual agreement with the injurer that – for whatever reason – are prohibitively high. Any third-party beneficiaries of substantial positive externalities from the injurer's actions should also face acceptably low transaction costs to reaching a contractual agreement with the injurer or the victims. (This way, these beneficiaries could offer to pay some of or all the damages so that the activity could continue at a socially optimal level.) Finally, there must be an adjudicative forum that can determine causation and – if appropriate – failure to abide by a duty of care, assess the harm, and enforce an award of damages, all with acceptably low administrative costs.

These characteristics are not present in most implementation scenarios. The injurers might be few or many. Due to the difficulty of attribution, a given injury – such as an extreme weather event – might not be able to be sufficiently causally linked to solar geoengineering. There could be many third parties that benefit from the reduced climate change. Even presuming that they would be willing to pay some or all of the damages in order to allow the solar geoengineering to continue, they would still face a collective action problem among themselves as to how to distribute the burden of paying. The International Court of Justice (ICJ) – a prominent possible forum for international adjudication – might not be able to determine possible negligence due to solar geoengineering's technical nature. More generally, states have been reluctant or unable to authorize international adjudicators (Trachtman 2008). Ultimately, liability is an inappropriate mechanism to compensate states who may be harmed by solar geoengineering deployment, at least in most feasible scenarios.

This creates a dilemma. Among scenarios of solar geoengineering deployment, those in which multiple countries implement and many other countries would face positive and negative impacts seem not unlikely. Yet in the absence of a compensation agreement, international tensions could rise, or solar geoengineering would not be undertaken or only at a suboptimal level. Although the prospects for a forum that is authorized to adjudicate such disputes and to enforce its decisions are dim, solar geoengineering might present sufficient incentives for states to agree to one. Even then, a traditional liability model would be challenged by the problems of attribution and numerous third-party beneficiaries.

This points toward a compensation mechanism that would be administrative, in which parties delegate to an institution the authority to govern rights with respect to a set of risky actions. Although this would have high administrative costs, these may be worth paying in certain circumstances. In general, justifications for potentially expensive administration include when the activity has substantial widespread positive external effects, when injurers may not have sufficient assets to pay damages, when harm is widely dispersed among victims, and when administrators have better information than the injurers and victims regarding what harm has occurred, who caused the harm, and what efficient precautions injurers and victims could take (Shavell 2007b). However, administration is rarely used internationally. The example closest to genuine administration is the governance of rights to minerals in the international seabed (UNCLOS, Articles 150–91).

These circumstances that favor administration could apply to solar geoengineering implementation. This would especially be the case in the complex scenarios in which various constellations of states caused climate change in the first place; engage in, endorse, oppose, benefit from, and are harmed by the solar geoengineering; and have some ability to pay. A hypothetical global administrator could – at least in principle – gather, assess, and share information regarding efficient precautions; identify the injurers, the victims, and the harm; determine the extent to which the harm was caused by solar geoengineering; punish injurers that failed to take these precautions; consider external benefits to third parties; collect funds for compensation, perhaps from those who benefitted and have ability to pay; and compensate victims that had taken appropriate precaution. If solar geoengineering provided global net benefits but harmed some actors, compensation could be provided by an international fund to distribute costs widely and equitably and to not unduly disincentivize those who might provide solar geoengineering. Those countries that benefit from solar geoengineering, those with high historical greenhouse gas emissions, and those with great financial assets could contribute the most.

A different approach to compensation for harm from solar geoengineering has been suggested by Joshua Horton and David Keith. If a central challenge is the tension between deploying states' assertions that deployment would reduce climate change impacts and some other states' claims that they would be harmed by it, then innovative international insurance mechanisms might be able to help resolve this through a risk management approach. In Horton's proposal, deploying states could offer insurance-like contracts to other states that would trigger payouts if specified parametric indices, such as temperature and rainfall, significantly differ from specified ranges (Horton and Keith 2019). If deployers' assertions about solar geoengineering's effectiveness are correct, then climate risks will be reduced and payouts will not be triggered. If resistant states' claims of harm are borne out, then they would be sufficiently compensated. Such parametric insurance might be able to overcome issues of attribution, victim's precaution, moral hazard, adverse selection, and high administrative costs. Some challenges remain. If the number of deploying states were great, then they could face a collective action problem regarding how to distribute costs among themselves. Furthermore, if deploying states were relatively poor or if they needed to make many large payouts, then they might not have the financial assets to underwrite the scheme.

A few final observations are in order regarding international compensation of harm from solar geoengineering implementation. First, this section did not explore possible harm from reducing or ending solar geoengineering. This is important because the possible sudden and sustained termination of solar geoengineering is among the most prominent concerns. Yet there is little precedent on liability arising from ending an ongoing benefit. This is part of a broader phenomenon in which the law and ethics treat negative and positive externalities differently (see Porat 2009; Dari-Mattiacci 2009).

Second, it is not immediately obvious whether compensation for harm would be beneficial. From a welfarist perspective, the administrative costs – including determining harm, the extent to which it could be attributed to solar geoengineering, whether injurers and victims had exercised appropriate precautions, and countries' appropriate contributions to compensation – would be expensive in terms of limited resources. Furthermore, trying to make these determinations could increase international tensions. Along these lines, legal scholar Gareth Davies argues that improved attribution of harm from solar geoengineering could be counterproductive by "lead-[ing] to an unprecedented growth in inter-state claims of responsibility and liability. Accepting such claims would hugely diminish sovereignty and state independence, as science gives states ever more opportunities to object to the internal polices of their neighbours" (Davies 2008, 634–5).

From a deontological view, various justice frameworks offer mixed assessments of compensation through administration. Compensatory justice would clearly be satisfied, and retribution could be achieved through punishing the injurers who failed to take care. In terms of distribution, if solar geoengineering's harm arose primarily from reducing the climate change that benefits some states, compensation might be inequitable. This is because relatively wealthy countries are expected to benefit from moderate climate change and thus could be relatively harmed by solar geoengineering's cooling, whereas poor ones will suffer under climate change and thus would likely benefit more from solar geoengineering. If so, then compensation for harm for solar geoengineering could end up flowing from poor to rich countries, at least in some scenarios.[10] To the extent that contributions to an international

[10] This does not mean that compensation would necessarily be inequitable or contrary to distributive justice. There are many salient factors besides states' relative wealth, such as their relative contributions to responses to climate change. Also, in a Coasian approach, Richard Tol suggests that countries' quasi-property right could alternatively be one to an optimally altered climate, and those that are

compensation fund are based on an ability-to-pay principle, then such inequitable outcomes would be less likely. Corrective justice would not be satisfied by a compensation fund and arguably not through any other feasible means of compensation in complex international situations due to the lack of a repaired relationship between injurer and victim.

The third observation is that it would be preferable to expand, if possible, the scope of harm that would be eligible for compensation (Reynolds 2015b, 208). As proposed, compensation would require that harm be attributed to solar geoengineering, at least to some extent, which would be a burdensome and uncertain process. Moreover, those countries that are not at risk from solar geoengineering may be reluctant to contribute to a compensation fund. Yet if the fund were to cover harm from all climatic changes, whether they are due to solar geoengineering or to anthropogenic greenhouse gas emissions, then the burden of attribution would be reduced and the base of international support broadened (Kellogg and Schneider 1974, 1170; Wong, Douglas, and Savulescu 2014). Further expanding eligible harm to include that from extreme weather phenomena or from all natural disasters would further those trends. Indeed, such a broad-based international compensation fund would likely be a more economically justified means for countries to pool their risks.

The final observation is that, as already noted, if both administration and compensation for harm are so rare in international law and international relations, then their manifestations in the case of solar geoengineering seem unlikely. At the same time, some scholars argue that contemporary environmental challenges require novel international institutions to develop and improve the governance of Earth systems (Biermann et al. 2012). Solar geoengineering may offer the basis to broaden and deepen such international cooperation (Davies 2010, 279).

12.5 SUMMARY AND CONCLUSION

Scholars and other observers regularly suggest compensation or liability for transboundary harm from solar geoengineering. In general, multilateral agreements rarely provide for compensation or liability. There could be a basis – at least in principle – found in customary international law for state liability for transboundary harm from solar geoengineering that was due to an act or omission contrary to international law. Regardless, liability or compensation for harm from solar geoengineering would face numerous conceptual and practical challenges regarding what harm to compensate, injurers' and victims' identities, mechanisms for securing compensation, and the reasons for compensation. While recognizing states' reluctance to consent to compensation, I suggest an international compensation fund in the case of large-scale outdoor research and offer some initial thoughts regarding deployment scenarios.

denied this could be compensated. Compensation would then generally go from rich countries to poor ones (Tol 2016).

13

A Path Forward

Since the rise of the solar geoengineering discourse, the questions of how it could and should be governed have loomed large (see Kellogg and Schneider 1974). Although until now this book has remained mostly descriptive and analytical, this final prescriptive chapter suggests further governance.[1] This begins by making my goals and approach explicit and by reviewing solar geoengineering's relevant characteristics that inform my suggestions. The proposals then follow, divided into three rough stages of small-scale research, large-scale research, and implementation (akin to those in Solar Radiation Management Governance Initiative 2011, 26; Keith 2013, 80–8).

13.1 GOALS, APPROACH, AND CHARACTERISTICS

My goal in proposing further governance for solar geoengineering is to maximize current and future humans' well-being (that is, welfare) in ways that are sustainable, consistent with widely shared norms such as the need for governance to be legitimate, and seemingly feasible. Like any set of multiple goals, these sometimes are in tension with one another. Well-being is meant to broadly encompass all about which people care, including the natural world. In addition, because conditions at one point in time shape those at subsequent ones, governance should aim not only to improve well-being at the present but also to facilitate future desirable conditions. Climate change is and solar geoengineering would be multigenerational – if not multicentennial – phenomena, and governance should consequently consider future generations and sustainability. Legitimacy recognizes the importance of democracy and the rule of law to maintaining a functioning and well-governed society. Feasibility is a minimal criterion because, in its absence, the apparent benefits of idealistic proposals that are highly unlikely to be adopted would dominate the suggestions. Patterns of past individual, state, and collective behavior offer a reasonable guide to future possibilities. One of the strongest feasibility constraints arises from the fact that people and institutions – including states – act to pursue

---

[1] Note that in Chapters 11 and 12, I offer policy suggestions regarding compensation and intellectual property.

their own diverse objectives given their limited resources. Proposals for governance should thus attempt to align actors' self-interests with increasing wider well-being. Greater international cooperation could make for more effective governance of solar geoengineering but might not be in states' own perceived interests, at least in the short term (see Jinnah 2018). For this reason, the second and third stages of proposed governance concerning large-scale research and deployment, which substantially rely on state activity, include some suggestions that rely less on international cooperation.

The characteristics of a given challenge will have a large bearing on suggesting and assessing governance (see Chapters 2 and 4). Climate change will have generally negative impacts on humans, other species, and the wider environment. These might be severe and will be most harmful to already vulnerable populations and ecosystems. Furthermore, there is a small chance of very large harm from climate change. Greenhouse gas emissions abatement, adaptation, and negative emissions technologies (NETs) can reduce climate change and its impacts but will be unable to eliminate them. Abatement and NETs pose difficult global collective action problems, and those countries that need to adapt the most tend to have the least capacity to do so. Importantly, these actions' beneficial effects will occur – for the most part – decades after their undertaking. These responses will continue to be substantially insufficient to prevent dangerous climate change and its impacts.

The current evidence is that solar geoengineering could greatly reduce climate change and the resulting harm. It appears to be technologically feasible, relatively inexpensive, rapid, and reversible in its direct climate effects. The first two characteristics imply that some states could implement it on their own and for their own net benefit, at least in principle (see Chapter 4). Its speed of action would allow it to manage short-term – that is, on the order of years to a couple of decades – climate risks in ways that abatement, adaptation, and NETs cannot. At the same time, solar geoengineering would pose genuine environmental risks and social challenges. Although which of these are most serious is a matter of judgment, I am most concerned about premature, uncoordinated, or excessive implementation; international disagreements regarding deployment decisions and responsibility for perceived negative impacts; and lessened efforts for other responses to climate change (see Chapter 3). For these reasons, and because it would imperfectly reduce climate change, solar geoengineering is not a solution to climate change and offers instead a possible complementary response option that poses a risk-risk trade-off with substantial uncertainty.

Additional, dedicated governance of solar geoengineering is warranted because it could help solar geoengineering proceed responsibly while reducing these risks and challenges. This does not mean that additional governance is *required*. The proposed technologies' development and use would proceed in some fashion in the absence of additional intentional guidance by states and other authoritative institutions. However, in the absence of governance that is specific to solar geoengineering, the risks and challenges would likely be greater. There would be less research, and what is undertaken would be less coordinated. Implementation would be more likely to be done with a weak knowledge base by a small number of countries contrary to the wishes of others, increasing international tensions. Deployment systems would be less redundant and resilient, increasing the chance of sudden and sustained termination.

There is already substantial governance of solar geoengineering, despite regular claims of a regulatory vacuum (for example, McKinnon 2019, 443). Existing national and subnational law would apply to small-scale activities that would not pose transboundary risks (see Chapter 9). Large-scale outdoor activities would implicate a wide range of international law, including treaties and custom; binding and nonbinding agreements; and in environmental, human rights, and other issue areas (see Chapters 5 through 8). Nonstate governance has apparently already influenced solar geoengineering activities and can continue to do so in ways that can be more flexible, open-ended, and better informed than legal governance and can lay a foundation for future legalized governance (see Chapters 10 and 11). Furthermore, at all scales, most actors have incentives, such as maintaining reputations, for responsible behavior. Together, this extant heterogeneous, polycentric governance is incomplete, fragmented, and inconsistent, but it is also adaptable and offers opportunities for learning (Galaz 2014; Reynolds 2018b).

Claims of a regulatory vacuum implicitly point to the absence of a binding, inforce international legal instrument that is specific to solar geoengineering. Yet this is not a problem, and in fact, it remains too early to develop one. Doing so under the present conditions of great scientific, social, and political uncertainty and low knowledge would lock-in rigid governance that might turn out to be ineffective or otherwise inappropriate. As research and development proceed, we will learn more about solar geoengineering as well as what we do – and do not – want from it. That is, solar geoengineering and its governance should co-evolve (Parker 2014; Parson 2017b). Meanwhile, a broad societal conversation as to whether developing and perhaps using solar geoengineering would be an acceptable means to reduce the risks of climate change is necessary. Moreover, a new dedicated international legal instrument is very unlikely in the near term. Ironically, demanding one only makes it more probable that solar geoengineering would proceed without appropriate additional governance.

Crafting governance of solar geoengineering will not be simple, and I highlight five key challenges here. First, one must define the activities to be governed. In general, solar geoengineering is the intentional modification of the Earth's radiative balance, excluding changes to greenhouse gases, and including the research and development activities that would inform and enable such interventions. However, the definition's boundaries for governance purposes remain unclear. For example, should intentionality – whether to modify energy fluxes or reduce climate change – always be a requisite? These definitional issues manifest differently at small- and large-scales. At the former, other research activities – such as those examining the interactions of aerosols, clouds, and climate – can resemble solar geoengineering research yet should generally not be subject to its additional regulation. On the other hand, solar geoengineering researchers could evade this regulation by misrepresenting their intentions. At large scales, research and deployment are difficult to distinguish. An actor who wished to implement might be able to avoid certain governance requirements by claiming to be conducting research.

Second, because climate change, solar geoengineering, and global politics are uncertain, there is a tension between developing governance early, before problems manifest, and doing so late, when more will be known.[2] Climate change impacts and efforts to reduce them via emissions abatement, adaptation, and NETs might be more or less severe than we presently believe. Solar geoengineering might turn out to be expensive, unacceptably risky, or politically rejected. It could consist of one or more of the currently leading techniques or something new and unexpected. For example, it is also not out of the question that, at some point in time, a combination of miniaturization and information technology could enable solar geoengineering through numerous small drones that would both be regional in effect and rapidly responsive to climatic conditions. If so, then solar geoengineering could develop to reduce extreme weather events, and distinctions among categories of activities would further blur. Moreover, we cannot know how widespread deployment capacity would be, which climatic conditions countries would prefer, and the value that states would place on having their preferred climate. Global politics could be cooperative or conflictual, and the distribution of states' relative power could have many forms from highly concentrated to widely dispersed. Likewise, the preferences of political leaders and laypersons regarding the environment, risk, and technology will continue to evolve. Because of this increasing uncertainty as one considers the more distant future, my suggestions for governance of the large-scale research stage and especially the implementation one are tentative.

However, this uncertainty should not paralyze decision-making. Research can reduce uncertainty, although some uncertainty will probably be irreducible. The governance of solar geoengineering should not only coevolve with its research and development but also remain adaptive to changing conditions and improved knowledge.[3] Although I consider here what I believe to be a middle range of possible futures, this range widens as we consider the further future. To be clear, although the future might be quite different than that described in this chapter, I assume in each section that the activities would, in fact, be undertaken without claiming that this would be certain or even to be expected.

---

[2] This is the Collingridge dilemma (Collingridge 1980).
[3] Adaptive governance is an established line of research and practice with respect to feedback from natural systems but is less so with respect to human ones (Folke et al. 2005; Bennear and Wiener 2019).

Third, I expect solar geoengineering to remain controversial, at least for some time, which presents challenges to developing governance. Its legitimacy will be particularly important. One implication is that governance processes should engage the public relatively early and deeply. Another is that, for the near future, state actors – which offers a basis of legitimacy – will probably remain reluctant to engage with the topic. Nonstate actors could consequently fill governing roles, offering effectiveness and expertise as bases of legitimacy (see Chapter 10). This need will be heightened by solar geoengineering's complexity and dynamism, indicating that traditional national regulators might have insufficient state-of-the-art knowledge. Furthermore, researchers could – especially early on – see themselves as a reputationally sensitive community of shared fate and have incentives to not only act responsibly but also monitor each other to maintain a social license to operate. A final implication of controversy, coupled with countries' possible suspicions of others' intentions, is that too strong a push for an early debate on international governance could result in a premature ban, taboo, or excessive regulation. There could be substantial costs to human well-being and sustainability of such undue restrictions.

Fourth, governance should not be based solely upon the reduction of aggregated physical risks of climate change and solar geoengineering and the maximization of any co-benefits. At the very least, impact assessments should give additional weight to those who are already disadvantaged. Moreover, decisions should be driven by people's preferences as expressed through legitimate channels. This reinforces the fact that informing and engaging with the public as well as diverse social science research will be essential. Nevertheless, maintaining a connection between governance and the public's preferences will be difficult, given solar geoengineering's global impacts, the world's diversity, and the complexity of how people perceive the environment, technology, and risk.

Finally, reliance upon law as a centerpiece of governance offers benefits and limitations. Perhaps the greatest advantage is law's legitimacy through states' centrality. At the same time, law is slow to change, whereas technology and technical knowledge often advance rapidly. Furthermore, the governance of large-scale solar geoengineering activities must in some ways rely on international law, which lacks centralized development, monitoring, and enforcement. Independent of the pros and cons, solar geoengineering will be researched and developed in a pre-existing governance landscape with numerous norms, rules, procedures, and institutions of diverse scale and legal character. In developing additional governance, these conditions should be acknowledged and even seen as an opportunity, not a barrier to be overcome (Nicholson, Jinnah, and Gillespie 2018).

13.2 SMALL-SCALE RESEARCH

The first of three rough stages of governance is that of small-scale solar geoengineering research. This includes indoor work as well as outdoor experiments whose impacts can be expected to be moderate. Here, the primary goals would probably be to reduce uncertainty regarding impacts and technical feasibility. Governance at this stage would be decentralized and heterogeneous. Developing governance will be constrained by uncertainty both in solar geoengineering's potential benefits and risks as well as in what we normatively want from it.

Like the other stages presented here, this one requires definitional boundaries. Activities that otherwise appear to be solar geoengineering research should be exempted based upon the researchers' reasonable stated intentions, but only if the expected impacts are *de minimus*. Although this raises the possibility of solar geoengineering research not being governed as such, this should be balanced with that of other research being subjected to unnecessary rules. If some genuine solar geoengineering research with *de minimus* expected impacts escapes additional oversight due to researchers' misleading statements, the consequences may be acceptable.

At the "upper" end, beyond which research would be large-scale, should be activities whose impacts are expected to be transboundary, widespread, long-lasting, or severe. The first criterion is important as this is where international law becomes salient. The other trio of terms is found in two international agreements.[4] A document related to the Convention on the Prohibition of Military or Any Other Hostile Use of Environmental Modification Techniques (ENMOD) defined them as:

> "widespread": encompassing an area on the scale of several hundred square kilometres; "long-lasting": lasting for a period of months, or approximately a season; "severe": involving serious or significant disruption or harm to human life, natural and economic resources or other assets. (ENMOD Understandings)

The upper boundary for small-scale research activities could use the expected radiative forcing – a measurement of the change in the Earth's energy balance – as one criterion to determine severity. For example, Parson and Keith suggest 0.01 watts per square meter, which is less than one percent of current anthropogenic radiative forcing due to greenhouse gases (Parson and Keith 2013).

Reducing the uncertainty regarding solar geoengineering's effectiveness, capability, costs, speed, reversibility, and risks is important and requires substantial investments. Although nonstate sources can contribute, state support is necessary. Especially as research moves outdoors, funders should consider allocating supplemental resources for compliance with governance as well as for insurance for compensating for potential harm. Furthermore, as research grows, funding would benefit from coordination within and among countries. This could take a range of forms and degrees of centralization including informal communication, a formalized consultation forum, and a genuine international program.

[4] The three words are in ENMOD itself (Article I.1) as well as in the approved amendment to the London Protocol to regulate marine geoengineering (Article. 1).

Governance should help ensure that solar geoengineering research should, as it proceeds, include three specific aspects. First, specialized clusters of scientists should aim to demonstrate either solar geoengineering's effectiveness and safety or its ineffectiveness and risks (Bipartisan Policy Center's Task Force on Climate Remediation Research 2011, 24). Second, efforts should explore diverse proposed technologies and approaches to help prevent technological lock-in. Third, international cooperation should be a priority, in order to capitalize on scientists' knowledge and skills worldwide, to foster socialization and build trust, and to avoid perceptions of rivalry or hostile activities. Moreover, international cooperation should not be mere sporadic collaboration and co-authorship but also include active recruitment, exchange programs, and capacity building in developing countries, which ideally could eventually develop their own research programs. The Solar Radiation Management Governance Initiative (SRMGI) is currently taking important steps in this direction.

Independent synthesis and assessment of results is also essential to reducing uncertainty. The Intergovernmental Panel on Climate Change (IPCC) appears qualified and authorized to do so and possesses the legitimacy that could help solar geoengineering integrate better into the broader climate change discourse.[5] However, because its members might be resistant to seriously considering solar geoengineering, David Victor argues that geoengineering is a poor match for the IPCC due to its open, weak, and consensus-oriented character (Victor 2008). One or more alternative forums – such as multiple national academies of science – might be warranted.

Much, if not most, of the environmental, safety, and other risks of small-scale solar geoengineering research can be effectively managed through existing national and subnational law (see Chapter 9). The industrialized countries, which are the most probable locations and funders of research, generally have robust regulation regarding air pollution, protection of vulnerable species and ecosystems, environmental assessment, public notification and participation, aviation, marine vessels and structures, weather modification, and liability for harm. Institutions that intend to conduct or support outdoor solar geoengineering research should undertake legislative analyses to confirm the extent to which existing mechanisms can appropriately govern environmental and other risks (Chhetri et al. 2018, 23). If there are gaps, then state or nonstate governing institutions should address them. In some cases, new national legislation or administrative regulations specific to solar geoengineering might be warranted.

---

[5] The IPCC's original mandate included reviewing "Possible response strategies to delay, limit or mitigate the impact of adverse climate change," although its own current principles for governing its work is limited to "options for adaptation and mitigation" (UN General Assembly Resolution on Protection of Global Climate for Present and Future Generations of Mankind; Intergovernmental Panel on Climate Change 2013). The latter might need to be amended to consider solar geoengineering.

Nevertheless, states should not be the sole source of governance of solar geoengineering. It would be unwise to fully rely upon them, either through domestic law or international cooperation, at least initially, due to the need for expertise, insufficient incentives for state actors to engage with the issue, the possibility of forum shopping, and normative uncertainty. Instead, in these circumstances, nonstate governance can offer adaptiveness, expertise, and an ability to operate across borders (see Chapter 10). Therefore, the further bottom-up development of norms and codes of conduct is important (Victor 2008; Parker 2014). Steps in this direction are evident in the various – and notably overlapping – sets of principles, particularly the Oxford ones (Rayner et al. 2013; Chhetri et al. 2018, 11). Most of the important features of the governance of small-scale sole geoengineering have been put forth in the proposed principles, including transparency; public participation; independent assessment; cooperation among researchers; monitoring, reporting, and verification; adaptive governance; reducing risk and maximizing benefit; and compensation for harm. Further development and crystallization of norms will require actual practice of solar geoengineering research, and their contours thus remain presently unclear. Nonstate governance could be linked to the proposed research commons (see Chapter 11), which could attract researchers, their institutions, funders, and publishers, bringing them into the fold of governance.

One central principle that warrants some elaboration is transparency, which is essential to responsible solar geoengineering research. Governance standards should include open publication of methods and results, including negative ones as well as disclosure of funding sources. Some of the precise details of transparency still need to be resolved. For example, should all research activities – including indoor modeling work – be publicly described before they are undertaken? To what extent should data be standardized (see Chapter 11)? Should researchers be required to publish in open access journals and to include laypersons' summaries? Are there any exceptions for confidentiality? Transparency could be linked to independent assessment and could be furthered through an online clearinghouse of researchers, projects, programs, and results (Craik 2015).

The principles for solar geoengineering research should, at some point in time, be gradually legalized; that is, their precision, delegation, and obligation should be increased (Abbott et al. 2000). The first process is one of providing greater specificity, through for example codes of conduct, to clarify the norms and to align expectations. The proposed Code of Conduct for Responsible Geoengineering Research is an important step in that direction, in part through helping keep nonstate governance aligned with existing international law (Hubert 2017). At the same time, its reliance upon international law as a basis also limits its utility, particularly for small-scale research. Some issues – such as the proper roles of commercial actors – remain unsettled and will need to be confronted. Legalization's second process – delegation – refers to authorizing governing actors distinct from the targets to develop, monitor, and enforce rules, standards, and norms. In other words, this would be a shift from self- to private regulation, meta-regulation, or – where possible and appropriate – state governance. Numerous nonstate actors, including research institutions, funders, professional societies, publishers, and nongovernmental organizations, could assume some delegated authority.[6] This points to the final process of legalization, that of obligation. At the very least, as a reputationally sensitive community of shared fate, solar geoengineering researchers have both the incentives and means to punish clear violators among them through naming, shaming, and exclusion. Furthermore, the potential delegated governing nonstate institutions listed above could offer rewards and punishments for those who comply with or contravene the nonbinding governance. During this legalization process, multiple sites of nonstate governance might develop. Ideally, any competition among them could foster those that are more effective, efficient, and just, instead of creating fragmented governance thickets.

A possible model for the nonstate governance of solar geoengineering's physical risks is independent, institutionally affiliated committees. These currently review research with recombinant DNA, animals, and human subjects. For example, the International Ethical Guidelines for Health-Related Research Involving Humans, the Declaration of Helsinki, US federal law, and European Union law each call for proposals for research with human subjects to be approved by the appropriate ethics committees before proceeding (Council for International Organizations of Medical Sciences and World Health Organization 2016; World Medical Association 2018; Common Rule, 45 CFR 46; Clinical Trials Directive; see Chapter 7). Analogous solar geoengineering research review boards could help ensure that experiments, projects, and programs proceed in a manner that minimizes risks relative to expected benefits and are consistent with widely shared norms. Their members should be independent from both the research and, to the extent possible, from the home institution. Research that presents negligible environmental impacts could be subject to an expedited review. Importantly, establishing the boards does not require state action. Instead, research institutions can act soon, and nonstate actors such as funders can demand these boards from the institutions that they support. Initially, these review boards could each operationalize general principles in their own diverse ways. Indeed, central coordination would not initially be needed and might not even be beneficial. Instead, the boards could experiment and learn from each other's approaches and experiences. However, as research scales up, greater coordination and harmonization – such as sharing more elaborate principles and best practices – would be warranted and increasingly feasible (Chhetri et al. 2018, 44–5). Here, professional societies, major research funders, public agencies, and nongovernmental organizations could participate.

With respect to the standards' substance, the emerging international governance of ocean fertilization provides a model that could, in some ways, be transposed to

---

[6] Whether governance by a research institution of its own researchers should be considered self-regulation or delegated private regulation depends upon the institution-researcher relationship.

outdoor solar geoengineering experiments that would be expected to have significant environmental impacts. In this, proponents are to demonstrate that a proposed activity would be legitimate scientific research; that it would fulfill its purpose; that its "rationale, goals, methods, scale, timings and locations as well as predicted benefits and risks" are justified; and that the researchers have the financial resources to carry it out adequately (Marine Geoengineering Amendment to the London Protocol). The proposal to conduct outdoor ocean fertilization research must be subject to an environmental impact assessment that includes, among other things, site selection and description, exposure assessment, risk characterization, and risk management. Furthermore, proposals should describe how the activity would be as encapsulated and reversible as possible; contain risk management plans, including regarding monitoring and contingencies for emergencies; consider alternatives to the suggested activity; and be peer reviewed. However, some caution is warranted in replicating this regulatory framework. No legitimate ocean fertilization research has been proposed since it was developed, implying that it might be too cumbersome.[7]

Growing solar geoengineering research would provoke social challenges. Most of these would not be related to any particular project or experiment and are thus not best managed through institutionally affiliated boards.[8] A wide social discourse is necessary and should ideally involve elected officials, state and intergovernmental bureaucrats, experts, thought leaders, and the general public. To some degree, this can be facilitated. Institutions that are active in solar geoengineering research should engage with the public, assess its opinions, and ensure that its members have appropriate opportunities to participate in decision-making. Indeed, the Carnegie Climate Geoengineering Governance Initiative (C2G2) and the SRMGI are currently initiating and carrying out international dialogues of growing breadth and depth. At the same time, some of this discourse must occur organically from the bottom-up. Although such engagement and conversations might not lead to consensus, it is important that its participants form their opinions, to the extent possible, based upon the best available evidence. Given solar geoengineering's present controversy, the public's low levels of awareness, and the popular media's frequently inaccurate and sensationalist coverage, active outreach will remain necessary.

Two related salient social challenges are that solar geoengineering research might unduly catalyze deployment and that a capable actor might prematurely implement it. To address these, this stage of small-scale research should have explicit predefined conditions under which research would end (Parson and Herzog 2016). These criteria should include a range of physical and social aspects such as the expected

---

[7] Another explanation is that previous ocean fertilization research has indicated little potential to reliably sequester carbon dioxide and relatively great environmental risks.
[8] In fact, US institutional review boards for human subjects research "should not consider possible long-range effects of applying knowledge gained in the research (for example, the possible effects of the research on public policy) as among those risks that fall within the purview of its responsibility" (45 CFR 46.101–409).

benefits and risks of further research as well as their distributions, the best understanding of solar geoengineering's potential, the presence of key characteristics of legitimate scientific research, support and opposition among the public and elites, the risks of climate change, and the apparent likelihood of sufficient emissions abatement, adaptation, and NETs.[9] At relatively small scales, establishing these breakpoint criteria and deciding whether they have been satisfied would be decentralized, such as at the level of research institutions or funders. As activities scale up, the criteria should be stricter and the breakpoints decisions should be more centrally managed, perhaps functionally linked to the assessment of research carried out by the IPCC or other bodies. How these criteria are formulated and how decision-making is informed would be essential to the overall research endeavor.

In principle, these breakpoint criteria could be coupled with a moratorium on solar geoengineering activities beyond a certain scale. A moratorium, which has been frequently proposed for solar geoengineering, can be thought of as a breakpoint that defaults to prohibition. However, a moratorium would face interrelated challenges of intertemporal credibility, legitimacy, and effectiveness (see Parker 2014; Chhetri et al. 2018, 11, 31). Adoption of a moratorium at an earlier time – such as when outdoor research remains small scale – initially seems to confer greater benefits of both making premature large-scale activities less likely and reassuring a potentially concerned public and other actors. At this stage, governance would remain decentralized, and those who could propose and implement a moratorium might be a few states, professional and scientific societies, research institutions, funders, and scientists. They might diverge on details, including the threshold of temporarily prohibited activities and the conditions for the moratorium's removal. Indeed, multiple moratoria are feasible. Independent of the number of moratoria, the developers of one could impose it credibly – more or less – on themselves, but whether others would accept it depends, among other things, on their perception of the developers' legitimacy to govern in this manner. Because of changing circumstances and knowledge as well as generational turnover, this credibility challenge becomes greater as a moratorium's length of time and of intended geographic reach increase. Would researchers at one time in one region of the world perceive as legitimate a moratorium that had been established much earlier in another region? What if the expected impacts of climate change had greatly increased in the meantime? What if various institutional leaders, scientists, and other experts disagreed as to whether a vague criterion, such as "an adequate scientific basis ... and appropriate consideration of the associated risks" had been met?[10] What if there were

---

[9] Regarding legitimate scientific research, the emerging regime under the Marine Geoengineering Amendment to the London Protocol bases its assessment of this upon whether a proposal would add to scientific knowledge, be subject to peer review, be published in a peer-reviewed outlet, and make data publicly available, as well as whether the researchers stand to directly financial gain from the research (Annex 5, Article 8).

[10] The quotation is from the 2010 CBD COP decision (X/33, paragraph 8(w)).

multiple moratoria and the lifting conditions for some of them had been met and others not?

To these components of governing small-scale research, I add three specific components, the first of which is the upstream development of a novel policy for intellectual property that is related to solar geoengineering. By extension, this will shape commercial actors' roles in small-scale research. This policy should be organized by nonstate actors in a bottom-up manner. In Chapter 11, I proposed four components of such as regime: the sharing of appropriately standardized research data, an intellectual property pledge community, monitoring of patenting activity, and clarification of states' intentions regarding compulsory licensing and march-in powers.

Second, the military's role in solar geoengineering should ideally be restricted. Some observers have claimed that the technologies could be used for hostile purposes or their operation subsumed by the military, perhaps even inevitably so (Horton and Reynolds 2016, 444–5). Although solar geoengineering's potential tactical utility is often overstated, the involvement of the military – particularly in hegemonic states – could undermine international public confidence and trust. At the same time, militaries possess useful equipment and knowledge regarding complex logistical operations at high altitudes and at sea. A complete rejection of any military role might be harmful. One possible boundary would be to allow military institutions to serve as secondary partners in research projects but not as primary ones or as funders and to require full disclosure of their involvement. Nevertheless, given governance's decentralized nature at this stage, it is unclear how such limitations could be adopted and effective.

Third, the governance of small-scale solar geoengineering should lay the foundation for that of responsible larger-scale activities. One way would be to educate and engage with the public, thought leaders, decision-makers, and other elites, as described. State and intergovernmental leaders should become informed so that their decision-making, at this time and in the future, would be based upon a more robust knowledge base and so that solar geoengineering integrates into the mainstream climate change debates in a balanced, responsible manner. C_2G_2 is doing this, especially with intergovernmental organizations. Such work should continue and expand, particularly to national policy-makers.

This raises the question of what intergovernmental organizations should do at this stage with respect to solar geoengineering. Many of them, especially those that are more advisory and facilitative in nature, such as UN Environment, World Meteorological Organization (WMO), the UN Education, Social and Cultural Organization (UNESCO), the International Maritime Organization (IMO), and the Food and Agriculture Organization, should undertake processes and establish subsidiary bodies, as appropriate, through which they could grow organizational knowledge, identifying capacities, and clarify some responsibilities. In fact, UN Environment's governing assembly has considered a resolution on geoengineering.

Those in which states' leaders could negotiate their positions, such as the UN General Assembly, its Security Council, and the UN Framework Convention on Climate Change (UNFCCC) Conference of the Parties (COP), should not directly engage with solar geoengineering too early, especially if the objective were a multilateral agreement. If they were to do so, then they might adopt poorly-informed policies and catalyze international polarization (Victor 2008, 331). There may be some exceptions, such as whether solar geoengineering research and development could be included as part of nationally determined contributions and global stocktaking under the Paris Agreement (see Chapter 6; Chhetri et al. 2018, 38–9; Nicholson, Jinnah, and Gillespie 2018).

A foundation for subsequent governance of large-scale activities could be laid by a high-level commission. Edward Parson has proposed a World Commission on Climate Engineering that would be authorized at a high level, such as by the UN; empowered by a broad, strong mandate; include diverse and distinguished members; and be supported with sufficient support resources. Such a body could offer a path forward in a controversial domain with great uncertainty. Importantly, this would not be directly connected to states yet probably would be widely perceived as legitimate. Notably, its mandate need not include the challenging and contentious task of developing specific rules and process for deployment but instead could "make a contribution simply by clarifying questions to be addressed, issues at stake, broad response options and factors militating for and against each" (Parson 2017a, 5; see also Chhetri et al. 2018; Nicholson, Jinnah, and Gillespie 2018, 327). In a similar vein, international relations scholar Joseph Nye points to a more state-affiliated group of governmental experts, such as that one regarding cybersecurity, to develop and help crystalize norms as well as lay a foundation for subsequent international cooperation (Nye 2018b).

13.3 LARGE-SCALE RESEARCH

The second stage of governance is that of large-scale outdoor solar geoengineering research, which would strive to further reduce uncertainty and to develop equipment, materials, and techniques for possible deployment. Governance would be characterized by managing expected transboundary impacts through rules and procedures that would be increasingly centralized, legalized, and promulgated by states and intergovernmental institutions. International law and international relations would be more salient as well.

"Large scale" encompasses outdoor experiments or tests that would go beyond small scale and expected to have transboundary, widespread, long-lasting, or severe impacts, as described in Section 13.2. The "upper" boundary is not sharply distinct from deployment but could still be drawn, albeit somewhat arbitrarily. For example, research could be limited to significant impacts up to some intensity (that is, radiative forcing). Intention should not be a criterion at this scale; any project that would change the Earth's radiative balance and be expected to have transboundary, widespread, long-lasting, or severe impacts should be governed as large-scale solar geoengineering. If future technologies enable effective regional solar geoengineering, then an additional spatial criterion might be needed.

Because new issues of international law and international relations as well as heightened demands for legitimacy would arise, states and intergovernmental organizations should at this stage assume more prominent if not leading roles in governance. At the very least, greater financial support from public sources for research would be needed. Furthermore, states, intergovernmental organizations and the parties of relevant multilateral agreements such as the UNFCCC, the Convention on Biological Diversity (CBD), and the UN Convention on the Law of the Sea (UNCLOS), should communicate and coordinate their actions and policies with respect to solar geoengineering governance in order to avoid undue fragmentation of governance. Moreover, the authorization of large-scale experiments should remain with the states. This process should comply with states' international legal commitments and obligations, including those under customary international law to prevent and reduce risks of transboundary harm. States should thus commit to prosecute, to the extent appropriate, any nonstate actor under their jurisdiction who conducts unauthorized large-scale solar geoengineering activities.

The centerpiece of my proposed governance of these large-scale activities is an intergovernmental institution, here called the Solar Geoengineering Organization (SGO) as a placeholder.[11] Although some of its functions would be possible through non-institutionalized international cooperation, these would probably be done less effectively or perceived as less legitimate in the absence of institutionalization. The first of the SGO's five core functions would be to facilitate research. This would include providing a forum for debating priorities; coordinating projects so that resources are used efficiently and outdoor experiments do not interfere with each other; fostering the growth and utilization of countries' research capacity, especially in developing ones; and standardizing and centralizing data sharing and other salient information. To help prevent technological lock-in, the SGO should support research that aims to identify solar geoengineering's limitations and risks and that explores nondominant proposed technologies and approaches. The SGO could assess and synthesize research, but there is a good argument that, to counter possible institutional momentum, these functions should be conducted by an institution that is independent from the SGO. Because this could conflict with the IPCC's work, the allocation of responsibilities between these organizations would need to be clarified.

Second, the SGO should help ensure that research is done responsibly. It could provide advice and models for national legislation that would be specific to solar

---

[11] This approach is inspired in part by the International Atomic Energy Agency (IAEA) due to similarities between nuclear technologies and solar geoengineering, and the IAEA's success in promoting the peaceful use of nuclear energy while preventing weapons proliferation (see Reynolds 2014a).

geoengineering. The SGO could also continue increasing the precision, delegation, and obligation of the research standards that arise from principles, codes of conduct, and the work of institutionally affiliated review boards. Among the standards' most important aspects would be impact assessment, monitoring, and reporting, each of which the institution could help operationalize. If some solar geoengineering methods and applications were regarded as unacceptably dangerous or contrary to widely shared norms, the SGO could facilitate processes through which these would be not pursued or even prohibited, likely through national laws or administrative regulations. Precision also points toward increasing specificity in the form of, for example, detailed best practices. Although greater precision can reduce risk, the standards should also remain adaptable because research projects would likely be diverse, knowledge would grow, and the social acceptability of risk would change. Delegation should include the refining, promulgation, and maintenance of the standards by the SGO and their implementation and enforcement by state and nonstate bodies. Greater obligation would be most clearly manifested in states' implementation of the standards, including by requiring that proposed research projects meet the standards. Likewise, the SGO could assume responsibility for managing the research commons, including its governance of patents related to solar geoengineering and, indirectly, of commercial actors.

The SGO's third function would be preventing premature escalation of large-scale research or deployment. This could be done through a continuation of the breakpoint criteria. However, in contrast with those for small-scale research, the rules could at this stage default to negative, at least in principle. In other words, research and development should arguably proceed only if certain criteria are met. This could be desirable beyond helping address environmental risks, "slippery slope," emissions abatement displacement, and public confidence. A moratorium – which is what such negative defaulting breakpoints would amount to – could, depending on its design, also catalyze greater precision, delegation, and obligation of governance and help build trust among countries that might be suspicious of others' intentions (Parson and Herzog 2016). The SGO might be able to resolve the problems of intertemporal credibility and legitimacy that a moratorium raises. It could assume the management of any previously developed moratoria and address related uncertainties. Yet even at the large-scale research stage, governing actors should be cautious with adopting a moratorium. Past moratoria such as those on commercial whaling, growing genetically modified crops, and mining in Antarctica ossified into hard-to-lift de facto prohibitions, even when they were developed by a single clear institution or mechanism (Bodansky 2013). Furthermore, like other strong regulations, a moratorium could have the perverse effect of being followed by only the responsible states and other actors, creating an opening for the less responsible ones to take the lead (Victor 2008). Ultimately, however, the states with the capacity to conduct large-scale solar geoengineering activities – which would be necessary for such a moratorium – might be unwilling to participate in one.

Another, perhaps more advantageous way in which the SGO could help prevent premature deployment would be through a nonproliferation mechanism. The reason for this is that, as with nuclear weapons, states might each simultaneously prefer to have the capacity to implement solar geoengineering and desire that few other states have such a capacity. Yet if all states pursue the former, then the latter would be clearly undermined. This would present a collective action problem. The international nuclear nonproliferation regime has largely resolved this by offering international assistance in developing the peaceful uses of nuclear power to those states that forego nuclear weapons. In the case of solar geoengineering, those states that commit to not developing deployment capacity could be offered a substantial role and voice, both in current research activities and in future implementation. Although this might call for a multilateral agreement, a less formal instrument might suffice. The necessary states might find a nonproliferation mechanism more appealing than a moratorium.

A final way to help prevent premature implementation is through monitoring solar geoengineering activity. For example, satellites and atmospheric sampling could detect changes in the Earth's radiative balance and observe high-altitude aircraft, while modeling may be able to help locate any source of undeclared activities. Such monitoring could also identify, and thus help prevent, secretive research programs. The SGO could coordinate and secure the requisite financial, technical, and other resources as well as ensure that a widely respected, independent institution is responsible for managing the actual monitoring work and publicly disseminating the results.

Related to preventing possible premature deployment is the question of how states might respond to revelations that one or a small number of actors was, in fact, undertaking solar geoengineering with global impacts. In principle, the SGO could provide a platform where states could have initial discussions regarding how they would react. However, such conversations may be better suited to an institution with an appropriate mandate and longer history in managing security issues, such as the UN Security Council.

The fourth core function of the SGO would be to maintain and foster international trust in and the perception of fairness of the solar geoengineering undertaking. This chapter has already suggested some activities that would do so, such as providing a forum for debating and establishing research priorities; facilitating transparency and data sharing, possibly through a research commons and a clearinghouse; preventing premature implementation; nonproliferation; international cooperation and capacity building, especially in developing countries; limiting the roles of the military; and monitoring. These should be continued. Ensuring that state and nonstate governance of major research programs and projects are consistent with existing international law, such as the customary obligation to reduce transboundary risks through, among other things, impact assessment, notification, and consultation, would also help maintain international trust. In some cases, the SGO could facilitate these procedural obligations, such as by providing a forum for international notification and consultation.

The SGO could also do more to foster international trust. For one thing, large-scale research would be a fairly expensive endeavor (although not when compared to aggressive emissions abatement, adaptation, and climate change damages), perhaps on the order of tens of billions of dollars annually. The SGO could offer a site where states negotiate the sharing of these costs based upon factors such as their ability to pay, historical greenhouse gas emissions, and expectations of benefits from solar geoengineering. Although this would be a collective action problem, the apparent large benefits-cost ratio implies that this could be managed. The SGO could also provide a forum where participating countries work to prevent and resolve disputes. For example, a requirement for participation could be that states commit to negotiate in good faith in the event of a conflict related to solar geoengineering. Related to that, and more speculatively, the SGO might be able to provide a home for a compensation mechanism, although this might require a multilateral agreement (see Chapter 10). Finally, it could liaise with and help coordinate the relevant activities of other intergovernmental bodies, including associated with relevant multilateral agreements, particularly the UNFCCC.

The SGO's final function would be to try to minimize emissions abatement displacement (see Chapter 3). The SGO's facilitation and coordination of research would help reduce uncertainties regarding solar geoengineering's potential, risks, and limitations. To help keep decision-making connected to the public's preferences, the SGO could ensure that public opinion studies and engagement exercises are undertaken. Furthermore, it could be the site of international linkage of solar geoengineering policies with those for emissions abatement and adaptation. For example, aggressive domestic abatement and adaptation actions could be prerequisites for membership (Parson 2014). However, this might have the pernicious effects of both encouraging unambitious abatement and adaptation targets as well as causing some states to remain outside the institution's governing framework.

This all raises the questions of the proposed intergovernmental institution's legal architecture, its relationship with existing international law, and states' participation in the SGO. Although its functions are largely coordination and facilitation, participation would entail some firm commitments, such as international cooperation, financial contributions to research and monitoring costs and a compensation fund (if appropriate), nonproliferation, good faith negotiation to resolve disputes, domestically governing large-scale research, and a possible moratorium or other breakpoint mechanism. Therefore, one or more new international legal instruments appear to be necessary to constitutionalize the institution and to outline participants' commitments and rights. The agreement(s) could resemble those of the International Atomic Energy Agency (IAEA), which was established by an initial Statute and subsequently oversaw almost twenty multilateral agreements on topics that include safety, liability, research, and technology transfer. Simplicity points to fewer agreements, while gradually building international trust and allowing states to opt-in and opt-out of agreements implies a greater number.

The development of a foundational multilateral agreement should occur under the auspices of an existing intergovernmental agency with broad membership and high perceived legitimacy. There are three clear candidates for this role. First, UN Environment has facilitated many such agreements. Second, the UN General Assembly enjoys both global participation and a nearly universal mandate.[12] Third, the UNFCCC bodies serve as the leading international forum for international discussions regarding reducing climate change and its risks.[13] Regardless, once established, the SGO should work to establish congenial relations with these other international institutions and avoid adversarial ones.

However, there are good – although not overwhelming – arguments for separating the governance of solar geoengineering from the UNFCCC architecture (Reynolds 2018a; see also Armeni and Redgwell 2015c). For one thing, the UNFCCC's objective is the stabilization of atmospheric greenhouse gas concentrations, which solar geoengineering would not address. For another, keeping the coordination and other governance of solar geoengineering activities at arm's length could reduce the displacement of emissions abatement and adaptation (although there is also a case for the reverse logic, in which coupled governance could minimize displacement). Furthermore, the UNFCCC may be so institutionally committed to emissions abatement and adaptation that it could be hostile to solar geoengineering. Finally, not only is it in some ways a politicized forum, but legal scholar Daniel Bodansky also notes that "the UNFCCC is seen as dysfunctional by many countries, and few trust its ability to make decisions" (Bodansky 2013, 550; but see Barrett 2010). However, if separate from the UNFCCC institutions, the SGO would need to cooperate closely with them, as some solar geoengineering activities would be governed to a limited extent by the UNFCCC and its protocols. For example, the building of capacity for solar geoengineering research in developing countries would be a form of technology transfer, and states may be able to include solar geoengineering research and development in their nationally determined contributions under the Paris Agreement.

An additional way in which the SGO could integrate with existing international law would be to help revive ENMOD, an agreement with widespread participation which prohibits the military or hostile use of environmental modification, implicitly including solar geoengineering (see Chapter 8). Although it has no standing bodies, its parties can call a meeting. The SGO could facilitate such meetings and encourage more states to ratify ENMOD to help ensure that solar geoengineering is used for only peaceful purposes.

---

[12] An alternative site to develop the foundational SGO agreement might be UNCLOS, whose objective includes protecting the marine environment (see Chapter 8).
[13] The WMO could also play an important role (see Chapter 5).

In terms of participation, a broad base would strengthen the SGO's international legitimacy, which would be critical given solar geoengineering's political contestation. Those states that would lack the capacity to undertake large-scale research and deployment appear to have little to lose through participation and something to gain. As proposed here, their greatest prices would be foregoing the development of implementation capacity under a potential nonproliferation mechanism and contributing to shared research and monitoring costs, while they could gain a voice and participation in international research as well as a means to seek compensation for demonstrated harm. The challenge would be attracting those countries that would have the relevant capacity, all of which would be necessary for the SGO to be genuinely effective. After all, at this stage the problem structure would largely be one of mutual restraint in order to prevent premature implementation (see Chapter 4). In terms of their costs, these states of deployment would commit to domestically governing research according to international standards, foregoing outdoor activities above the threshold of any moratorium, helping build capacity in developing countries, negotiating in good faith to resolve disputes, contributing to a possible compensation fund, and funding monitoring and research, possibly disproportionately so. In turn, they would gain in three ways. First, a nonproliferation agreement would limit the number of countries with the capacity to implement solar geoengineering. This would not only help stabilize international relations but also implicitly maintain or even increase these capable states' relative political power. Second, their participation in the SGO could reduce any international opposition to their solar geoengineering research activities, which are likely to be controversial. Third, those countries with implementation capacity would benefit through their restraint from moving forward too rapidly.

I concede that my suggestions for the governance of large-scale solar geoengineering research could be too state-centric. It might be that such international cooperation will be too difficult, or at least the costs in terms of international political capital too great. An approach that relies less on national leadership and cooperation, yet indirectly accountable to states, could be sufficiently effective. For example, a high-level commission that had been authorized by the UN, as introduced in Section 13.2, might have enough legitimacy, influence, and flexibility to guide governance through the challenges described in this section. Nevertheless, my sense is that transboundary climatic impacts that are significant – even if not substantial – would be too contentious and close to state's core interests for them to relegate it to an international institution in which they do not have direct say.

13.4 IMPLEMENTATION

The final stage of governance is that of the possible deployment of solar geoengineering. Here, there would be substantial transboundary and presumably global climatic effects. The purposes of governance would now include preventing, managing, and resolving international disputes; encouraging cooperation regarding deployment; and ensuring that any implementation is done in ways that are close to states' preferences and that are consistent with widely shared norms. The boundaries of this stage would be any solar geoengineering activity that exceeds large-scale research by, for example, having an expected impact on the Earth's radiative balance beyond a certain magnitude.

States would be central to the governance of deployment. Nevertheless, they could likely benefit by establishing or empowering an intergovernmental decision-making institution through a multilateral agreement. To be effective, it should aim to count as participants all states with the capacity, international political clout, and willingness to implement solar geoengineering (or counter–solar geoengineering) in a sustained manner (Virgoe 2009; Benedick 2011; Barrett 2014). These presently number perhaps one or two dozen. The institution should also try to attract any other states with the relative power and willingness to retaliate in other issue areas in response to solar geoengineering activities with which they disagree. This would be a handful of great powers, most or all of which would also be capable states (see Chapter 4). I collectively call these two groups the "target states," whose participation would be an objective – but not a requirement – of establishing an institution for decision-making regarding solar geoengineering deployment. Ultimately, the core challenge is thus making it in target states' perceived self-interests to relinquish some of their sovereign authority.

An important first question is whether states capable of sustained deployment would, to be a member of the institution, need to commit to refraining from deploying solar geoengineering outside of or contrary to the institution's decision-making process. An affirmative answer might be tempting, but requiring this could cause some of them to not join. These resistant target states might claim that they intend to generally abide by the institution's decisions but are unwilling to commit. Moreover, superpowers and states that are weakly integrated into the national order might see little negative consequences – such as reputational costs or retaliation – to taking this somewhat ambivalent position. My sense is that the international community would be better served by ensuring that as many target states as possible participate, even at the "cost" of weaker commitments. To that end, the founding agreement's parties should not prohibit unauthorized solar geoengineering implementation but instead use language with reduced force. For example, an article could provide that participants "should" refrain from undertaking deployment-scale solar geoengineering activities that have not been endorsed by the decision-making institution. Others could call on all deploying states to publicly notify the international community prior to solar geoengineering activities and to report on them afterwards; to coordinate among themselves; to take other states' interests into account in decision-making; to comply with prevailing best practices, including prior impact assessment; to try in good faith to resolve international disputes; and to not use solar geoengineering for military, hostile, or otherwise intentionally harmful purposes. The agreement could even include an explicit right for parties to deploy solar geoengineering, perhaps limited to safeguarding their citizens from clear and substantial threats (as implied by Barrett 2014, 11). It could reiterate existing obligations under international environmental law, specifically the customary law of the prevention of transboundary harm through means such as prior assessment, notification, and consultation. In other words, the agreement and the institution would not aim to prevent all uni- and minilateral deployment but instead to "cabin" it (Bodansky 2013, 549). This is a compromise that all target states might be willing to make.

There are reasons that an agreement should exclude or – more likely – limit the roles of some nontarget states. Wide participation, especially with a requirement of unanimity or a large supermajority, can cause decision-making to be slow, unduly conservative, and vulnerable to gridlock. These outcomes can arise due to objections of one of a handful of states that lie on the fringes of the international order. Recognizing this, some target states would be reluctant to join an institution with such decision-making architecture. At the same time, broad participation and stringent voting rules would increase the institution's perceived legitimacy and could help distribute costs and other burdens more widely. A potential middle ground could be two tiers of decision-making among countries' leaders. Like at the UN, a general assembly could be open to all states and agree to nonbinding resolutions, while an executive committee of target states plus, if necessary, a handful of representative nontarget ones would make operational decisions. The rules for decision-making within the executive committee would be critical. One approach would be to strive for consensus and, in its absence, to agree to vote, perhaps with a supermajority threshold.

In addition to two tiers of states, there is another axis of decision-making. Decisions regarding deployment would vary in their resolution, that is, their degree of detail. Low resolution decisions would address general goals, less technical issues, long temporal scales, and large spatial scales. For example, should solar geoengineering be used to reduce climate change and its risks? Is the objective to maintain present climatic conditions, to return to preindustrial ones, or to slow the rate of climate change? Do the interests of the northern and southern hemispheres and of the high and low latitudes diverge, and if so, could multiple objectives be balanced? How should solar geoengineering be altered once atmospheric greenhouse gas concentrations have begun to decline due to net negative emissions (that is, through abatement and NETs)? Should solar geoengineering be phased out due to a return to acceptable greenhouse gas concentrations or changes in preferences? States will probably insist on retaining decision-making authority for these matters. In contrast, high resolution decisions would address specific goals, more technical issues, brief temporal scales, and regional and local scales. For example, in the cases of stratospheric aerosol injection or cirrus cloud thinning, what types, quantities, locations, and timings of injected material would most likely achieve the general goals? If the weather had been globally warmer or cooler than expected for a few years, is this an indication that these parameters should be modified or is it mere climatic "noise"?

If there was evidence that anomalous precipitation at the continental scale had been caused by solar geoengineering, could that be reduced, and if so how? These sorts of matters should be delegated to an expert body. In its short-term management of solar geoengineering deployment, the expert body would be able to respond to feedback on various time scales ranging from seasonal to decadal (MacMartin and Kravitz 2019; see also Chris 2015).

Because national leaders could be tempted to interfere in these decisions for political reasons, there should be institutional firewalls to shield the expert body from undue influence from the executive committee and other political actors. In these ways, monetary policy offers a useful analogy, in which politicians set general goals that are implemented by boards of economists and other experts. Although this proposed arrangement might raise concerns of technocracy to some, it is within the current bounds of experts' roles in democratic societies, albeit at a global scale (Parson 2015; Horton et al. 2018). Nevertheless, the distinction of responsibilities between those of the state-led general assembly and executive body, on the one hand, and the expert body, on the other, would not necessarily be clear. For example, solar geoengineering might be able to reduce impending extreme weather events such as hurricanes and droughts. Or after such an event, a participating country might claim that it had been harmed by solar geoengineering and demand immediate changes to the parameters of implementation. These would be high-resolution decisions, yet responses to them appear to go beyond the expert body's justified remit.

A multilateral agreement that would establish such an institutionalized means to make solar geoengineering implementation decisions might need to include assurances of compensation for harm or other concessions. One reason for this is that it might be necessary to obtain the consent, or to end the opposition, of target states that believe that solar geoengineering would cause them more harm than benefit (see Chapter 10). Another related reason is that even those necessary target states that would not expect net harm might hold out their support to try to obtain a greater share of the welfare surplus. If so, then side payments might be necessary to expand participation. Distinguishing between demands for compensation for harm and those for mere side payment might not be possible, as those in the latter group would likely portray themselves as belonging to the former. Regardless, a multilateral agreement could include a compensation fund for harmed states. As with that at the research stage, contributions should ideally be based upon ability to pay, historical greenhouse gas emissions, and expectation of benefits from solar geoengineering, as described in Chapter 10.

An intergovernmental institution to make decisions regarding solar geoengineering deployment could have a handful of additional responsibilities, some of which are common with those of the SGO. First, assessment of solar geoengineering deployment would be critical to help ensure that decisions are made on a sound scientific basis. An independent assessment body should regularly report to the institution, but its core activities should be kept at arm's length from the latter. Second, the agreement should include a mechanism to share financial and other burdens in ways that are considered fair. Third, its bodies should, to the extent possible, cooperate and be integrated with other relevant international institutions, include those for climate change in general, weather, the environment, oceans, food, and international security.[14] In some cases, these institutions should ex ante delineate their respective responsibilities, whereas in others it would be preferable to do so on a case-by-case basis. Fourth, the agreement and institution should, again to the extent possible, seek to prevent, minimize, and resolve international conflicts related to solar geoengineering. As in the SGO, a requirement for participating – and thus having a voice in implementation decisions – could be to commit to negotiate in good faith and to try to peacefully resolve disputes with other potentially affected parties. Finally, the institution should help prevent dramatic changes in the deployment regime, especially sudden and sustained termination. One way would be to ensure that the systems, equipment, supplies, relationships, and knowledge necessary for implementation would be redundant and secure from disruption. Another would be to develop a plan to reduce and, if appropriate, phase out solar geoengineering activities.

How a multilateral agreement to launch such an international institution would come about depends on the previous stage of large-scale research. If it were to proceed as I describe, then the SGO could be the vehicle for this and perhaps even evolve into the decision-making institution. Alternatively, it could arise from other international climate change institutions such as through a solar geoengineering protocol to the UNFCCC. Alternatively, other international bodies, including the UN General Assembly, the WMO, UN Environment, or even the UNCLOS parties, could catalyze an agreement. A final option is that the UN General Assembly and Security Council assume the roles of the proposed institution's general assembly and executive committee.

My suggestions thus far have been of optimistic but arguably realistic international cooperation. However, a legalized intergovernmental institution with operational decision-making authority might not be necessary or feasible. To the extent that monetary policy offers an instructive analogy to solar geoengineering, as noted above, then the lack of highly centralized decision-making there is notable. The most important institution in that domain, the International Monetary Fund, neither sets global monetary policy nor dictates its members' policies (apart from conditional loans). Instead, it monitors economic conditions, conducts research, provides a forum for sharing information, coordinates, offers advice, helps build capacity, and responds to crises. Perhaps such a model would be more appropriate for an international solar geoengineering institution. Furthermore, states'

---

[14] For an analysis of the needs of solar geoengineering governance with respect to potential conflicts and security, see Parson and Ernst (2013).

preferences regarding deployment might not diverge so greatly, which would reduce the benefits of deep cooperation. Likewise, the costs of cooperation to form a legalized decision-making institution in terms of limited negotiating resources, political capital, and the risk of forcing unnecessary divisive debates might be too great. And credible threats of counter–solar geoengineering could foster mutual restraint by decreasing states' incentives to use solar geoengineering earlier or to a greater degree than the wishes of the international community, or at least those of the states capable of sustained implementation.

If so, then an intergovernmental institution with the functions and the mandate of the SGO might suffice. That is, it could continue to facilitate research and monitoring; help ensure that solar geoengineering is done responsibly by assisting states with their domestic legislation and other regulation, developing and maintaining nonbinding governance instruments, and encouraging the prohibition of specific solar geoengineering methods and activities; prevent undue and premature escalation of deployment through, among other means, a nonproliferation mechanism; maintain and foster international trust and the perception of fairness by providing a forum for international discussions and consultation, limiting the roles of the military, facilitating transparency and procedural obligations; help share costs and other burdens; coordinate the relevant activities of other intergovernmental bodies; and try to minimize emissions abatement displacement.

Yet states might not cooperate, especially if international relations are fragmented. In the absence of both an international decision-making body with sufficient participation by target states and a coordinating agency in the model of the International Monetary Fund, existing international law, legal institutions, and norms could offer minimal and perhaps sufficient governance. First, all states are bound by the customary international law of preventing transboundary environmental harm (see Chapter 5). Any state that planned to deploy solar geoengineering would be obligated to practice due diligence, undertake all appropriate measures to reduce the risk, perform an environmental impact assessment, require prior authorization for any nonstate actors' solar geoengineering, notify and consult with potentially affected states, inform the public, and develop plans for possible emergencies. Second, under the UNFCCC, when states undertake activities to reduce climate change risks, they are to minimize adverse effects including those on the economy, public health, and the environment (Article 4.1(f)). Furthermore, if solar geoengineering were considered adaptation, then under the Paris Agreement it may not threaten food production and must be done "with a view to contributing to sustainable development" (Articles 2.1(b), 7.1). Third, parties to the UNCLOS have several commitments regarding protecting the marine environment, which would be affected by solar geoengineering (see Chapter 8). Fourth, ENMOD prohibits the hostile or military use of solar geoengineering and has fairly widespread participation (Article I.1). Finally, states could utilize traditional international diplomacy and negotiation to address potential disputes. In this, the UN Security Council could act as a backstop forum of last resort under its remit to maintain international peace and security.

Independent of the degree of formalized cooperation, nonlegal norms, including unwritten ones, of international behavior can have compliance pull. Economist and strategist Thomas Schelling – who was the first social scientist to write on solar geoengineering – dedicated his Nobel Prize lecture to the power of taboos and other norms in preventing nuclear war (Schelling 2006). It might seem remarkable that mere norms would have substantial influence on decision-making in arguably the most consequential of international affairs. However, it is precisely in such high-stakes issue areas that states are often less willing to make explicit commitments. Norms can thus fill a void, take hold for solar geoengineering, and play a substantial role in shaping states' behavior. After all, assuming that international law can guide their decision-making in novel domains puts the cart before the horse. Instead, practice gives rise to norms, which in turn can crystalize into international law.[15]

13.5 SUMMARY AND CONCLUSION

Geographer Mike Hulme asserts that solar geoengineering is "*ungovernable* because there is no plausible and legitimate process for deciding who sets the world's temperature" (Hulme 2014, xii; italics in original). I disagree. The legitimate and effective international governance of solar geoengineering is plausible. In this chapter, I have described possible governance that could increase current and future humans' well-being in ways that are sustainable, consistent with widely shared norms, and feasible. To be clear, I am not claiming that the future will unfold in this way. Importantly, the world might be much more conflictual. If so, then solar geoengineering – like numerous other phenomena with transboundary impacts – would have substantially suboptimal outcomes. Likewise, solar geoengineering's risks might outweigh its expected benefits, or society might reject it for noninstrumental reasons.

---

[15] Consider the examples of humanitarian intervention and international cyber conflict (Buchanan 2003; Nye 2018a).

14

Conclusion

Earth may have entered a new geological epoch. Some scientists have proposed that we are now in the Anthropocene, in which humans are a dominant force in shaping the planet (Waters et al. 2018).[1] The proposal has upset some environmentalists who are concerned that the designation would foster acceptance of humanity's impact on the natural world and apathy regarding possible steps to limit and reduce it. They are disconcerted that a declaration of the Anthropocene would give humanity license to increase our intervention in and impacts on natural systems. In contrast, other environmentalists respond that we should acknowledge and even embrace our responsibilities to manage the natural world. They argue that some conscious and responsible control of natural systems is the only feasible path to sustainable human flourishing while preserving valued species and ecosystems. If humanity is indeed crossing the planetary boundaries that offer a "safe operating space for humanity," then powerful technologies could help us manage ourselves, our impacts on the natural world, and the boundaries themselves (Rockström et al. 2009; Galaz 2014, 101–2). For example, we exceeded the boundary with respect to the Earth's nitrogen cycle decades ago, but the Haber–Bosch process empowers humans to control the majority of the cycle and feed billions of people (Morton 2018).

These contrasting visions of humans' ideal role as either treading as lightly as possible on the Earth or as stewards of managed nature is a long-standing divide within environmentalism, stretching in some form at least back to the contrast between John Muir's preservationism and Gifford Pinchot's conservationism. This difference – as well as the symptomatic debates over solar geoengineering – can be viewed from several angles. One way is the philosophical relationship between what is and what should be. Environmentalism is based upon a wise respect for the natural world. This respect is sometimes used to derive norms by appealing to nature, in which the natural is presumptively good, and the unnatural presumptively bad. Such logic is often criticized as based upon the naturalistic fallacy, pointing out that the natural includes much suffering and that the unnatural includes many

---

[1] In fact, one of the earliest to propose the designation was Paul Crutzen, who later wrote an essay that helped crack the solar geoengineering taboo (Crutzen 2002; Crutzen 2006).

sources of well-being and joy.[2] Another way is to examine environmentalism's ends and means. To some, the goal is to ensure that people can thrive sustainably in a world that still possesses biological diversity and intact ecosystems. Means still matter: we should achieve this goal in ways that are right, both to each other and to nature. Yet for other environmentalists, these means – being just and reducing humanity's natural footprint – *are* the goals. The third view regards the politics of change. Ultimately, this division in environmentalism in general and that concerning solar geoengineering specifically can be understood based on worldviews, risk perception, tenuous political coalitions, and sunk political capital (see Reynolds 2017; Chapter 3).

I go further with environmentalism's sometimes problematic politics of change. Some strains within the movement are progressive, welcoming and even advocating for change, innovation, and ambition. Although environmentalism is presently associated with the political left, that was not so before the mid-twentieth century. I suggest that much of contemporary environmentalism is actually conservative, in that it is skeptical of change, new technologies, and large expertise-driven undertakings and – in some ways – seeks to return society to an earlier ideal.

Solar geoengineering lies, in many ways, astride these divisions in environmentalism, and its controversy should thus not be surprising. The problem that I see is not that solar geoengineering is contentious. It should be, as it is a dangerous, unpleasant proposal. My concern is that its discourse is unduly driven by intuition, ideology, and pre-existing conclusions instead of empiricism and rationality. Both popular and academic writing foregrounds solar geoengineering's risks and limitation while neglecting the evidence that it could reduce climate change and its risk. This is exacerbated by a "curious asymmetry" in which experts assess the various possible responses to climate change, especially emissions abatement and solar geoengineering, with significantly different standards (Heyward and Rayner 2015, 2212). Ideas that are supported only weakly or have been debunked nevertheless persist (Reynolds, Parker, and Irvine 2016). Scholars – even natural scientists – and other observers are reluctant to clearly present a balance of evidence and to openly discuss solar geoengineering, lest they be perceived as part of "a freak show in otherwise serious discussions of climate science and policy" (Victor 2008, 323).

I opened Chapter 2 by stating that climate change is arguably the most important and difficult environmental problem that presently confronts global society. It poses serious risks to people, other species, and ecosystems, especially the most vulnerable of these. Emissions abatement, adaptation, and negative emissions technologies (NETs) can – and should – be aggressively pursued to reduce climate change and its risks. However, these responses have been insufficient and will, in all likelihood, continue to be so. There is substantial and growing evidence that solar geoengineering could effectively reduce climate change; that it could manage climate risks in ways that the other response options could not; that it would have relatively low

---

[2] See the well-known exchange between Krieger (1973) and Tribe (1974).

direct costs of deployment; that it would be rapid, reversible in its direct effects; and that it would be technologically feasible. While it poses genuine physical risks and social challenges, these increasingly appear manageable.

The malignment and the possible dismissal of solar geoengineering is a gamble that actual abatement, adaptation, and NETs will be enough to prevent dangerous climate change and the resulting negative impacts. Both thirty years of international efforts and the underlying problem structures suggest that these responses will remain woefully inadequate (see Chapter 4). Notably, those who are making this gamble are not those who would face the negative consequences of losing it. Resistance to solar geoengineering arises largely from those who live in industrialized countries, while vulnerable people in developing ones (as well as other species and ecosystems) would pay the price of inadequate action. The critics of solar geoengineering also exist in the present, while future generations would suffer from climate change. This misalignment between who chooses and who bears the consequences is undermining effective management of climate risks.

This is not an enthusiastic endorsement of solar geoengineering as the solution. In fact, problems as complex as anthropogenic climate change do not have solutions; they are merely managed. It is instead a call for us to be explicit about our goals and to rationally assess the evidence of the possible means to achieve them. Given the stakes, solar geoengineering should be taken seriously, including through governance that facilitates its responsible research, development, and – if warranted – implementation.

Legal Sources

1. INTERNATIONAL TREATIES

Additional Protocol to the American Convention on Human Rights in the Area of Economic, Social and Cultural Rights, 28 ILM 156 (1989). Adopted November 17, 1988, in force November 16, 1999.

African Charter on Human and Peoples' Rights, 21 ILM 52 (1982). Adopted June 1, 1981, in force October 21, 1986.

Agreement on Trade-Related Aspects of Intellectual Property Rights, 33 ILM 1197 (1994). Adopted April 15, 1994, in force January 1, 1995.

Amendment to the London Protocol to Regulate the Placement of Matter for Ocean Fertilization and Other Marine Geoengineering Activities (Marine Geoengineering Amendment), Report of the Thirty-Fifth Consultative Meeting and the Eighth Meeting of Contracting Parties, October 14–18, 2013, UN Doc. LC 35/15 (2013). Adopted on October 18, 2013, not in force.

American Convention on Human Rights, 9 ILM 99 (1969). Adopted November 22, 1969, in force July 18, 1978.

Annex V to the Protocol on Environmental Protection to the Antarctic Treaty Area Protection and Management, Recommendation adopted by the XVIth Antarctic Treaty Consultative Meeting, October 17, 1991, in force May 24, 2002.

Annex VI to the Protocol on Environmental Protection to the Antarctic Treaty, on Liability Arising from Environmental Emergencies, Recommendation adopted by the XXVIIIth Antarctic Treaty Consultative Meeting, June 14, 2005, not in force.

Antarctic Treaty, 402 UNTS 71 (1961). Adopted December 1, 1959, in force June 23, 1961.

Basel Protocol on Liability and Compensation for Damage Resulting from Transboundary Movements of Hazardous Wastes and Their Disposal, www.basel.int/Countries/StatusofRatifications/TheProtocol/tabid/1345/Default.aspx. Adopted December 10, 1999, not in force.

Charter of Fundamental Rights of the European Union, 2012 OJ (C 326) 2. Adopted October 2, 2000, in force December 1, 2009 (as part of the Treaty of Lisbon Amending the Treaty on European Union and the Treaty Establishing the European Community).

Charter of the Organization of American States, 119 UNTS 3 (1952). Adopted April 30, 1948, in force December 13, 1951.

Charter of the United Nations, 1 UNTS xvi (1945). Adopted June 26, 1945, in force October 24, 1945.

Convention for the Protection of Human Rights and Fundamental Freedoms (European Convention on Human Rights), 213 UNTS 222 (1955). Adopted November 4, 1950, in force September 3, 1953.

Convention for the Protection of the Ozone Layer (Vienna Convention), 26 ILM 1529 (1987). Adopted March 22, 1985, in force September 22, 1988.

Convention on Access to Information, Public Participation in Decision-Making and Access to Justice in Environmental Matters (Aarhus Convention), 38 ILM 517 (1999). Adopted June 25, 1998, in force October 29, 2001.

Convention on Biological Diversity (CBD), 31 ILM 818 (1992). Adopted June 5, 1992, in force December 29, 1993.

Convention on Civil Liability for Damage Resulting from Activities Dangerous to the Environment (Lugano Convention), 32 ILM 1228 (1993). Adopted June 21, 1993, not in force.

Convention on Civil Liability for Nuclear Damage (Vienna Nuclear Liability Convention), 2 ILM 727 (1963). Adopted May 29, 1963, in force November 12, 1977.

Convention on Damage Caused by Foreign Aircraft to Third Parties on the Surface (Rome Convention), 310 UNTS 181 (1958). Adopted October 7, 1952, in force February 4, 1958.

Convention on Environmental Impact Assessment in a Transboundary Context (Espoo Convention), 30 ILM 802 (1971). Adopted February 25, 1991, in force June 27, 1997.

Convention on International Civil Aviation (Chicago Convention), 15 UNTS 295 (1948). Adopted December 7, 1944, in force April 4, 1947.

Convention on International Liability for Damage Caused by Space Objects (Space Liability Convention), 961 UNTS 187 (1975). Adopted March 29, 1972, in force September 1, 1972.

Convention on International Trade in Endangered Species of Wild Fauna and Flora, 12 ILM 1085 (1973). Adopted March 2, 1973, in force July 1, 1975.

Convention on Long-Range Transboundary Air Pollution (CLRTAP), 18 ILM 1442 (1979). Adopted November 13, 1979, in force March 16, 1983.

Convention on Nature Protection and Wildlife Preservation in the Western Hemisphere, 161 UNTS 193 (1953). Adopted October 12, 1940, in force May 1, 1942.

Convention on Offences and Certain Other Acts Committed on Board Aircraft, 2 ILM 1042 (1963). Adopted September 14, 1963, in force December 4, 1969.

Convention on Supplementary Compensation, 36 ILM 1473 (1997). Adopted September 12, 1997, in force April 15, 2015.

Convention on the Elimination of All Forms of Discrimination against Women, 19 ILM 33 (1980). Adopted December 18, 1979, in force September 3, 1981.

Convention on the Prevention of Marine Pollution by Dumping of Wastes and Other Matter (London Convention), 11 ILM 1294 (1972). Adopted November 13, 1972, in force August 30, 1975.

Convention on the Prohibition of Military or Any Other Hostile Use of Environmental Modification Techniques (ENMOD), 16 ILM 88 (1978). Adopted December 10, 1976, in force May 18, 1977.

Convention on the Rights of Persons with Disabilities, 2515 UNTS 3 (2008). Adopted December 13, 2006, in force May 3, 2008.

Convention on the Rights of the Child, 28 ILM 1456 (1989). Adopted November 20, 1989, in force September 2, 1990.

Convention on Third Party Liability in the Field of Nuclear Energy (Paris Nuclear Liability Convention), 2 ILM 685 (1963). Adopted July 29, 1960, in force April 1, 1968.

Council of Europe Convention on Human Rights and Biomedicine, 36 ILM 817 (1997). Adopted April 4, 1997, in force December 1, 1999.

International Convention for the Prevention of Pollution from Ships (MARPOL), 12 ILM 1319 (1973). Adopted November 2, 1973. Amended by the Protocol Relating to the Convention for the Prevention of Pollution from Ships, 17 ILM 546 (1978). Adopted February 17, 1978, in force October 2, 1983.

International Convention on Civil Liability for Oil Pollution Damage (CLC 1969), 9 ILM 45 (1970). Adopted November 29, 1969, in force June 19, 1975.

International Convention on Civil Liability for Oil Pollution Damage (CLC 1992), 1956 UNTS 255 (1997). Adopted November 27, 1992, in force May 3, 1996.

International Convention on Liability and Compensation for Damage in Connection with the Carriage of Hazardous and Noxious Substances by Sea (HNS Convention), 35 ILM 1415 (1996). Adopted May 3, 1996, not in force.

International Covenant on Civil and Political Rights (ICCPR), 6 ILM 368 (1967). Adopted December 16, 1966, in force March 23, 1976.

International Covenant on Economic, Social and Cultural Rights (ICESCR), 6 ILM (1967) 360. Adopted December 16, 1966, in force January 3, 1976.

Joint Protocol Relating to the Application of the Vienna Convention and the Paris Convention (Joint Protocol), 1672 UNTS 293 (1992). Adopted September 21, 1988, in force April 27, 1992.

Nagoya-Kuala Lumpur Supplementary Protocol on Liability and Redress to the Cartagena Protocol on Biosafety, 50 ILM 105 (2011). Adopted October 15, 2010, in force March 5, 2018.

Paris Agreement, 55 ILM 743 (2016). Adopted December 12, 2015, in force November 4, 2016.

Protocol of 1992 to Amend the International Convention on the Establishment of an International Fund for Compensation for Oil Pollution Damage, 1971 (FUND92), 35 ILM 1406 (1996). Adopted November 27, 1992, in force May 30, 1996.

Protocol of 1997 to Amend the International Convention for the Prevention of Pollution from Ships, 1973, as Modified by the Protocol of 1978 Relating Thereto (Annex VI), Adopted by the International Conference of Parties to the MARPOL Convention, UN Doc. MP/CONF.3/34, September 26, 1997, in force May 19, 2005.

Protocol on Civil Liability and Compensation for Damage Caused by the Transboundary Effects of Industrial Accidents on Transboundary Waters to the 1992 Convention on the Protection and Use of Transboundary Watercourses and International Lakes and to the 1992 Convention on the Transboundary Effects of Industrial Accidents, UN Doc. ECE/MP.WAT/11–ECE/CP.TEIA/9. Adopted May 21, 2003, not in force.

Protocol on Further Reduction of Sulphur Emissions (Oslo Protocol), 33 ILM 1542 (1994). Adopted June 13, 1994, in force 5 August 1998.

Protocol on Pollutant Release and Transfer Registers to the Convention on Access to Information, Public Participation in Decision-Making and Access to Justice in Environmental Matters (Kiev Protocol), 2626 UNTS 119 (2009). Adopted May 21, 2003, in force October 8, 2009.

Protocol on Strategic Environmental Assessment to the Convention on Environmental Impact Assessment in a Transboundary Context, 2685 UNTS 140 (2010). Adopted May 21, 2003, in force July 11, 2010.

Protocol on Substances that Deplete the Ozone Layer (Montreal Protocol), 26 ILM 1550 (1987). Adopted September 16, 1987, in force January 1, 1989.

Protocol on the Reduction of Sulphur Emissions (Helsinki Protocol), 27 ILM 707 (1988). Adopted July 8, 1985, in force September 2, 1987.

Protocol to the 1979 Convention on Long-Range Transboundary Air Pollution to Abate Acidification Eutrophication and Ground-Level Ozone (Gothenburg Protocol), 2319 UNTS 81 (2005). Adopted November 30, 1999, in force May 17, 2005.

Protocol to the Antarctic Treaty on Environmental Protection (including Annexes I to IV) (Madrid Protocol), 30 ILM 1461 (1991). Adopted October 4, 1991, in force January 14, 1998.

Protocol to the Framework Convention on Climate Change (Kyoto Protocol), 37 ILM 22 (1998). Adopted December 10, 1997, in force February 16, 2005.

Protocol to the London Dumping Convention (London Protocol), 36 ILM 7 (1997). Adopted November 7, 1996, in force March 24, 2006.

Statute of the International Atomic Energy Agency, 276 UNTS 3 (1957). Adopted October 23, 1956, in force July 29, 1957.

Treaty on Principles Governing the Activities of States in the Exploration and Use of Outer Space, Including the Moon and Other Celestial Bodies (Outer Space Treaty), 6 ILM 386 (1967). Adopted December 19, 1966, in force October 10, 1967.

United Nations Convention on the Law of the Sea (UNCLOS), 21 ILM 1261 (1982). Adopted December 10, 1982, in force November 16, 1994.

United Nations Framework Convention on Climate Change (UNFCCC), 31 ILM 851 (1992). Adopted May 9, 1992, in force March 21, 1994.

2. INTERNATIONAL CASES

Certain Activities Carried out by Nicaragua in the Border Area (Costa Rica v. Nicaragua), Compensation Owed by the Republic of Nicaragua to the Republic of Costa Rica, ICJ Judgement of February 2, 2018, www.icj-cij.org/files/case-related/150/150-20180202-JUD-01-00-EN.pdf.

Certain Activities Carried out by Nicaragua in the Border Area and Construction of a Road in Costa Rica along the San Juan River (Costa Rica/Nicaragua), (2015) ICJR Reports 665.

Gabčíkovo-Nagymaros Project (Hungary/Slovakia), (1997) ICJ Reports 7.

Legality of the Use by a State of Nuclear Weapons (Advisory Opinion), (1996) ICJ Reports 226.

Pulp Mills on the River Uruguay (Argentina v. Uruguay), (2010) ICJ Reports 14.

Responsibilities and Obligations of States Sponsoring Persons and Entities with Respect to Activities in the Area (Advisory Opinion), (2011) International Tribunal for the Law of the Sea, Case No. 17, February 1, 2011.

The Environment and Human Right, Advisory Opinion, OC-23/17, Inter-American Court of Human Rights, (ser. A) No. 23 (November 15, 2017).

3. INTERNATIONAL LEGAL DOCUMENTS

Convention on Biological Diversity. Tenth Meeting of the Conference of Parties, *Decision X/33. Biodiversity and Climate Change*, UN Doc. UNEP/CBD/COP/DEC/X/33 (2010).

Convention on Biological Diversity. Eleventh Meeting of the Conference of Parties, *Decision XI/20. Climate-Related Geoengineering*, UN Doc. UNEP/CBD/COP/DEC/XI/20 (2012).

Convention on Biological Diversity. Thirteenth Meeting of the Conference of Parties, *Decision XIII/14. Climate-Related Geoengineering*, UN Doc. UNEP/CBD/COP/DEC/XIII/14 (2016).

Declaration of the United Nations Conference on the Human Environment (Stockholm Declaration), Report of the United Nations Conference on the Human Environment, UN Doc. A/CONF.48/14/Rev.1 (1972).

Geoengineering and Its Governance. Resolution for Consideration at the 4th United Nations Environment Assembly. (2019) http://jreynolds.org/wp-content/uploads/2019/03/2019UNEAres.pdf.

Human Rights Council. Report of the Office of the United Nations High Commissioner for Human Rights on the Relationship between Climate Change and Human Rights, UN Doc. A/HRC/10/61 (2009).

Human Rights Council. Report of the Special Rapporteur in the Field of Cultural Rights: The Right to Enjoy the Benefits of Scientific Progress and Its Applications, UN Doc A/HRC/20/26 (2012).

Human Rights Council. Report of the Special Rapporteur on the Issue of Human Rights Obligations Relating to the Enjoyment of a Safe, Clean, Healthy and Sustainable Environment, UN Doc. A/HRC/31/52 (2016).

Human Rights Council. Report of the Special Rapporteur on the Issue of Human Rights Obligations Relating to the Enjoyment of a Safe, Clean, Healthy and Sustainable Environment, Framework Principles on Human Rights and the Environment, UN Doc. A/HRC/37/59 (2018).

Informal Single Negotiating Text, Part III, Third United Nations Conference on the Law of the Sea, UN Doc. A/CONF.62/WP.8/Part III (1975).

International Law Commission. Draft Articles on Prevention of Transboundary Harm from Hazardous Activities, UN General Assembly Official Records, UN Doc. A/56/10 (2001).

International Law Commission. Draft Articles on Responsibility of States for Internationally Wrongful Acts, UN General Assembly Official Records, UN Doc. A/56/10 (2001).

International Law Commission. Draft Principles on the Allocation of Loss in The Case of Transboundary Harm Arising out of Hazardous Activities, UN Doc. A/61/10 (2006).

International Law Commission. Protection of the Atmosphere: Texts and Titles of Draft Guidelines and Preamble Adopted by the Drafting Committee on First Reading, UN Doc. A/CN.4/L.909 (2018).

Office of the High Commissioner for Human Rights, Committee on Economic, Social and Cultural Rights. Substantive Issues Arising in the Implementation of the International Covenant on Economic, Social and Cultural Rights: General Comment 12 (Twentieth Session, 1999) The Right to Adequate Food (Art. 11), UN Doc E/C.12/1999/5 (1999).

Provisions for Co-operation between States in Weather Modification, UN Environment Programme Decision 8/7/A, UN Doc. UNEP/GC.8/7 (1980).

Rio Declaration on Environment and Development, 31 ILM 874, adopted June 14, 1992.

Understandings regarding the Convention (ENMOD), Report of Conference of the Committee on Disarmament, Vol. 1, UN Doc. A/31/27 (1976).

United Nations Educational, Scientific and Cultural Organization and International Council for Science. Declaration on Science and the Use of Scientific Knowledge, adopted by the World Conference on Science (July 1, 1999).

United Nations Educational, Scientific and Cultural Organization. Universal Declaration on Bioethics and Human Rights, adopted by the General Conference, Thirty-Third Session (October 19, 2005).

United Nations Framework Convention on Climate Change. Conference of the Parties' Sixteenth Session, *Decision 1/CP.16. The Cancun Agreements: Outcome of the Work of the Ad Hoc Working Group on Long-term Cooperative Action under the Convention*, UN Doc. FCCC/CP/2010/7/Add.1 (2010).

United Nations Framework Convention on Climate Change. Conference of the Parties' Twenty-First Session, *Adoption of the Paris Agreement* in Decision 1 of the COP 21 Decisions, UN Doc. FCCC/CP/2015/10/Add.1 (2015).

United Nations Security Council. Resolution 2349 (2017) Adopted by the Security Council at Its 7911th Meeting, on March 31, 2017, UN Doc. S/RES/2349 (2017).

United Nations Security Council. Resolution 2408 (2018) Adopted by the Security Council at Its 8215th Meeting, on March 27, 2018, UN Doc. S/RES/2408 (2018)

United Nations Security Council. *Statement by the President of the Security Council*, UN Doc. S/PRST/2011/15 (2011).

United Nations General Assembly. Universal Declaration of Human Rights (UDHR), December 10, 1948, UN Doc. A/RES/217(III) (1948).

United Nations General Assembly. Right to Exploit Freely Natural Wealth and Resources, December 21, 1952, UN Doc. A/RES/626(VII) (1952).

United Nations General Assembly. Declaration on Permanent Sovereignty over Natural Resources, December 14, 1962, UN Doc. A/RES/1803(XVII) (1962).

United Nations General Assembly. Protection of Global Climate for Present and Future Generations of Mankind, December 6, 1988, UN Doc. A/RES/43/53 (1988).

United Nations General Assembly. United Nations Sustainable Development Goals, September 25, 2015, UN Doc. A/RES/70/1 (2015).

Vienna Declaration and Programme of Action, adopted by the World Conference on Human Rights, UN Doc. A/CONF.157/23 (1993).

4. EUROPEAN UNION LEGISLATION

Directive 2001/20/EC of the European Parliament and of the Council of 4 April 2001 on the approximation of the laws, regulations and administrative provisions of the Member States relating to implementation of good clinical practice in the conduct of clinical trials on medicinal products for human use (Clinical Trials Directive), 2001 OJ (L 121) 34.

5. US FEDERAL LEGISLATION

Act to Prevent Pollution from Ships, 33 USC 1905–1915.
Clean Air Act, 42 USC 7401–7671.
Endangered Species Act, 16 USC 1531–1544.
Federal Policy for the Protection of Human Subjects (Common Rule), 45 CFR part 46.
Marine Debris Act, 33 USC 1951–1958.
Marine Mammal Protection Act, 16 USC 703–712.
Marine Protection, Research and Sanctuaries Act (Ocean Dumping Act), 16 USC 1431–1445, 16 USC 1447–1447, 33 USC 1401–1445, 33 USC 2801–2805.
Migratory Bird Treaty Act, 16 USC 703–712.
National Environmental Policy Act (NEPA), 42 USC 4321–4347.
Weather Modification Reporting Act, 15 USC 330.

6. US STATE LEGISLATION

Rhode Island, "An Act Relating to Health and Safety – Geoengineering," H 7655, 2014 session (2014).
Rhode Island, "An Act Relating to Health and Safety – Geoengineering," H 5480, 2015 session (2015).
Rhode Island, "An Act Relating to Health and Safety – Geoengineering," H 7578, 2016 session (2016).
Rhode Island, "An Act Relating to Health and Safety – Geoengineering," H 5607, 2017 session (2017).
Rhode Island, "An Act Relating to Health and Safety – The Geoengineering Act of 2017," H 6011, 2017 session (2017).
Rhode Island, House Resolution, H 6011 Substitute A, 2017 session (2017).

7. US CASES

Defenders of Wildlife v. Lujan, 911 F.2d 117 (8th Cir. 1990).
Environmental Defense Fund, Inc. v. Massey, 772 F. Supp. 1296 (DDC 1991).
Hartford Fire Insurance Co. v. California, 509 US 764 (1993).
Lujan v. Defenders of Wildlife, 504 US 555 (1992).
Massachusetts v. Environmental Protection Agency, 549 US 497 (2007).
Natural Resources Defense Council v. US Department of Navy, No. CV-01–07781, 2002 WL 32095131 (CD Cal. September 17, 2002).
Steele v. Bulova Watch Co., Inc., 344 US 280 (1952).
United States v. Aluminum Co. of America (Alcoa), 148 F.2d 416 (2d Cir. 1945).

8. NATIONAL LEGAL DOCUMENTS

Council on Environmental Quality. Guidance on NEPA Analyses for Transboundary Impacts (1997).

Estado Plurinacional de Bolivia. Submission to Joint Workshop of Experts on Geoengineering, Lima, Peru, June 20–22 (2011).

Great Britain Department of Energy and Climate Change. Government Response to the House of Commons Science and Technology Committee 5th Report of Session 2009–10: The Regulation of Geoengineering.

House of Commons (UK), Science and Technology Committee, The Regulation of Geoengineering, Fifth Report of Session 2009–10, HC 221 (2010).

Bibliography

Abate, Randall S. 2006. "Dawn of a New Era in the Extraterritorial Application of US Environmental Statutes: A Proposal for an Integrated Judicial Standard Based on the Continuum of Context." *Columbia Journal of Environmental Law* 31 no. 1:87–137, https://heinonline.org/HOL/P?h=hein.journals/cjel31&i=93.

Abbott, Kenneth W., Robert O. Keohane, Andrew Moravcsik, Anne-Marie Slaughter, and Duncan Snidal. 2000. "The Concept of Legalization." *International Organization* 54 no. 3:401–19, DOI:10.1162/002081800551271.

Adelman, Sam. 2017. "Geoengineering: Rights, Risks and Ethics." *Journal of Human Rights and the Environment* 8 no. 1:119–38, DOI:10.4337/jhre.2017.01.06.

Adger, Neil W., Juan M. Pulhin, Jon Barnett, Geoffrey D. Dabelko, Grete K. Hovelsrud, Marc Levy, et al. 2014. "Human Security." In *Climate Change 2014: Impacts, Adaptation, and Vulnerability: Contribution of Working Group II to the Fifth Assessment Report of the Intergovernmental Panel on Climate Change*, edited by Christopher B. Field, Vicente R. Barros, David Jon Dokken, Katharine J. Mach, Michael D. Mastrandrea, T. Eren Bilir. et al., 755–92. New York: Cambridge University Press, www.ipcc.ch/pdf/assessment-report/ar5/wg2/WGIIAR5-Chap12_FINAL.pdf.

Alexander, Lisa V., Simon K. Allen, Nathaniel L. Bindoff, François-Marie Bréon, John A. Church, Ulrich Cubasch et al. 2013. "Summary for Policymakers." In *Climate Change 2013: The Physical Science Basis: Working Group I Contribution to the Fifth Assessment Report of the Intergovernmental Panel on Climate Change*, edited by Thomas F. Stocker, Q. Dahe, Gian-Kasper Plattner, Melinda M. B. Tignor, Simon K. Allen, Judith Boschung et al., 3–29. New York: Cambridge University Press, DOI:10.1017/CBO9781107415324.004.

Allen, Myles, Mustafa Babiker, Yang Chen, Heleen de Coninck, Sarah Connors, Renée van Diemen et al. 2018. "Summary for Policymakers." In *Global Warming of 1.5 °C: An IPCC Special Report on the Impacts of Global Warming of 1.5 °C above Pre-Industrial Levels and Related Global Greenhouse Gas Emission Pathways, in the Context of Strengthening the Global Response to the Threat of Climate Change, Sustainable Development, and Efforts to Eradicate Poverty*, edited by Valérie Masson-Delmotte, Panmao Zhai, Hans-Otto Pörtner, Debra Roberts, Jim Skea, Priyadarshi R. Shukla et al., 3–26. Intergovernmental Panel on Climate Change. www.ipcc.ch/sr15/chapter/summary-for-policy-makers/.

Alley, R. B., J. Marotzke, W. D. Nordhaus, J. T. Overpeck, D. M. Peteet, R. A. Pielke et al. 2003. "Abrupt Climate Change." *Science* 299 no. 5615:2005–10, DOI:10.1126/science.1081056.

American Law Institute. 2009. *Restatement of the Law Third: Liability for Physical and Emotional Harm*. Vol. 1. St. Paul: American Law Institute.

American Meteorological Society Council. 2009. "AMS Policy Statement on Geoengineering the Climate System." Accessed March 8, 2019. www.ametsoc.org/index.cfm/ams/about-ams/ams-statements/statements-of-the-ams-in-force/geoengineering-the-climate-system/.

Anderson, J. W. 1997. *The Kyoto Protocol on Climate Change: The Negotiations over Global Warming.* Washington, DC: Resources for the Future.

Anshelm, Jonas, and Anders Hansson. 2014. "Battling Promethean Dreams and Trojan Horses: Revealing the Critical Discourses of Geoengineering." *Energy Research & Social Science* 2:135–44, DOI:10.1016/j.erss.2014.04.001.

Armeni, Chiara, and Catherine Redgwell. 2015a. "Geoengineering under National Law: A Case Study of Germany." *Climate Geoengineering Governance Working Paper* 24, accessed March 8, 2019. www.geoengineering-governance-research.org/perch/resources/workingpaper24armeniredgwellgermany-1.pdf.

—2015b. "Geoengineering under National Law: A Case Study of the United Kingdom." *Climate Geoengineering Governance Working Paper* 23, accessed March 8, 2019. www.geoengineering-governance-research.org/perch/resources/workingpaper23armeniredgwelltheukcombine.pdf.

—2015c. "International Legal and Regulatory Issues of Climate Geoengineering Governance: Rethinking the Approach." *Climate Geoengineering Governance Working Paper* 21, accessed March 8, 2019. www.geoengineering-governance-research.org/perch/resources/workingpaper21armeniredgwelltheinternationalcontextrevise-.pdf.

Asilomar Scientific Organizing Committee. 2010. *The Asilomar Conference Recommendations on Principles for Research into Climate Engineering Techniques.* Washington, DC: Climate Institute, http://jreynolds.org/wp-content/uploads/2018/05/MacCracken-2010-The-Asilomar-Conference-Recomm.pdf.

Baatz, Christian. 2016. "Can We Have It Both Ways? On Potential Trade-Offs between Mitigation and Solar Radiation Management." *Environmental Values* 25 no. 1:29–49, DOI:10.3197/096327115X14497392134847.

Banerjee, Bidisha. 2011. "The Limitations of Geoengineering Governance in a World of Uncertainty." *Stanford Journal of Law, Science, and Policy* 4 no. 1:15–36, https://law.stanford.edu/publications/the-limitations-of-geoengineering-governance-in-a-world-of-uncertainty/.

Barrett, Scott. 1994. "Self-Enforcing International Environmental Agreements." *Oxford Economic Papers* 46 no. 4:878–94, www.jstor.org/stable/2663505.

—2007. *Why Cooperate? The Incentive to Supply Global Public Goods.* Oxford: Oxford University Press.

—2008. "The Incredible Economics of Geoengineering." *Environmental and Resource Economics* 39 no. 1:45–54, DOI:10.1007/s10640-007-9174-8.

—2010. Written Statement, In *Geoengineering: Parts I, II, and III. Hearing before the Committee on Science and Technology, House of Representatives, One Hundred Eleventh Congress, First Session and Second Session, November 5, 2009, February 4, 2010, and March 18, 2010,* 312–20, Washington, DC: US Government Printing Office. www.govinfo.gov/content/pkg/CHRG-111hhrg53007/pdf/CHRG-111hhrg53007.pdf.

—2014. "Solar Geoengineering's Brave New World: Thoughts on the Governance of an Unprecedented Technology." *Review of Environmental Economics and Policy* 8 no. 2:249–69, DOI:10.1093/reep/reu011.

Baum, Seth D., Timothy M. Maher, Jr., and Jacob Haqq-Misra. 2012. "Double Catastrophe: Intermittent Stratospheric Geoengineering Induced by Societal Collapse." *Environment, Systems and Decisions* 33 no. 1:168–80, DOI:10.1007/s10669-012-9429-y.

Bellamy, Rob, Jason Chilvers, Naomi E. Vaughan, and Timothy M. Lenton. 2012. "A Review of Climate Geoengineering Appraisals." *Wiley Interdisciplinary Reviews: Climate Change* 3 no. 6:597–615, DOI:10.1002/wcc.197.

Bellamy, Rob, and Peter Healey. 2018. "'Slippery Slope' or 'Uphill Struggle'? Broadening out Expert Scenarios of Climate Engineering Research and Development." *Environmental Science & Policy* 83:1–10, DOI:10.1016/j.envsci.2018.01.021.

Benedick, Richard Elliot. 1998. *Ozone Diplomacy: New Directions in Safeguarding the Planet.* Cambridge, MA: Harvard University Press.

—2011. "Considerations on Governance for Climate Remediation Technologies: Lessons from the 'Ozone Hole.'" *Stanford Journal of Law, Science, and Policy* 4 no. 1:6–9, https://law.stanford.edu/publications/considerations-on-governance-for-climate-remediation-technologies-lessons-from-the-ozone-hole/.

Bennear, Lori S., and Jonathan B. Wiener. 2019. "Adaptive Regulation: Instrument Choice for Policy Learning over Time." Working paper, https://www.hks.harvard.edu/sites/default/files/centers/mrcbg/files/Regulation%20-%20adaptive%20reg%20-%20Bennear%20Wiener%20on%20Adaptive%20Reg%20Instrum%20Choice%202019%2002%2012%20clean.pdf.

Bickel, J. Eric, and Lee Lane. 2010. "Climate Engineering." In *Smart Solutions to Climate Change: Comparing Costs and Benefits*, edited by Bjorn Lomborg, 9–51. Cambridge: Cambridge University Press, DOI:10.1017/CBO9780511779015.002.

Biermann, F., K. Abbott, S. Andresen, K. Bäckstrand, S. Bernstein, M. M. Betsill et al. 2012. "Navigating the Anthropocene: Improving Earth System Governance." *Science* 335 no. 6074:1306–7, DOI:10.1126/science.1217255.

Bipartisan Policy Center's Task Force on Climate Remediation Research. 2011. *Geoengineering: A National Strategic Plan for Research on the Potential Effectiveness, Feasibility, and Consequences of Climate Remediation Technologies.* Washington, DC: Bipartisan Policy Center, https://bipartisanpolicy.org/library/task-force-climate-remediation-research/.

Birnie, Patricia W., Alan E. Boyle, and Catherine Redgwell. 2009. *International Law and the Environment.* Oxford: Oxford University Press.

Black, Henry Campbell. 1910. *A Law Dictionary Containing Definitions of the Terms and Phrases of American and English Jurisprudence, Ancient and Modern.* St. Paul, MN: West.

Blackstock, Jason J., Nigel Moore, and Clarisse Kehler Siebert. 2011. "Engineering the Climate: Research Questions and Policy Implications." *UNESCO-SCOPE-UNEP Policy Brief* 14, United Nations Educational, Scientific and Cultural Organization, http://unesdoc.unesco.org/images/0021/002144/214496e.pdf.

Bodansky, Daniel. 1996. "May We Engineer the Climate?" *Climatic Change* 33 no. 3:309–21, DOI:10.1007/bf00142579.

—2000. "What's So Bad about Unilateral Action to Protect the Environment?" *European Journal of International Law* 11 no. 2:339–47, DOI:10.1093/ejil/11.2.339.

—2011. "Governing Climate Engineering: Scenarios for Analysis." *The Harvard Project on Climate Agreements Discussion Paper* 11-47, Cambridge, MA: Harvard Kennedy School, DOI:10.2139/ssrn.1963397.

—2012. "What's in a Concept? Global Public Goods, International Law, and Legitimacy." *European Journal of International Law* 23 no. 3:651–68, DOI:10.1093/ejil/chs035.

—2013. "The Who, What, and Wherefore of Geoengineering Governance." *Climatic Change* 121 no. 3:539–51, DOI:10.1007/s10584-013-0759-7.

Bodle, Ralph. 2010. "Geoengineering and International Law: The Search for Common Legal Ground." *Tulsa Law Review* 46 no. 2:305–22, https://digitalcommons.law.utulsa.edu/tlr/vol46/iss2/4.

Boucher, Olivier, David Randall, Paulo Artaxo, Christopher Bretherton, Graham Feingold, Piers Forster et al. 2013. "Clouds and Aerosols." In *Climate Change 2013: The Physical Science Basis: Contribution of Working Group I to the Fifth Assessment Report of the Intergovernmental Panel on Climate Change*, edited by Thomas F. Stocker, Dahe Qin, Gian-Kasper Plattner, Melinda. M.B. Tignor, Simon K. Allen, Judith Boschung et al., 571–657. Cambridge: Cambridge University Press, DOI:10.1017/CBO9781107415 324.016.

Boucher, Olivier, Piers M. Forster, Nicolas Gruber, Minh Ha-Duong, Mark G. Lawrence, Timothy M. Lenton et al. 2014. "Rethinking Climate Engineering Categorization in the Context of Climate Change Mitigation and Adaptation." *Wiley Interdisciplinary Reviews: Climate Change* 5 no. 1:23–35, DOI:10.1002/wcc.261.

Boyle, A. E. 2005. "Globalising Environmental Liability: The Interplay of National and International Law." *Journal of Environmental Law* 17 no. 1:3–26, DOI:10.1093/envlaw/eqi001.

Boyle, Alan. 2012. "Law of the Sea Perspectives on Climate Change." *The International Journal of Marine and Coastal Law* 27 no. 4:831–8, DOI:10.1163/15718085-12341244.

Brent, Kerryn. 2018. "Solar Radiation Management Geoengineering and Strict Liability for Ultrahazardous Activities." In *Global Environmental Change and Innovation in International Law*, edited by Cameron S. G. Jefferies, Neil Craik, Sara L. Seck, and Tim Stephens, 161–79. Cambridge: Cambridge University Press, DOI:10.1017/9781108526081.010.

Brent, Kerryn, Jeffrey McGee, and Amy Maguire. 2015. "Does the 'No-Harm' Rule Have a Role in Preventing Transboundary Harm and Harm to the Global Atmospheric Commons from Geoengineering?" *Climate Law* 5 no. 1:35–63, DOI:10.1163/18786561-00501007.

Briggs, Chad. 2018. "Is Solar Geoengineering a US National Security Risk?" In *Geoengineering Our Climate? Ethics, Politics, and Governance*, edited by Jason J. Blackstock and Sean Low, 178–81. London: Earthscan.

Brownlie, Ian. 1985. "The Rights of Peoples in Modern International Law." *Bulletin of the Australian Society of Legal Philosophy* 9 no. 2:104–19, http://classic.austlii.edu.au/au/journals/AUSocLegPhilB/1985/7.html.

Brownsword, Roger. 2008. *Rights, Regulation, and the Technological Revolution*. Oxford: Oxford University Press.

Bruintjes, Roelof. 2015. *Report of the Expert Team on Weather Modification Meeting, Phitsanulok, Thailand, 17–19 March 2015*. World Meteorological Organization, www.wmo.int/pages/prog/arep/wwrp/new/documents/WMO_expert_mtg_Phisanulok_2015_report_FINAL.pdf.

Buchanan, Allen. 2003. "Reforming the International Law of Humanitarian Intervention." In *Humanitarian Intervention: Ethical, Legal and Political Dilemmas*, edited by J. L. Holzgrefe and Robert O. Keohane, 130–74. Cambridge: Cambridge University Press, DOI:10.1017/CBO9780511494000.005.

Buck, Holly Jean. 2012. "Geoengineering: Re-Making Climate for Profit or Humanitarian Intervention?" *Development and Change* 43 no. 1:253–70, DOI:10.1111/j.1467-7660.2011.01744.x.

—2013. "Climate Engineering: Spectacle, Tragedy, or Solution? A Content Analysis of News Media Framing." In *Interpretive Approaches to Global Climate Governance. (De) Constructing the Greenhouse*, edited by Chris Methmann, Delf Rothe, and Benjamin Stephen, 166–80. London: Routledge.

Budyko, Mikhail I. 1977. *Climatic Changes*. Translated by AGU Translation Board. Washington, DC: American Geophysical Union.

Bunzl, Martin. 2011. "Geoengineering Harms and Compensation." *Stanford Journal of Law, Science, and Policy* 4 no. 1:69–75, https://law.stanford.edu/publications/geoengineering-harms-and-compensation/.

Burns, Elizabeth T., Jane A. Flegal, David W. Keith, Aseem Mahajan, Dustin Tingley, and Gernot Wagner. 2016. "What Do People Think When They Think about Solar Geoengineering? A Review of Empirical Social Science Literature, and Prospects for Future Research." *Earth's Future* 4 no. 11:536–42, DOI:10.1002/2016EF000461.

Burns, William C. G. 2011. "Climate Geoengineering: Solar Radiation Management and Its Implications for Intergenerational Equity." *Stanford Journal of Law, Science, and Policy* 4 no. 1:37–55, https://law.stanford.edu/publications/climate-geoengineering-solar-radiation-management-and-its-implications-for-intergenerational-equity/.

—2016. "The Paris Agreement and Climate Geoengineering Governance: The Need for a Human Rights-Based Component." *CIGI Paper* 111, Centre for International Governance Innovation, www.cigionline.org/publications/paris-agreement-and-climate-geoengineering-governance-need-human-rights-based.

Burton, Ian. 1994. "Deconstructing Adaptation ... and Reconstructing." *Delta* 5 no. 1:14–15.

Bykvist, Krister. 2009. "Preference Formation and Intergenerational Justice." In *Intergenerational Justice*, edited by Axel Gosseries and Lukas H. Meyer, 301–23. Oxford: Oxford University Press, DOI:10.1093/acprof:oso/9780199282951.001.0001.

Cairns, Rose C. 2014. "Climate Geoengineering: Issues of Path-Dependence and Socio-Technical Lock-In." *Wiley Interdisciplinary Reviews: Climate Change* 5 no. 5:649–61, DOI:10.1002/wcc.296.

—2016. "Climates of Suspicion: 'Chemtrail' Conspiracy Narratives and the International Politics of Geoengineering." *The Geographical Journal* 182 no. 1:70–84, DOI:10.1111/geoj.12116.

Cairns, Rose C., and Paul Nightingale. 2014. "The Security Implications of Geoengineering: Blame, Imposed Agreement and the Security of Critical Infrastructure." *Climate Geoengineering Governance Working Paper* 18, www.geoengineering-governance-research.org/perch/resources/workingpaper18nightingalecairnssecurityimplications.pdf.

Callies, Daniel Edward. 2018. "The Slippery Slope Argument against Geoengineering Research." *Journal of Applied Philosophy*. Online ahead of print:1–13, DOI:10.1111/japp.12345.

Campos, Nauro F., and Francesco Giovannoni. 2007. "Lobbying, Corruption and Political Influence." *Public Choice* 131 no. 1:1–21, DOI:10.1007/s11127-006-9102-4.

Carrico, Amanda R., Heather Barnes Truelove, Michael P. Vandenbergh, and David Dana. 2015. "Does Learning about Climate Change Adaptation Change Support for Mitigation?" *Journal of Environmental Psychology* 41:19–29, DOI:10.1016/j.jenvp.2014.10.009.

Cascio, Jamais. 2009. *Hacking the Earth: Understanding the Consequences of Geoengineering*. www.lulu.com/spotlight/openthefuture.

Chalecki, Elizabeth L., and Lisa L. Ferrari. 2018. "A New Security Framework for Geoengineering." *Strategic Studies Quarterly* 12 no. 2:82–106, www.airuniversity.af.edu/Portals/10/SSQ/documents/Volume-12_Issue-2/Chalecki_Ferrari.pdf.

Chavez, Anthony E. 2015. "Exclusive Rights to Saving the Planet: The Patenting of Geoengineering Inventions." *Northwestern Journal of Technology and Intellectual Property* 13 no. 1: article 1, https://scholarlycommons.law.northwestern.edu/njtip/vol13/iss1/1/.

Chemnick, Jean. 2019. "U.S. Blocks U.N. Resolution on Geoengineering." *E&E News*, March 15, 2019, www.scientificamerican.com/article/u-s-blocks-u-n-resolution-on-geoen gineering/.

Chhetri, Netra, Dan Chong, Ken Conca, Richard Falk, Alexander Gillespie, Aarti Gupta et al. 2018. *Governing Solar Radiation Management*. Forum for Climate Engineering Assessment, DOI:10.17606/M6SM17.

Chris, Robert. 2015. *Systems Thinking for Geoengineering Policy: How to Reduce the Threat of Dangerous Climate Change by Embracing Uncertainty and Failure*. London: Earthscan.

Cohen, Alma, and Peter Siegelman. 2010. "Testing for Adverse Selection in Insurance Markets." *Journal of Risk and Insurance* 77 no. 1:39–84, DOI:10.1111/j.1539-6975.2009.01337.x.

Coleman, Jules L. 1995. "The Practice of Corrective Justice." *Arizona Law Review* 37 no. 1:15–31, https://digitalcommons.law.yale.edu/fss_papers/4210/.

Collingridge, David. 1980. *The Social Control of Technology*. Milton Keynes, UK: Open University Press.

Colyvan, Mark, Damian Cox, and Katie Steele. 2010. "Modelling the Moral Dimension of Decisions." *Noûs* 44 no. 3:503–29, DOI:10.1111/j.1468-0068.2010.00754.x.

Contreras, Jorge L. 2015. "Patent Pledges." *Arizona State Law Journal* 47 no. 3:543–608, http://arizonastatelawjournal.org/2015/12/14/patent-pledges/.

Corner, Adam, and Nick Pidgeon. 2014. "Like Artificial Trees? The Effect of Framing by Natural Analogy on Public Perceptions of Geoengineering." *Climatic Change* 130 no. 3:425–38, DOI:10.1007/s10584-014-1148-6.

Corry, Olaf. 2017. "The International Politics of Geoengineering: The Feasibility of Plan B for Tackling Climate Change." *Security Dialogue* 48 no. 4:297–315, DOI:10.1177/0967010617704142.

Council for International Organizations of Medical Sciences, and World Health Organization. 2016. *International Ethical Guidelines for Health-Related Research Involving Humans*. Geneva: Council for International Organizations of Medical Sciences, https://cioms.ch/wp-content/uploads/2017/01/WEB-CIOMS-EthicalGuidelines.pdf.

Craik, Neil. 2015. "International EIA Law and Geoengineering: Do Emerging Technologies Require Special Rules?" *Climate Law* 5 no. 2–4:111–41, DOI:10.1163/18786561-00504002.

Craik, Neil, Jason J. Blackstock, and Anna-Maria Hubert. 2013. "Regulating Geoengineering Research through Domestic Environmental Protection Frameworks: Reflections on the Recent Canadian Ocean Fertilization Case." *Carbon & Climate Law Review* 7 no. 2:117–24, DOI:10.21552/CCLR/2013/2/253.

Crutzen, Paul J. 2002. "Geology of Mankind." *Nature* 415 no. 6867:23, DOI:10.1038/415023a.

—2006. "Albedo Enhancement by Stratospheric Sulfur Injections: A Contribution to Resolve a Policy Dilemma?" *Climatic Change* 77 no. 3:211–20, DOI:10.1007/s10584-006-9101-y.

Curry, Charles L., Jana Sillmann, David Bronaugh, Kari Alterskjaer, Jason N. S. Cole, Duoying Ji et al. 2014. "A Multi-Model Examination of Climate Extremes in an Idealized Geoengineering Experiment." *Journal of Geophysical Research: Atmospheres* 119 no. 7:3900–23, DOI:10.1002/2013JD020648.

Dalby, Simon. 2015. "Geoengineering: The Next Era of Geopolitics?" *Geography Compass* 9 no. 4:190–201, DOI:10.1111/gec3.12195.

—2017. "Anthropocene Formations: Environmental Security, Geopolitics and Disaster." *Theory, Culture & Society* 34 no. 2–3:233–52, DOI:10.1177/0263276415598629.

Dari-Mattiacci., Giuseppe. 2009. "Negative Liability." *The Journal of Legal Studies* 38 no. 1:21–59, DOI:10.1086/596197.

Dave, Dhaval, and Robert Kaestner. 2009. "Health Insurance and Ex Ante Moral Hazard: Evidence from Medicare." *International Journal of Health Care Finance and Economics* 9 no. 4:367–90, DOI:10.1007/s10754-009-9056-4.

Davies, Gareth. 2008. "Law and Policy Issues of Unilateral Geoengineering: Moving to a Managed World." In *Select Proceedings of the European Society of International Law*, edited by Hélène Ruiz Fabri, Rüdiger Wolfrum, and Jana Gogolin, 627–40. Oxford: Hart.

—2010. "Framing the Social, Political, and Environmental Risks and Benefits of Geoengineering: Balancing the Hard-to-Imagine against the Hard-to-Measure." *Tulsa Law Review* 46 no. 2:261–82, https://digitalcommons.law.utulsa.edu/tlr/vol46/iss2/2.

—2013. "Privatisation and De-Globalisation of the Climate." *Carbon & Climate Law Review* 7 no. 3:187–93, DOI:10.21552/CCLR/2013/3/263.

Davis, William Daniel. 2009. "What Does "Green" Mean?: Anthropogenic Climate Change, Geoengineering, and International Environmental Law." *Georgia Law Review* 43 no. 3:901–51, https://heinonline.org/HOL/P?h=hein.journals/geolr43&i=919.

de Coninck, Heleen, Aromar Revi, Babiker Mustafa, Paolo Bertoldi, Marcos Buckeridge, Anton Cartwright et al. 2018. "Strengthening and Implementing the Global Response." In *Global Warming of 1.5 °C: An IPCC Special Report on the Impacts of Global Warming of 1.5 °C above Pre-Industrial Levels and Related Global Greenhouse Gas Emission Pathways, in the Context of Strengthening the Global Response to the Threat of Climate Change, Sustainable Development, and Efforts to Eradicate Poverty*, edited by Valérie Masson-Delmotte, Panmao Zhai, Hans-Otto Pörtner, Debra Roberts, Jim Skea, Priyadarshi R. Shukla et al., 313–42. Intergovernmental Panel on Climate Change. www.ipcc.ch/sr15/chapter/4-0/.

Dellink, Rob, Jean Chateau, Elisa Lanzi, and Bertrand Magné. 2016. "Long-Term Economic Growth Projections in the Shared Socioeconomic Pathways." *Global Environmental Change* 42:200–14, DOI:10.1016/j.gloenvcha.2015.06.004.

Dilling, Lisa, and Rachel Hauser. 2013. "Governing Geoengineering Research: Why, When and How?" *Climatic Change* 121 no. 3:553–65, DOI:10.1007/s10584-013-0835-z.

Dobbs, Dan, Paul Hayden, and Ellen Bublick. 2015. *Hornbook on Torts*. St. Paul, MN: West.

Drabiak-Syed, Katherine. 2010. "Lessons from Havasupai Tribe V. Arizona State University Board of Regents: Recognizing Group, Cultural, and Dignity Harms as Legitimate Risks Warranting Integration into Research Practice." *Journal of Health & Biomedical Law* 6 no. 2:175–225, https://heinonline.org/HOL/P?h=hein.journals/jhbio6&i=183.

Du, Haomiao. 2017. *An International Legal Framework for Geoengineering: Managing the Risks of an Emerging Technology*. London: Routledge.

Dykema, John A., David W. Keith, James G. Anderson, and Debra Weisenstein. 2014. "Stratospheric Controlled Perturbation Experiment: A Small-Scale Experiment to Improve Understanding of the Risks of Solar Geoengineering." *Philosophical Transactions of the Royal Society A: Mathematical, Physical and Engineering Sciences* 372 no. 2031: article 20140059, DOI:10.1098/rsta.2014.0059.

Eastham, Sebastian D., Debra K. Weisenstein, David W. Keith, and Steven R. H. Barrett. 2018. "Quantifying the Impact of Sulfate Geoengineering on Mortality from Air Quality and UV-B Exposure." *Atmospheric Environment* 187: 424–34, DOI:10.1016/j.atmosenv.2018.05.047.

Eckersley, Robyn. 2017. "Geopolitan Democracy in the Anthropocene." *Political Studies* 65 no. 4:983–99, DOI:10.1177/0032321717695293.

Edenhofer, Ottmar, Ramon Pichs-Madruga, Youba Sokona, Christopher Field, Vicente Barros, Thomas F. Stocker et al. 2012. *IPCC Expert Meeting on Geoengineering, Lima, Peru, 20–22 June 2011 Meeting Report*. Potsdam, Germany: IPCC Working Group III Technical Support Unit, Potsdam Institute for Climate Impact Research, https://wg1.ipcc.ch/publications/supportingmaterial/EM_GeoE_Meeting_Report_final.pdf.

Emmerling, Johannes, and Massimo Tavoni. 2017a. "Geoengineering and Climate Change Mitigation: Trade-Offs and Synergies as Foreseen by Integrated Assessment Models." In *Climate Justice and Geoengineering: Ethics and Policy in the Atmospheric Anthropocene*, edited by Christopher J. Preston, 175–88. London: Rowman & Littlefield.

—2017b. "Quantifying Non-Cooperative Climate Engineering." *Fondazione Eni Enrico Mattei Working Paper* 058–2017, Fondazione Eni Enrico Mattei, DOI:10.2139/ssrn.3090312.

Engineering and Physical Sciences Research Council. 2012. "SPICE Update." https://epsrc.ukri.org/newsevents/news/spiceupdateoct/.

Evans, Laurel, Taciano L. Milfont, and Judy Lawrence. 2014. "Considering Local Adaptation Increases Willingness to Mitigate." *Global Environmental Change* 25:69–75, DOI:10.1016/j.gloenvcha.2013.12.013.

Fairbrother, Malcolm. 2016. "Geoengineering, Moral Hazard, and Trust in Climate Science: Evidence from a Survey Experiment in Britain." *Climatic Change* 139 no. 3–4:477–89, DOI:10.1007/s10584-016-1818-7.

Faure, Michael, and Marjan Peeters, eds. 2011. *Climate Change Liability*. Cheltenham, UK: Edward Elgar.

Field, Christopher B., Vicente R. Barros, Katharine J. Mach, Michael D. Mastrandrea, Maarten K. van Aalst, W. Neil Adger et al. 2014. "Technical Summary." In *Climate Change 2014: Impacts, Adaptation, and Vulnerability: Contribution of Working Group II to the Fifth Assessment Report of the Intergovernmental Panel on Climate Change*, edited by Christopher B. Field, Vicente R. Barros, David Jon Dokken, Katharine J. Mach, Michael D. Mastrandrea, T. Eren Bilir et al., 35–94. Cambridge: Cambridge University Press, DOI:10.1017/CBO9781107415379.004.

Fleming, James R. 2007. "The Climate Engineers." *Wilson Quarterly*, Spring 2007, 46–60, http://archive.wilsonquarterly.com/essays/climate-engineers.

Foley, Rider W., David H. Guston, and Daniel Sarewitz. 2018. "Towards the Anticipatory Governance of Geoengineering." In *Geoengineering Our Climate? Ethics, Politics and Governance*, edited by Jason J. Blackstock and Sean Low, 223–43. London: Earthscan.

Folke, Carl, Thomas Hahn, Per Olsson, and Jon Norberg. 2005. "Adaptive Governance of Social-Ecological Systems." *Annual Review of Environment and Resources* 30 no. 1:441–73, DOI:10.1146/annurev.energy.30.050504.144511.

Foray, Dominique. 2004. *The Economics of Knowledge*. Cambridge, MA: MIT Press.

Fortune. 2018. "Global 500." http://fortune.com/global500/list/.

Forum for Climate Engineering Assessment. n.d., accessed March 8, 2019. "Who We Are." http://ceassessment.org/who-we-are/.

Fountain, Henry. 2012. "A Rogue Climate Experiment Has Ocean Experts Outrages." *New York Times*, October 19, 2012, A1, www.nytimes.com/2012/10/19/science/earth/iron-dumping-experiment-in-pacific-alarms-marine-experts.html.

Fragnière, Augustin, and Stephen M. Gardiner. 2016. "Why Geoengineering Is Not 'Plan B'" In *Climate Justice and Geoengineering*, edited by Christopher J. Preston, 15–32. London: Rowman & Littlefield.

Franckx, Erik, ed. 2001. *Vessel-Source Pollution and Coastal State Jurisdiction: The Work of the ILA Committee on Coastal State Jurisdiction Relating to Marine Pollution (1991–2000)*. The Hague: Kluwer Law International.

Frank, Veronica. 2007. *The European Community and Marine Environmental Protection in the International Law of the Sea: Implementing Global Obligations at the Regional Level.* Leiden: Martinus Nijhoff.

Frumhoff, Peter C., and Jennie C. Stephens. 2018. "Towards Legitimacy of the Solar Geoengineering Research Enterprise." *Philosophical Transactions of the Royal Society A: Mathematical, Physical and Engineering Sciences* 376 no. 2119: article 20160459, DOI:10.1098/rsta.2016.0459.

Galaz, Victor. 2014. *Global Environmental Governance, Technology and Politics: The Anthropocene Gap*. Cheltenham, UK: Edward Elgar.

Gardiner, Stephen M. 2010. "Is 'Arming the Future' with Geoengineering Really the Lesser Evil? Some Doubts about the Ethics of Intentionally Manipulating the Climate System." In *Climate Ethics: Essential Readings*, edited by Stephen M. Gardiner, Simon Caney, Dale Jamieson, and Henry Shue, 284–312. Oxford: Oxford University Press.

Gardiner, Stephen M., and Augustin Fragnière. 2018. "The Tollgate Principles for the Governance of Geoengineering: Moving Beyond the Oxford Principles to an Ethically More Robust Approach." *Ethics, Policy & Environment* 21 no. 2:143–74, DOI:10.1080/21550085.2018.1509472.

Ghosh, Arunabha. 2018. "Environmental Institutions, International Research Programmes, and Lessons for Geoengineering Research." In *Geoengineering Our Climate? Ethics, Politics, and Governance*, edited by Jason J. Blackstock and Sean Low, 199–213. London: Earthscan.

Ginzky, Harald, and Robyn Frost. 2014. "Marine Geo-Engineering: Legally Binding Regulation under the London Protocol." *Carbon & Climate Law Review* 8 no. 2:82–96, https://cclr.lexxion.eu/article/cclr/2014/2/284

Glienke, Susanne, Peter J. Irvine, and Mark G. Lawrence. 2015. "The Impact of Geoengineering on Vegetation in Experiment G1 of the Geoengineering Model Intercomparison Project (GeoMIP)." *Journal of Geophysical Research: Atmospheres* 120 no. 19:10196–213, DOI:10.1002/2015JD024202.

Goeschl, Timo, Daniel Heyen, and Juan Moreno-Cruz. 2013. "The Intergenerational Transfer of Solar Radiation Management Capabilities and Atmospheric Carbon Stocks." *Environmental and Resource Economics* 56 no. 1:85–104, DOI:10.1007/s10640-013-9647-x.

Goeschl, Timo, and Tobias Pfrommer. 2015. "Learning by Negligence: Torts, Experimentation, and the Value of Information," *Department of Economics Discussion Paper* 598, University of Heidelberg, DOI:10.11588/heidok.00019197.

Gore, Al. 1992. *Earth in the Balance: Ecology and the Human Spirit*. Boston: Houghton Mifflin.

Govindasamy, Bala, and Ken Caldeira. 2000. "Geoengineering Earth's Radiation Balance to Mitigate CO_2-Induced Climate Change." *Geophysical Research Letters* 27 no. 14:2141–4, DOI:10.1029/1999GL006086.

Graham, John D., and Jonathan Baert Wiener, eds. 1995. *Risk vs. Risk: Tradeoffs in Protecting Health and the Environment*. Cambridge, MA: Harvard University Press.

Great Barrier Reef Foundation. 2018. "Reef 'Sun Shield' Trials Show Promise to Prevent Coral Bleaching." www.barrierreef.org/latest/news/reef-sun-shield-trials-show-promise-to-prevent-coral-bleaching.

Green, Jessica F. 2014. *Rethinking Private Authority: Agents and Entrepreneurs in Global Environmental Governance*. Princeton: Princeton University Press.

Gunningham, Neil, Robert A. Kagan, and Dorothy Thornton. 2004. "Social License and Environmental Protection: Why Businesses Go beyond Compliance." *Law and Social Inquiry* 29 no. 2:307–41, DOI:10.1111/j.1747-4469.2004.tb00338.x.

Guzman, Andrew T. 2008. *How International Law Works: A Rational Choice Theory*. Oxford: Oxford University Press.

Hale, Benjamin. 2012. "The World That Would Have Been: Moral Hazard Arguments against Geoengineering." In *Engineering the Climate: The Ethics of Solar Radiation Management*, edited by Christopher J. Preston, 113–32. Lanham, MD: Lexington.

Halstead, John. 2018. "Stratospheric Aerosol Injection Research and Existential Risk." *Futures* 102:63–77, DOI:10.1016/j.futures.2018.03.004.

Hamilton, Clive. 2013. *Earthmasters: The Dawn of the Age of Climate Engineering*. New Haven: Yale University Press.

Hansson, Sven Ove. 2006. "Informed Consent out of Context." *Journal of Business Ethics* 63 no. 2:149–54, DOI:10.1007/s10551-005-2584-z.

Harding, Anthony, and Juan B. Moreno-Cruz. 2016. "Solar Geoengineering Economics: From Incredible to Inevitable and Half-Way Back." *Earth's Future* 4 no. 12:569–77, DOI:10.1002/2016EF000462.

Harvard's Solar Geoengineering Research Program. n.d., accessed March 8, 2019. "About Us." https://geoengineering.environment.harvard.edu/about.

Hauser, Rachel. 2013. "Using Twentieth-Century US Weather Modification Policy to Gain Insight into Global Climate Remediation Governance Issues." *Weather, Climate, and Society* 5 no. 2:180–93, DOI:10.1175/WCAS-D-11-00011.1.

Herter, Christian. 1971. "Memorandum of C. Herter, Special Assistant to the Secretary of State for Environmental Affairs." In *Administration of the National Environmental Policy Act: Hearings before the Subcommittee on Fisheries and Wildlife Conservation of the Committee on Merchant Marine and Fisheries, House of Representatives, Ninety-First Congress, Second Session*, 551. Washington, DC: US Government Printing Office.

Hester, Tracy. 2018. "Liability and Compensation." In *Climate Engineering and the Law: Regulation and Liability for Solar Radiation Management and Carbon Dioxide Removal*, edited by Michael B. Gerrard and Tracy Hester, 224–68. Cambridge: Cambridge University Press, DOI:10.1017/9781316661864.005.

Heutel, Garth, Juan Moreno-Cruz, and Katharine Ricke. 2016. "Climate Engineering Economics." *Annual Review of Resource Economics* 8 no. 3:99–118, DOI:10.3386/w21711.

Heutel, Garth, Juan Moreno-Cruz, and Soheil Shayegh. 2018. "Solar Geoengineering, Uncertainty, and the Price of Carbon." *Journal of Environmental Economics and Management* 87:24–41, DOI:10.1016/j.jeem.2017.11.002.

Heyen, Daniel. 2016. "Strategic Conflicts on the Horizon: R&D Incentives for Environmental Technologies." *Climate Change Economics* 7 no. 4: article 1650013, DOI:10.1142/s2010007816500135.

Heyen, Daniel, Thilo Wiertz, and Peter James Irvine. 2015. "Regional Disparities in SRM Impacts: The Challenge of Diverging Preferences." *Climatic Change* 133 no. 4:557–63, DOI:10.1007/s10584-015-1526-8.

Heyward, Clare. 2013. "Situating and Abandoning Geoengineering: A Typology of Five Responses to Dangerous Climate Change." *PS: Political Science and Politics* 46 no. 1:23–7, DOI:10.1017/S1049096512001436.

———. 2014. "Benefiting from Climate Geoengineering and Corresponding Remedial Duties: The Case of Unforeseeable Harms." *Journal of Applied Philosophy* 31 no. 4:405–19, DOI:10.1111/japp.12075.

Heyward, Clare, and Steve Rayner. 2015. "Uneasy Expertise: Geoengineering, Social Science, and Democracy in the Anthropocene." In *Policy Legitimacy, Science and Political Authority: Knowledge and Action in Liberal Democracies*, edited by Michael Heazle and John Kane, 101–21. London: Earthscan.

Honegger, Matthias, Henry Derwent, Nicholas Harrison, Axel Michaelowa, and Stefan Schäfer. 2018. *Carbon Removal and Solar Geoengineering: Potential Implications for Delivery of the Sustainable Development Goals*. New York: Carnegie Climate Geoengineering Governance Initiative, www.c2g2.net/geoeng-sdgs/.

Horton, Joshua B. 2011. "Geoengineering and the Myth of Unilateralism: Pressures and Prospects for International Cooperation." *Stanford Journal of Law, Science, and Policy* 4 no. 1:56–69, https://law.stanford.edu/publications/geoengineering-and-the-myth-of-unilateralism-pressures-and-prospects-for-international-cooperation/.

———. 2015. "The Emergency Framing of Solar Geoengineering: Time for a Different Approach." *The Anthropocene Review* 2 no. 2:147–51, DOI:10.1177/2053019615579922.

Horton, Joshua B., and David W. Keith. 2019. "Solar Geoengineering and Parametric Insurance: A Proposal to Facilitate Agreement on Deployment." *Climate Policy*, forthcoming.

Horton, Joshua B., Andrew Parker, and David Keith. 2015. "Liability for Solar Geoengineering: Historical Precedents, Contemporary Innovations, and Governance Possibilities." *New York University Environmental Law Journal* 22 no. 3:225–73, www.nyuelj.org/wp-content/uploads/2015/02/Horton_READY_FOR_WEBSITE.pdf.

Horton, Joshua B., and Jesse L. Reynolds. 2016. "The International Politics of Climate Engineering: A Review and Prospectus for International Relations." *International Studies Review* 18 no. 3:438–61, DOI:10.1093/isr/vivo13.

Horton, Joshua B., Jesse L. Reynolds, Holly Jean Buck, Daniel Callies, Stefan Schäfer, David W. Keith, and Steve Rayner. 2018. "Solar Geoengineering and Democracy." *Global Environmental Politics* 18 no. 3:5–24, DOI:10.1162/glep_a_00466.

Houston, David J., and Lilliard E. Richardson. 2007. "Risk Compensation or Risk Reduction? Seatbelts, State Laws, and Traffic Fatalities." *Social Science Quarterly* 88 no. 4:913–36, DOI:10.1111/j.1540-6237.2007.00510.x.

Hubert, Anna-Maria. 2011. "The New Paradox in Marine Scientific Research: Regulating the Potential Environmental Impacts of Conducting Ocean Science." *Ocean Development and International Law* 42 no. 4:329–55, DOI:10.1080/00908320.2011.619368.

———. 2017. "Code of Conduct for Responsible Geoengineering Research." www.ucalgary.ca/grgproject/files/grgproject/revised-code-of-conduct-for-geoengineering-research-2017-hubert.pdf.

Hubert, Anna-Maria, and David Reichwein. 2015. "An Exploration of a Code of Conduct for Responsible Scientific Research Involving Geoengineering: Introduction, Draft Articles and Commentaries." *IASS Working Paper*, Institute for Advanced Sustainability Studies, DOI:10.2312/iass.2015.013.

Hulme, Mike. 2009. *Why We Disagree about Climate Change: Understanding Controversy, Inaction and Opportunity*. Cambridge: Cambridge University Press.

———. 2014. *Can Science Fix Climate Change? A Case against Climate Engineering*. Cambridge: Polity.

Huttunen, Suvi, Emmi Skytén, and Mikael Hildén. 2015. "Emerging Policy Perspectives on Geoengineering: An International Comparison." *The Anthropocene Review* 2 no. 1:14–32, DOI:10.1177/2053019614557958.

Institute of Medicine, National Academy of Sciences, and National Academy of Engineering. 1992. *Policy Implications of Greenhouse Warming: Mitigation, Adaptation, and the Science Base*. Washington, DC: National Academy Press, DOI:10.17226/1605

Integrated Assessment of Geoengineering Proposals. 2014. "Views about Geoengineering: Key Findings from Public Discussion Groups." http://iagp.ac.uk/sites/default/files/Views%20about%20geoengineering%20IAGP.pdf.

Intellectual Ventures. 2009. "Intellectual Ventures' Answers about Geoengineering," Press release, October 23, 2009. www.intellectualventures.com/buzz/press-releases/intellectual-ventures-answers-about-geoengineering.

Intergovernmental Panel on Climate Change. 2013. "Principles Governing IPCC Work." www.ipcc.ch/site/assets/uploads/2018/09/ipcc-principles.pdf.

International Association Synthetic Biology. 2009. The IASB Code of Conduct for Best Practices in Gene Synthesis. Cambridge, MA: International Association Synthetic Biology, http://op.bna.com.s3.amazonaws.com/hl.nsf/r%3FOpen%3Djaqo-7xqpnr.

International Civil Aviation Organization. 2008. *Environmental Protection, Volume II: Aircraft Engine Emissions*. 3rd ed. Vol. 2. Montréal: International Civil Aviation Organization.

International Council on Human Rights Policy. 2011. *Beyond Technology Transfer: Protecting Human Rights in a Climate-Constrained World*. Geneva: International Council on Human Rights Policy.

International Energy Agency. 2015. *Energy and Climate Change: World Energy Outlook Special Report*. Paris: International Energy Agency www.iea.org/publications/freepublications/publication/WEO2015SpecialReportonEnergyandClimateChange.pdf.

International Energy Agency, and International Renewable Energy Agency. 2017. *Perspectives for the Energy Transition: Investment Needs for a Low-Carbon Energy System*, www.irena.org/publications/2017/Mar/Perspectives-for-the-energy-transition-Investment-needs-for-a-low-carbon-energy-system.

International Institute for Sustainable Development. 1998. "Report of Cop-4." *Earth Negotiations Bulletin*, November 16, 1998, http://enb.iisd.org/download/asc/enb1297e.txt.

Ipsos MORI. 2010. *Experiment Earth? Report on a Public Dialogue on Geoengineering*. Ipsos MORI, www.ipsos.com/sites/default/files/publication/1970-01/sri_experiment-earth-report-on-a-public-dialogue-on-geoengineering_sept2010.pdf.

Irvine, Peter, Kerry Emanuel, Jie He, Larry W. Horowitz, Gabriel Vecchi, and David Keith. 2019. "Halving Warming with Idealized Solar Geoengineering Moderates Key Climate Hazards." *Nature Climate Change*. Online ahead of print: 1–7, DOI:10.1038/s41558-019-0398-8.

Irvine, Peter J., Ben Kravitz, Mark G. Lawrence, and Helene Muri. 2016. "An Overview of the Earth System Science of Solar Geoengineering." *Wiley Interdisciplinary Reviews: Climate Change* 7 no. 6:815–33, DOI:10.1002/wcc.423.

Izrael, Y., V. Zakharov, N. Petrov, A. Ryaboshapko, V. Ivanov, A. Savchenko et al. 2009. "Field Studies of a Geo-Engineering Method of Maintaining a Modern Climate with Aerosol Particles." *Russian Meteorology and Hydrology* 34 no. 10:635–8, DOI:10.3103/s106837390910001x.

Jaeger, Jill. 1988. *Developing Policies for Responding to Climatic Change: A Summary of the Discussions and Recommendations of the Workshops Held in Villach (28 September-2*

October 1987) and Bellagio (9–13 November 1987) under the Auspices of the Beijer Institute, Stockholm. World Meteorological Organization, United Nations Environment Programme, https://library.wmo.int/pmb_ged/wmo-td_225_en.pdf.

Jamieson, Dale. 1996. "Ethics and Intentional Climate Change." *Climatic Change* 33 no. 3:323–36, DOI:10.1007/bf00142580.

Jinnah, Sikina. 2018. "Why Govern Climate Engineering? A Preliminary Framework for Demand-Based Governance." *International Studies Review* 20 no. 2:272–82, DOI:10.1093/isr/viy022.

Johnson, Eric E. 2008. "The Black Hole Case: The Injunction against the End of the World." *Tennessee Law Review* 76 no. 4:819–908, https://heinonline.org/HOL/P?h=hein.journals/tenn76&i=829.

Jones, Gregory N. 1991. "Weather Modification: The Continuing Search for Rights and Liabilities." *Brigham Young University Law Review* no. 2:1163–99, https://digitalcommons.law.byu.edu/lawreview/vol1991/iss2/9/.

Kahan, Dan M., Hank Jenkins-Smith, Tor Tarantola, Carol L. Silva, and Donald Braman. 2015. "Geoengineering and Climate Change Polarization: Testing a Two-Channel Model of Science Communication." *Annals of American Academy of Political and Social Science* 658 no. 1:192–222, DOI:10.1177/0002716214559002.

Kane, Sally, and Jason F. Shogren. 2000. "Linking Adaptation and Mitigation in Climate Change Policy." *Climatic Change* 45 no. 1:75–102, DOI:10.1023/a:1005688900676.

Kates, Robert W. 1997. "Climate Change 1995: Impacts, Adaptations, and Mitigation." *Environment: Science and Policy for Sustainable Development* 39 no. 9:29–33, DOI:10.1080/00139159709604767.

Keith, David. 2013. *A Case for Climate Engineering*. Cambridge, MA: Boston Review.

Keith, David, and John Dykema. 2018. "Why We Chose Not to Patent Solar Geoengineering Technologies." https://keith.seas.harvard.edu/blog/why-we-chose-not-patent-solar-geoengineering-technologies.

Keith, David W. 2000. "Geoengineering the Climate: History and Prospect." *Annual Review of Energy and the Environment* 25 no. 1:245–84, DOI:10.1146/annurev.energy.25.1.245.

—— 2010. "Photophoretic Levitation of Engineered Aerosols for Geoengineering." *Proceedings of the National Academy of Sciences* 107 no. 38:16428–31, DOI:10.1073/pnas.1009519107.

Keith, David W., and Hadi Dowlatabadi. 1992. "A Serious Look at Geoengineering." *EOS* 73 no. 27:289, 92–93, DOI:abs/10.1029/91EO00231.

Keith, David W., and Peter J. Irvine. 2016. "Solar Geoengineering Could Substantially Reduce Climate Risks: A Research Hypothesis for the Next Decade." *Earth's Future* 4 no. 11:549–59, DOI:10.1002/2016EF000465.

Keith, David W., Gernot Wagner, and Claire L. Zabel. 2017. "Solar Geoengineering Reduces Atmospheric Carbon Burden." *Nature Climate Change* 7 no. 9:617–19, DOI:10.1038/nclimate3376.

Keith, David W., Debra K. Weisenstein, John A. Dykema, and Frank N. Keutsch. 2016. "Stratospheric Solar Geoengineering without Ozone Loss." *Proceedings of the National Academy of Sciences* 113 no. 52:14910–14, DOI:10.1073/pnas.1615572113.

Kellogg, William W., and Stephen H. Schneider. 1974. "Climate Stabilization: For Better or for Worse?" *Science* 186 no. 4170:1163–72, DOI:10.1126/science.186.4170.1163.

Kempner, Joanna, Jon F. Merz, and Charles L. Bosk. 2011. "Forbidden Knowledge: Public Controversy and the Production of Nonknowledge." *Sociological Forum* 26 no. 3:475–500, DOI:10.1111/j.1573-7861.2011.01259.x.

Keohane, Robert O. 1984. *After Hegemony: Cooperation and Discord in the World Political Economy*. Princeton: Princeton University Press.
—2015. "The Global Politics of Climate Change: Challenge for Political Science." *PS: Political Science & Politics* 48 no. 1:19–26, DOI:10.1017/S1049096514001541.
Keohane, Robert O., and Joseph S. Jr. Nye. 2011. *Power and Interdependence*. 4th ed. Boston, MA: Longman.
Klepper, Gernot, and Wilfried Rickels. 2014. "Climate Engineering: Economic Considerations and Research Challenges." *Review of Environmental Economics and Policy* 8 no. 2:270–89, DOI:10.1093/reep/reu010.
Klimont, Z., S. J. Smith, and J. Cofala. 2013. "The Last Decade of Global Anthropogenic Sulfur Dioxide: 2000–2011 Emissions." *Environmental Research Letters* 8 no. 1:014003, DOI:10.1088/1748-9326/8/1/014003.
Kössler, Georg P. 2012. *Geo-Engineering: Gibt Es Wirklich Einen Plan(Eten) B?* Berlin: Heinrich Böll Foundation, www.boell.de/de/content/geo-engineering-gibt-es-wirklich-einen-planeten-b.
Kravitz, Ben, Douglas G. MacMartin, Alan Robock, Philip J. Rasch, Katharine L. Ricke, Jason N. S. Cole et al. 2014. "A Multi-Model Assessment of Regional Climate Disparities Caused by Solar Geoengineering." *Environmental Research Letters* 9 no. 7:074013, DOI:10.1088/1748-9326/9/7/074013.
Krieger, Martin H. 1973. "What's Wrong with Plastic Trees? Rationales for Preserving Rare Natural Environments Involve Economic, Societal, and Political Factors." *Science* 179 no. 4072:446–55, DOI:10.1126/science.179.4072.446.
Kruger, Tim. 2018. "A Commentary on the Oxford Principles." In *Geoengineering Our Climate? Ethics, Politics and Governance*, edited by Jason J. Blackstock and Sean Low, 190–8. London: Earthscan.
Lasswell, Harold D. 1936. *Politics: Who Gets What, When, How*. New York: McGraw-Hill.
Latham, John, Alan Gadian, Jim Fournier, Ben Parkes, Peter Wadhams, and Jack Chen. 2014. "Marine Cloud Brightening: Regional Applications." *Philosophical Transactions of the Royal Society A: Mathematical, Physical and Engineering Sciences* 372 no. 2031: article 20140053, DOI:10.1098/rsta.2014.0053.
Lederer, Markus, and Judith Kreuter. 2018. "Organising the Unthinkable in Times of Crises: Will Climate Engineering Become the Weapon of Last Resort in the Anthropocene?" *Organization* 25 no. 4:472–90, DOI:10.1177/1350508418759186.
Lefale, Penehuro F., and Cheryl Lea Anderson. 2018. "Climate Engineering and Small Island States: Panacea or Catastrophe?" In *Geoengineering Our Climate? Ethics, Politics and Governance*, edited by Jason J. Blackstock and Sean Low, 159–63. London: Earthscan.
Lefeber, René. 2012. "Climate Change and State Responsibility." In *International Law in the Era of Climate Change*, edited by Rosemary Rayfuse and Shirley V. Scott, 321–49. Cheltenham, UK: Edward Elgar, DOI:10.4337/9781781006085.
Lempert, Robert J., and Don Prosnitz. 2011. *Governing Geoengineering Research: A Political and Technical Vulnerability Analysis of Potential Near-Term Options*. Santa Monica, CA: RAND Corporation, www.rand.org/pubs/technical_reports/TR846.html.
Lessig, Lawrence. 2009. *Code: And Other Laws of Cyberspace*. 2nd ed. New York: Basic Books.
Liebowitz, Stan J., and Stephen E. Margolis. 2013. "The Troubled Path of the Lock-In Movement." *Journal of Competition Law and Economics* 9 no. 1:125–52, DOI:10.1093/joclec/nhs034.

Lin, Albert. 2013. "Does Geoengineering Present a Moral Hazard?" *Ecology Law Quarterly* 40 no. 3:673–712, DOI:10.15779/Z38JP1J.

—2016. "The Missing Pieces of Geoengineering Research Governance." *Minnesota Law Review* 100 no. 6:2509–76, www.minnesotalawreview.org/wp-content/uploads/2016/08/Lin_Online.pdf.

—2018. "US Law." In *Climate Engineering and the Law*, edited by Michael B. Gerrard and Tracy Hester, 154–223. Cambridge: Cambridge University Press, DOI:10.1017/9781316661864.004.

Lin, Jolene. 2014. "Geoengineering: An ASEAN Position." In *Adaptation to Climate Change: ASEAN and Comparative Experiences*, edited by Kheng-Lian Koh, Ilan Kelman, Robert Kibugi, and Rose-Liza Eisma Osorio, 191–212. Singapore: World Scientific, DOI:10.1142/9789814689748_0008.

Lind, Michael. 2013. "Democracy, Hacked." *New York Times*, February 10, 2013, BR11, www.nytimes.com/2013/02/10/books/review/the-future-six-drivers-of-global-change-by-al-gore.html.

Liu, Zhe, and Ying Chen. 2013. "Geoengineering: Ethical Considerations and Global Governance." *Chinese Journal of Urban and Environmental Studies* 01 no. 01: article 1350006, DOI:10.1142/S2345748113500061.

Lloyd, Ian D., and Michael Oppenheimer. 2014. "On the Design of an International Governance Framework for Geoengineering." *Global Environmental Politics* 14 no. 2:45–63, DOI:10.1162/GLEP_a_00228.

Lockley, Andrew, and D'Maris Coffman. 2016. "Distinguishing Morale Hazard from Moral Hazard in Geoengineering." *Environmental Law Review* 18 no. 3:194–204, DOI:10.1177/1461452916659830.

Lohmann, Ulrike, and Blaž Gasparini. 2017. "A Cirrus Cloud Climate Dial?" *Science* 357 no. 6348:248–9, DOI:10.1126/science.aan3325.

Long, Jane C. S., and Dane Scott. 2013. "Vested Interests and Geoengineering Research." *Issues in Science and Technology* 29 no. 3:45–52, https://issues.org/long-4/.

Long, Jane C. S., and John G. Shepherd. 2014. "The Strategic Value of Geoengineering Research." In *Global Environmental Change*, edited by Bill Freedman, 757–70. Dordrecht, The Netherlands: Springer, DOI:10.1007/978-94-007-5784-4_24.

Lord, Richard, Silke Goldberg, Lavanya Rajamani, and Jutta Brunnée, eds. 2011. *Climate Change Liability: Transnational Law and Practice*. Cambridge: Cambridge University Press.

Maas, Achim, and Irina Comardicea. 2013. "Climate Gambit: Engineering Climate Security Risks?" In *Backdraft: The Conflict Potential of Climate Change Adaptation and Mitigation*, edited by Geoffrey D. Dabelko, Lauren Herzer, Schuyler Null, Meaghan Parker, and Russell Sticklor, 37–48. Washington, DC: Woodrow Wilson International Center for Scholars.

Maas, Achim, and Jürgen Scheffran. 2012. "Climate Conflicts 2.0? Climate Engineering as Challenge for International Peace and Security." *Sicherheit und Frieden* 30 no. 4:193–200.

MacCracken, Michael. 2006. "Geoengineering: Worthy of Cautious Evaluation?" *Climatic Change* 77 no. 3:235–43, DOI:10.1007/s10584-006-9130-6.

MacMartin, Douglas G., and Ben Kravitz. 2019. "The Engineering of Climate Engineering." *Annual Review of Control, Robotics, and Autonomous Systems* 2: online ahead of print: 1–23, DOI:10.1146/annurev-control-053018-023725.

MacMartin, Douglas G., Katharine L. Ricke, and David W. Keith. 2018. "Solar Geoengineering as Part of an Overall Strategy for Meeting the 1.5°C Paris Target."

Philosophical Transactions of the Royal Society A: Mathematical, Physical and Engineering Sciences 376 no. 2119: article 20160454, DOI:10.1098/rsta.2016.0454.

Mahoney, James. 2000. "Path Dependence in Historical Sociology." *Theory and Society* 29 no. 4:507–48, DOI:10.1023/a:1007113830879.

Maler, Karl-Göran. 1989. "The Acid Rain Game." In *Studies in Environmental Science*, edited by H. Folmer and E. van Ierland, 231–52. Amsterdam: Elsevier, DOI:10.1016/S0166-1116(08)70035-9.

Manoussi, Vassiliki, and Anastasios Xepapadeas. 2015. "Cooperation and Competition in Climate Change Policies: Mitigation and Climate Engineering When Countries Are Asymmetric." *Environmental and Resource Economics* 66 no. 4:605–27, DOI:10.1007/s10640-015-9956-3.

Marine Cloud Brightening for the Great Barrier Reef. n.d., accessed January 22, 2019. www.savingthegreatbarrierreef.org/.

Marine Cloud Brightening for the Great Barrier Reef. n.d. "Intellectual Property." accessed January 22, 2019. www.savingthegreatbarrierreef.org/intellectual-property/.

Markusson, Nils, Franklin Ginn, Navraj Singh Ghaleigh, and Vivian Scott. 2014. "'In Case of Emergency Press Here': Framing Geoengineering as a Response to Dangerous Climate Change." *Wiley Interdisciplinary Reviews: Climate Change* 5 no. 2:281–90, DOI:10.1002/wcc.263.

Marshall, Michael. 2012. "Independent Geoengineer's Ocean Field Test Condemned." *New Scientist*, October 17, 2012, www.newscientist.com/article/dn22390-independent-geoengineers-ocean-field-test-condemned/.

McClellan, Justin, David W. Keith, and Jay Apt. 2012. "Cost Analysis of Stratospheric Albedo Modification Delivery Systems." *Environmental Research Letters* 7 no. 3:034019, DOI:10.1088/1748-9326/7/3/034019.

McCormack, Caitlin G., Wanda Born, Peter J. Irvine, Eric P. Achterberg, Tatsuya Amano, Jeff Ardron et al. 2016. "Key Impacts of Climate Engineering on Biodiversity and Ecosystems, with Priorities for Future Research." *Journal of Integrative Environmental Sciences* 13 no. 2–4:103–28, DOI:10.1080/1943815X.2016.1159578.

McDonald, Matt. 2018. "Climate Change and Security: Towards Ecological Security?" *International Theory* 10 no. 2:153–80, DOI:10.1017/S1752971918000039.

McKenzie, Richard B., and Gordon Tullock. 1981. *The New World of Economics: Explorations into the Human Experience*. Homewood, IL: R. D. Irwin.

McKinnon, Catriona. 2019. "Sleepwalking into Lock-In? Avoiding Wrongs to Future People in the Governance of Solar Radiation Management Research." *Environmental Politics* 28 no. 3: 441–59, DOI:10.1080/09644016.2018.1450344.

McLaren, Duncan. 2016. "Mitigation Deterrence and the 'Moral Hazard.'" *Earth's Future* 4 no. 12:596–602, DOI:10.1002/2016EF000445.

McNutt, Marcia. 2016. "Climate Intervention: Possible Impacts on Global Security and Resilience." *Engineering* 2 no. 1:50–1, DOI:10.1016/j.eng.2016.01.015.

Mercer, A. M., David W. Keith, and J. D. Sharp. 2011. "Public Understanding of Solar Radiation Management." *Environmental Research Letters* 6 no. 4: article 044006, DOI:10.1088/1748-9326/6/4/044006.

Merges, Robert P., and Richard R. Nelson. 1990. "On the Complex Economics of Patent Scope." *Columbia Law Review* 90 no. 4:839–916, DOI:10.2307/1122920.

Merk, Christine, and Gert Pönitzsch. 2017. "The Role of Affect in Attitude Formation toward New Technologies: The Case of Stratospheric Aerosol Injection." *Risk Analysis* 37 no. 12:2289–304, DOI:10.1111/risa.12780.

Merk, Christine, Gert Pönitzsch, Carola Kniebes, Katrin Rehdanz, and Ulrich Schmidt. 2015. "Exploring Public Perceptions of Stratospheric Sulfate Injection." *Climatic Change* 130 no. 2:299–312, DOI:10.1007/s10584-014-1317-7.

Merk, Christine, Gert Pönitzsch, and Katrin Rehdanz. 2016. "Knowledge about Aerosol Injection Does Not Reduce Individual Mitigation Efforts." *Environmental Research Letters* 11 no. 5: article 054009, DOI:10.1088/1748-9326/11/5/054009.

Merk, Christine, Gert Pönitzsch, and Katrin Rehdanz. 2019. "Do Climate Engineering Experts Display Moral-Hazard Behaviour?" *Climate Policy* 19 no. 2: 231–43, DOI:10.1080/14693062.2018.1494534.

Michaelson, Jay. 1998. "Geoengineering: A Climate Change Manhattan Project." *Stanford Environmental Law Journal* 17 no. 1:73–140, https://heinonline.org/HOL/P?h=hein.journals/staevj17&i=91.

Millard-Ball, Adam. 2012. "The Tuvalu Syndrome: Can Geoengineering Solve Climate's Collective Action Problem?" *Climatic Change* 110 no. 3–4:1047–66, DOI:10.1007/s10584-011-0102-0.

Moreno-Cruz, Juan B. 2015. "Mitigation and the Geoengineering Threat." *Resource and Energy Economics* 41:248–63, DOI:10.1016/j.reseneeco.2015.06.001.

Moreno-Cruz, Juan B., and David W. Keith. 2013. "Climate Policy under Uncertainty: A Case for Solar Geoengineering." *Climatic Change* 121 no. 3:431–44, DOI:10.1007/s10584-012-0487-4.

Morgan, M. Granger, Robert R. Nordhaus, and Paul Gottlieb. 2013. "Needed: Research Guidelines for Solar Radiation Management." *Issues in Science and Technology* 29 no. 3:37–44, https://issues.org/morgan-3/.

Moriyama, Ryo, Masahiro Sugiyama, Atsushi Kurosawa, Kooiti Masuda, Kazuhiro Tsuzuki, and Yuki Ishimoto. 2016. "The Cost of Stratospheric Climate Engineering Revisited." *Mitigation and Adaptation Strategies for Global Change* 22 no. 8:1207–28, DOI:10.1007/s11027-016-9723-y.

Morrow, David R. 2014a. "Why Geoengineering Is a Public Good, Even If It Is Bad." *Climatic Change* 123 no. 2:95–100, DOI:10.1007/s10584-013-0967-1.

—2014b. "Ethical Aspects of the Mitigation Obstruction Argument against Climate Engineering Research." *Philosophical Transactions of the Royal Society A: Mathematical, Physical and Engineering Sciences* 372 no. 2031: article 20140062, DOI:10.1098/rsta.2014.0062.

—2014c. "Starting a Flood to Stop a Fire? Some Moral Constraints on Solar Radiation Management." *Ethics, Policy and Environment* 17 no. 2:123–38, DOI:10.1080/21550085.2014.926056.

Morrow, David R., Robert E. Kopp, and Michael Oppenheimer. 2009. "Toward Ethical Norms and Institutions for Climate Engineering Research." *Environmental Research Letters* 4 no. 4: article 045106, DOI:10.1088/1748-9326/4/4/045106.

—2013. "Political Legitimacy in Decisions about Experiments in Solar Radiation Management." In *Climate Change Geoengineering*, edited by William C. G. Burns and Andrew L. Strauss, 146–67. New York: Cambridge University Press, DOI:10.1017/CBO9781139161824.009.

Morton, Oliver. 2015. *The Planet Remade: How Geoengineering Could Change the World.* Princeton: Princeton University Press.

—2018. "Nitrogen Geoengineering." In *Geoengineering Our Climate? Ethics, Politics, and Governance*, edited by Jason J. Blackstock and Sean Low, 34–7. London: Earthscan.

Mueller, Dennis C. 2003. *Public Choice III.* Cambridge: Cambridge University Press.

Nash, Jonathan Remy. 2009. "The Curious Legal Landscape of the Extraterritoriality of US Environmental Laws." *Virginia Journal of International Law* 50 no. 4:997–1020, https://heinonline.org/HOL/P?h=hein.journals/vajint50&i=1007.

National Research Council, Committee on Geoengineering Climate. 2015a. *Climate Intervention: Carbon Dioxide Removal and Reliable Sequestration.* Washington, DC: National Academies Press, DOI:10.17226/18805.

—2015b. *Climate Intervention: Reflecting Sunlight to Cool Earth.* Washington, DC: National Academies Press, DOI:10.17226/18988.

Necheles, Ella, Lizzie Burns, and David Keith. 2018. "Funding for Solar Geoengineering from 2008 to 2018." https://geoengineering.environment.harvard.edu/blog/funding-solar-geoengineering.

Nicholson, Simon, Sikina Jinnah, and Alexander Gillespie. 2018. "Solar Radiation Management: A Proposal for Immediate Polycentric Governance." *Climate Policy* 18 no. 3:322–34, DOI:10.1080/14693062.2017.1400944.

Nordhaus, William D. 2008. *A Question of Balance: Weighing the Options on Global Warming Policies.* New Haven, CT: Yale University Press.

Nye, Joseph S. 2018a. "Normative Restraints on Cyber Conflict." *Cyber Security: A Peer-Reviewed Journal* 1 no. 4:331–42, www.ingentaconnect.com/content/hsp/jcs/2018/00000001/00000004/art00006.

—2018b. "Notes on Insights from Other Regimes: Cyber." In *Governance of the Deployment of Solar Geoengineering*, 55–9. Harvard Project on Climate Agreements. www.belfercenter.org/index.php/publication/governance-deployment-solar-geoengineering.

Oldham, P., B. Szerszynski, J. Stilgoe, C. Brown, B. Eacott, and A. Yuille. 2014. "Mapping the Landscape of Climate Engineering." *Philosophical Transactions of the Royal Society* A: *Mathematical, Physical and Engineering Sciences* 372 no. 2031: article 20140065, DOI:10.1098/rsta.2014.0065.

Olson, Robert L. 2011. *Geoengineering for Decision Makers.* Washington, DC: Woodrow Wilson International Center for Scholars, www.wilsoncenter.org/publication/geoengineering-for-decision-makers.

Owen, Richard. 2014. "Solar Radiation Management and the Governance of Hubris." In *Geoengineering of the Climate System*, edited by R. E. Hester and R. M. Harrison, 212–48. Cambridge: Royal Society of Chemistry, DOI:10.1039/9781782621225-00212.

Parker, A., J. B. Horton, and D. W. Keith. 2018. "Stopping Solar Geoengineering through Technical Means: A Preliminary Assessment of Counter-Geoengineering." *Earth's Future* 6 no. 8:1058–65, DOI:10.1029/2018EF000864.

Parker, Andy. 2014. "Governing Solar Geoengineering Research as It Leaves the Laboratory." *Philosophical Transactions of the Royal Society* A: *Mathematical, Physical and Engineering Sciences* 372 no. 2031: article 20140173, DOI:10.1098/rsta.2014.0173.

Parker, Andy, and Peter J. Irvine. 2018. "The Risk of Termination Shock from Solar Geoengineering." *Earth's Future* 6 no. 3:456–67, DOI:10.1002/2017EF000735.

Parkes, B., A. Challinor, and K. Nicklin. 2015. "Crop Failure Rates in a Geoengineered Climate: Impact of Climate Change and Marine Cloud Brightening." *Environmental Research Letters* 10 no. 8: article 084003, DOI:10.1088/1748-9326/10/8/084003.

Parson, Edward A. 2003. *Protecting the Ozone Layer: Science and Strategy.* Oxford: Oxford University Press.

—2014. "Climate Engineering in Global Climate Governance: Implications for Participation and Linkage." *Transnational Environmental Law* 3 no. 1:89–110, DOI:10.1017/S2047102513000496.

—2015. "Expertise and Evidence in Public Policy: In Defence of (a Little) Technocracy." In *A Subtle Balance: Expertise, Evidence, and Democracy in Public Policy and Governance,*

1970–2010, edited by Edward A. Parson, 42–50. Montreal: McGill-Queen's University Press.
—2017a. "Starting the Dialogue on Climate Engineering Governance: A World Commission." *Fixing Climate Governance Series Policy Brief* 8, Centre for International Governance Innovation, www.cigionline.org/publications/starting-dialo gue-climate-engineering-governance-world-commission.
—2017b. "Climate Policymakers and Assessments Must Get Serious about Climate Engineering." *Proceedings of the National Academy of Sciences* 114 no. 35:9227–30, DOI:10.1073/pnas.1713456114.
Parson, Edward A., and Lia N. Ernst. 2013. "International Governance of Climate Engineering." *Theoretical Inquiries in Law* 14 no. 1:307–38, DOI:10.1515/til-2013-015.
Parson, Edward A., and Megan M. Herzog. 2016. "Moratoria for Global Governance and Contested Technology: The Case of Climate Engineering." *UCLA Public Law & Legal Theory Series*, University of California, Los Angeles School of Law, https://escholarship.org/uc/item/2c28w2tn.
Parson, Edward A., and David W. Keith. 2013. "End the Deadlock on Governance of Geoengineering Research." *Science* 339 no. 6125:1278–9, DOI:10.1126/science.1232527.
Parthasarathy, Shobita, Christopher Avery, Nathan Hedberg, Jessie Mannisto, and Molly Maguire. 2010. "A Public Good? Geoengineering and Intellectual Property." 10–11, University of Michigan Science, Technology, and Public Policy Program, http://jrey nolds.org/wp-content/uploads/2019/03/Parthasarathy-2010-A-Public-Good.pdf.
Pasztor, Janos, Cynthia Scharf, and Kai-Uwe Schmidt. 2017. "How to Govern Geoengineering?" *Science* 357 no. 6348:231, DOI:10.1126/science.aan6794.
Peltzman, Sam. 1975. "The Effects of Automobile Safety Regulation." *Journal of Political Economy* 83 no. 4:677–725, DOI:10.2307/1830396.
Perry, Ronen. 2005. "The Role of Retributive Justice in the Common Law of Torts: A Descriptive Theory." *Tennessee Law Review* 73 no. 2:177–236, https://heinonline.org/HOL/P?h=hein.journals/tenn73&i=185.
Pew Research Center. n.d. "Political Issue Priorities." accessed March 13, 2019. www.people-press.org/topics/political-issue-priorities/.
Pfrommer, Tobias. 2018. "A Model of Solar Radiation Management Liability." *Department of Economics Discussion Paper*, University of Heidelberg, DOI:10.11588/heidok.00023978.
Pfrommer, Tobias, Timo Goeschl, Alexander Proelss, Martin Carrier, Johannes Lenhard, Henrike Martin, Ulrike Niemeier, and Hauke Schmidt. 2019. "Establishing Causation in Climate Litigation: Admissibility and Reliability." *Climatic Change* 152 no. 1: 67–84, DOI:10.1007/s10584-018-2362-4.
Pidgeon, Nick, Karen Parkhill, Adam Corner, and Naomi Vaughan. 2013. "Deliberating Stratospheric Aerosols for Climate Geoengineering and the SPICE Project." *Nature Climate Change* 3 no. 5:451–7, DOI:10.1038/nclimate1807.
Pielke, Roger A., Jr. 1998. "Rethinking the Role of Adaptation in Climate Policy." *Global Environmental Change* 8 no. 2:159–70, DOI:10.1016/s0959-3780(98)00011-9.
—2010. *The Climate Fix: What Scientists and Politicians Won't Tell You about Global Warming*. New York: Basic Books.
Pielke, Roger, Jr., Gwyn Prins, Steve Rayner, and Daniel Sarewitz. 2007. "Lifting the Taboo on Adaptation." *Nature* 445 no. 7128:597–8, DOI:10.1038/445597a.
Porat, Ariel. 2009. "Private Production of Public Goods: Liability for Unrequested Benefits." *Michigan Law Review* 108 no. 2:189–227, https://heinonline.org/HOL/P?h=hein.jour nals/mlr108&i=193.

Posner, Eric A., and Alan O. Sykes. 2013. *Economic Foundations of International Law*. Cambridge, MA: Belknap.

Preston Christopher J., ed. 2017. *Climate Justice and Geoengineering: Ethics and Policy in the Atmospheric Anthropocene*. London: Rowman & Littlefield.

Proctor, Jonathan, Solomon Hsiang, Jennifer Burney, Marshall Burke, and Wolfram Schlenker. 2018. "Estimating Global Agricultural Effects of Geoengineering Using Volcanic Eruptions." *Nature* 560 no. 7719:480–3, DOI:10.1038/s41586-018-0417-3.

PwC. 2018. "The Low Carbon Economy Index 2018." www.pwc.co.uk/services/sustainability-climate-change/insights/low-carbon-economy-index.html.

Qu, Jingwen, and Emilson Caputo Delfino Silva. 2015. "Strategic Effects of Future Environmental Policy Commitments: Climate Change, Solar Radiation Management and Correlated Air Pollutants." *Journal of Environmental Management* 151:22–32, DOI:10.1016/j.jenvman.2014.11.033.

Quaas, Johannes, Martin F. Quaas, Olivier Boucher, and Wilfried Rickels. 2016. "Regional Climate Engineering by Radiation Management: Prerequisites and Prospects." *Earth's Future* 4 no. 12:618–25, DOI:10.1002/2016EF000440.

Quaas, Martin F., Johannes Quaas, Wilfried Rickels, and Olivier Boucher. 2017. "Are There Reasons against Open-Ended Research into Solar Radiation Management? A Model of Intergenerational Decision-Making under Uncertainty." *Journal of Environmental Economics and Management* 84:1–17, DOI:10.1016/j.jeem.2017.02.002.

Rabitz, Florian. 2016. "Going Rogue? Scenarios for Unilateral Geoengineering." *Futures* 84 no. A:98–107, DOI:10.1016/j.futures.2016.11.001.

—2019. "Governing the Termination Problem in Solar Radiation Management." *Environmental Politics* 28 no. 3: 502–22, DOI:10.1080/09644016.2018.1519879.

Rahman, A. Atiq, Paulo Artaxo, Asfawossen Asrat, and Andy Parker. 2018. "Developing Countries Must Lead on Solar Geoengineering Research." *Nature* 556 no. 7699:22–4, DOI:10.1038/d41586-018-03917-8.

Rayner, Steve. 1991. "The Greenhouse Effect in the US: The Legacy of Energy Abundance." In *Energy Policies and the Greenhouse Effect*, edited by Michael Grubb, 233–78. Dartmouth: Royal Institute of International Affairs.

—2015. "To Know or Not to Know? A Note on Ignorance as a Rhetorical Resource in Geoengineering Debates." In *Routledge International Handbook of Ignorance Studies*, edited by Matthias Gross and Linsey McGoey, 308–17. London: Routledge.

Rayner, Steve, Clare Heyward, Tim Kruger, Nick Pidgeon, Catherine Redgwell, and Julian Savulescu. 2013. "The Oxford Principles." *Climatic Change* 121 no. 3:499–512, DOI:10.1007/s10584-012-0675-2.

Rayner, Steve, Catherine Redgwell, Julian Savulescu, Nick Pidgeon, and Tim Kruger. 2010. "Memorandum Submitted by Tim Kruger et al (Geo 07a), the Regulation of Geoengineering, Science and Technology Committee, House of Commons (UK)." https://publications.parliament.uk/pa/cm200910/cmselect/cmsctech/221/10011316.htm.

Rees, Jospeh. 1994. *Hostages of Each Other: The Transformation of Nuclear Safety since Three Mile Island*. Chicago: University of Chicago Press.

Reichwein, David, Anna-Maria Hubert, Peter J. Irvine, Francois Benduhn, and Mark G. Lawrence. 2015. "State Responsibility for Environmental Harm from Climate Engineering." *Climate Law* 5 no. 2–4:142–81, DOI:10.1163/18786561-00504003.

Revelle, Roger, Wallace Broecker, Harmon Craig, C. D. Keeling, and J. Smagorinsky. 1965. "Atmospheric Carbon Dioxide." In *Restoring the Quality of Our Environment*, President's Science Advisory Committee, 111–33. Washington, DC: US Government Printing Office.

Reynolds, Jesse L. 2011. "The Regulation of Climate Engineering." *Law, Innovation and Technology* 3 no. 1:113–36, DOI:10.5235/175799611796399821.
—2014a. "The International Regulation of Climate Engineering: Lessons from Nuclear Power." *Journal of Environmental Law* 26 no. 2:269–89, DOI:10.1093/jel/equ006.
—2014b. "Response to Svoboda and Irvine." *Ethics, Policy & Environment* 17 no. 2:183–5, DOI:10.1080/21550085.2014.926080.
—2015a. "A Critical Examination of the Climate Engineering Moral Hazard and Risk Compensation Concern." *The Anthropocene Review* 2 no. 2:174–91, DOI:10.1177/2053019614554304.
—2015b. "An Economic Analysis of Liability and Compensation for Harm from Large-Scale Field Research in Solar Climate Engineering." *Climate Law* 5 no. 2–4:182–209.
—2017. "Solar Climate Engineering, Law, and Regulation." In *The Oxford Handbook of Law, Regulation, and Technology*, 799–822. Oxford: Oxford University Press, DOI:10.1093/oxfordhb/9780199680832.013.71.
—2018a. "Why the UNFCC and CBD Should Refrain from Regulating Solar Climate Engineering." In *Geoengineering Our Climate? Ethics, Politics and Governance*, edited by Jason J. Blackstock and Sean Low, 137–41. London: Earthscan.
—2018b. "Governing Experimental Responses: Negative Emissions Technologies and Solar Climate Engineering." In *Governing Climate Change: Polycentricity in Action?*, edited by Andrew Jordan, Dave Huitema, Harro Van Asselt, and Johanna Forster, 285–302. Cambridge: Cambridge University Press, DOI:10.1017/9781108284646.017.
—2018c. "International Law." In *Climate Engineering and the Law: Regulation and Liability for Solar Radiation Management and Carbon Dioxide Removal*, edited by Michael B. Gerrard and Tracy Hester, 57–153. Cambridge: Cambridge University Press, DOI:10.1017/9781316661864.003.
Reynolds, Jesse L., Jorge L. Contreras, and Joshua D. Sarnoff. 2017. "Solar Climate Engineering and Intellectual Property: Toward a Research Commons." *Minnesota Journal of Law, Science & Technology* 18 no. 1:1–110, https://scholarship.law.umn.edu/mjlst/vol18/iss1/1/.
—2018. "Intellectual Property Policies for Solar Geoengineering." *Wiley Interdisciplinary Reviews: Climate Change* 9 no. 2: article e512, DOI:10.1002/wcc.512.
Reynolds, Jesse L., and Floor Fleurke. 2013. "Climate Engineering Research: A Precautionary Response to Climate Change?" *Carbon & Climate Law Review* 7 no. 2:101–7, DOI:10.21552/CCLR/2013/2/251.
Reynolds, Jesse L., Andy Parker, and Peter Irvine. 2016. "Five Solar Geoengineering Tropes That Have Outstayed Their Welcome." *Earth's Future* 4 no. 12:562–8, DOI:10.1002/2016EF000416.
Reynolds, Jesse L., and Gernot Wagner. 2018. "Governance of Highly Decentralized Nonstate Actors: The Case of Solar Geoengineering." *Belfer Center for Science and International Affairs Discussion Paper*, Harvard Kennedy School, Harvard, www.belfercenter.org/publication/governance-highly-decentralized-nonstate-actors-case-solar-geoengineering.
Ricke, Katharine L., Juan B. Moreno-Cruz, and Ken Caldeira. 2013. "Strategic Incentives for Climate Geoengineering Coalitions to Exclude Broad Participation." *Environmental Research Letters* 8 no. 1:article 014021, DOI:10.1088/1748-9326/8/1/014021.
Robertson, John A. 1977. "The Scientist's Rights to Research: A Constitutional Analysis." *Southern California Law Review* 51 no. 6:1203–79, https://scholarship.law.umn.edu/mjlst/vol18/iss1/1/.
Robock, Alan. 2008. "20 Reasons Why Geoengineering May Be a Bad Idea." *Bulletin of the Atomic Scientists* 64 no. 2:14–18, DOI:10.2968/064002006.

Robock, Alan, Allison Marquardt, Ben Kravitz, and Georgiy Stenchikov. 2009. "Benefits, Risks, and Costs of Stratospheric Geoengineering." *Geophysical Research Letters* 36 no. 19: article L19703, DOI:10.1029/2009gl039209.

Rockström, Johan, Will Steffen, Kevin Noone, Åsa Persson, F. Stuart Chapin, Eric F. Lambin et al. 2009. "A Safe Operating Space for Humanity." *Nature* 461 no. 7263:472–5, DOI:10.1038/461472a.

Russell, Lynn M., Armin Sorooshian, John H. Seinfeld, Bruce A. Albrecht, Athanasios Nenes, Lars Ahlm et al. 2013. "Eastern Pacific Emitted Aerosol Cloud Experiment." *Bulletin of the American Meteorological Society* 94 no. 5:709–29, DOI:10.1175/BAMS-D-12-00015.1.

Samset, B. H., M. Sand, C. J. Smith, S. E. Bauer, P. M. Forster, J. S. Fuglestvedt et al. 2018. "Climate Impacts from a Removal of Anthropogenic Aerosol Emissions." *Geophysical Research Letters* 45 no. 2:1020–9, DOI:10.1002/2017GL076079.

Sandler, Todd. 1997. *Global Challenges: An Approach to Environmental, Political, and Economic Problems*. Cambridge: Cambridge University Press.

—2017. "Collective Action and Geoengineering." *The Review of International Organizations* 13 no. 1:105–25, DOI:10.1007/s11558-017-9282-3.

Sandman, Peter M. 1993. *Responding to Community Outrage: Strategies for Effective Risk Communication*. Farifax, VA: American Industrial Hygiene Association.

Sands, Philippe, and Jacqueline Peel. 2012. *Principles of International Environmental Law*. Cambridge: Cambridge University Press.

Saxler, Barbara, Jule Siegfried, and Alexander Proelss. 2015. "International Liability for Transboundary Damage Arising from Stratospheric Aerosol Injections." *Law, Innovation and Technology* 7 no. 1:112–47, DOI:10.1080/17579961.2015.1052645.

Schäfer, Stefan, Mark Lawrence, Harald Stelzer, Wanda Born, Sean Low, Asbjørn Aaheim et al. 2015. *The European Transdisciplinary Assessment of Climate Engineering (EUTRACE): Removing Greenhouse Gases from the Atmosphere and Reflecting Sunlight Away from Earth*. Potsdam, Germany: European Transdisciplinary Assessment of Climate Engineering, www.iass-potsdam.de/sites/default/files/files/eutrace_report_digital_second_edition_0.pdf.

Scheffran, Jürgen. 2013. "Energy, Climate Change and Conflict: Securitization of Migration, Mitigation and Geoengineering." In *International Handbook of Energy Security*, edited by Hugh Dyer and Maria Julia Trombetta, 319–44. Cheltenham, UK: Edward Elgar, DOI:10.4337/9781781007907.

Schelling, Thomas C. 1983. "Climatic Change: Implications for Welfare and Policy." In *Changing Climate: Report of the Carbon Dioxide Assessment Committee*, edited by William A. Nierenberg, Peter G. Brewer, Lester Machta, William D. Nordhaus, Roger R. Revelle, Thomas C. Schelling et al., 449–97. Washington, DC: National Academy Press, DOI:10.17226/18714.

—1996. "The Economic Diplomacy of Geoengineering." *Climatic Change* 33 no. 3:303–7, DOI:10.1007/bf00142578.

—2006. "An Astonishing Sixty Years: The Legacy of Hiroshima." *The American Economic Review* 96 no. 4:929–37, DOI:10.1257/aer.96.4.929.

Schellnhuber, Hans Joachim. 2011. "Geoengineering: The Good, the MAD, and the Sensible." *Proceedings of the National Academy of Sciences* 108 no. 51:20277–8, DOI:10.1073/pnas.1115966108.

Schipper, E. Lisa F. 2006. "Conceptual History of Adaptation in the UNFCCC Process." *Review of European Community and International Environmental Law* 15 no. 1:82–92, DOI:10.1111/j.1467-9388.2006.00501.x.

Schneider, Stephen H. 1996. "Geoengineering: Could – or Should – We Do It?" *Climatic Change* 33 no. 3:291–302, DOI:10.1007/bf00142577.

Schrijver, Nico. 2008. *Sovereignty over Natural Resources: Balancing Rights and Duties*. Cambridge: Cambridge University Press.

Schütte, Georg. 2014. "Speech by State Secretary Dr Georg Schütte, Federal Ministry of Education and Research, at the International Conference on 'Climate Engineering – Critical Global Discussions' of the Institute for Advanced Sustainability Studies Berlin, 18 August 2014." www.bmbf.de/pub/reden/Rede_StSchuette_IASS__Konferenz_18_08_ _engl.pdf.

Science Media Center. 2012. "Expert Reaction to Decision Not to Launch the 1 km Balloon as Part of the SPICE Geoengineering Research Project," Press release, May 16, 2012. www .sciencemediacentre.org/expert-reaction-to-decision-not-to-launch-the-1km-balloon-as-part-of-the-spice-geoengineering-research-project-2/.

Scotchmer, Suzanne. 1991. "Standing on the Shoulders of Giants: Cumulative Research and the Patent Law." *Journal of Economic Perspectives* 5 no. 1:29–41, DOI:10.1257/jep.5.1.29.

Scott, Karen N. 2013. "International Law in the Anthropocene: Responding to the Geoengineering Challenge." *Michigan Journal of International Law* 34 no. 2:309–58, https://repository.law.umich.edu/mjil/vol34/iss2/2/.

Seitz, Russell. 2011. "Bright Water: Hydrosols, Water Conservation and Climate Change." *Climatic Change* 105 no. 3:365–81, DOI:10.1007/s10584-010-9965-8.

"Selected International Legal Materials." 1990. *American University Journal of International Law and Policy* 5 no. 2:513–634, https://digitalcommons.wcl.american.edu/auilr/vol5/iss2/11/.

Seneviratne, Sonia I., Steven J. Phipps, Andrew J. Pitman, Annette L. Hirsch, Edouard L. Davin, Markus G. Donat et al. 2018. "Land Radiative Management as Contributor to Regional-Scale Climate Adaptation and Mitigation." *Nature Geoscience* 11 no. 2:88–96, DOI:10.1038/s41561-017-0057-5.

Shavell, Steven. 2007a. *Economic Analysis of Accident Law*. Cambridge, MA: Harvard University Press.

—2007b. "Liability for Accidents." In *Handbook of Law and Economics*, edited by A. Mitchell Polinsky and Steven Shavell, 139–82. Amsterdam: North-Holland.

Shepherd, John, Ken Caldeira, Joanna Haigh, David Keith, Brian Launder, Georgina Mace et al. 2009. *Geoengineering the Climate: Science, Governance and Uncertainty*. London: The Royal Society, https://royalsociety.org/topics-policy/publications/2009/geoengineer ing-climate/.

Sjöberg, Lennart. 2000. "Factors in Risk Perception." *Risk Analysis* 20 no. 1:1–12, DOI:10.1111/ 0272-4332.00001.

Smith, Wake, and Gernot Wagner. 2018. "Stratospheric Aerosol Injection Tactics and Costs in the First 15 Years of Deployment." *Environmental Research Letters* 13 no. 12: 124001, DOI:10.1088/1748-9326/aae98d.

Solar Radiation Management Governance Initiative. 2011. *Solar Radiation Management: The Governance of Research*. Solar Radiation Management Governance Initiative, www .srmgi.org/files/2016/02/SRMGI.pdf.

SPICE. n.d., accessed March 14, 2019. "What Is Geoengineering?" www.spice.ac.uk/about-us/ geoengineering/.

Stilgoe, Jack. 2015. *Experiment Earth: Responsible Innovation in Geoengineering*. London: Earthscan.

—2018. "Geoengineering." In *Companion to Environmental Studies*, edited by Noel Castree, Mike Hulme, and James D. Proctor, 679–83. London: Routledge.

Stilgoe, Jack, Matthew Watson, and Kirsty Kuo. 2013. "Public Engagement with Biotechnologies Offers Lessons for the Governance of Geoengineering Research and Beyond." *PLoS Biology* 11 no. 11: article e1001707, DOI:10.1371/journal.pbio.1001707.

Stirling, Andy. 2014. "Emancipating Transformations: From Controlling 'the Transition' to Culturing Plural Radical Progress." *Climate Geoengineering Governance Working Paper* 12, www.geoengineering-governance-research.org/perch/resources/workingpaper12stirlingemancipatingtransformations.pdf.

Stott, Peter A., Nikolaos Christidis, Friederike E. L. Otto, Ying Sun, Jean-Paul Vanderlinden, Geert Jan van Oldenborgh et al. 2016. "Attribution of Extreme Weather and Climate-Related Events." *Wiley Interdisciplinary Reviews: Climate Change* 7 no. 1:23–41, DOI:10.1002/wcc.380.

Sugiyama, Masahiro, and Taishi Sugiyama. 2010. *Interpretation of CBD COP10 Decision on Geoengineering, SERC Discussion Paper Serc10013*. Socio-Economic Research Center, Central Research Institute of Electric Power Industry, https://criepi.denken.or.jp/jp/serc/research_re/download/10013dp.pdf.

Sunstein, Cass R. 2005. *Laws of Fear: Beyond the Precautionary Principle*. Cambridge: Cambridge University Press.

Sütterlin, Bernadette, and Michael Siegrist. 2017. "Public Perception of Solar Radiation Management: The Impact of Information and Evoked Affect." *Journal of Risk Research* 20 no. 10:1292–307, DOI:10.1080/13669877.2016.1153501.

Svoboda, Toby. 2017. *The Ethics of Climate Engineering: Solar Radiation Management and Non-Ideal Justice*. London: Routledge.

Svoboda, Toby, Holly Jean Buck, and Pablo Suarez. 2019. "Forum: Climate Engineering and Human Rights." *Environmental Politics* 28 no. 3: 397–416, DOI:10.1080/09644016.2018.1448575.

Svoboda, Toby, and Peter J. Irvine. 2014. "Ethical and Technical Challenges in Compensating for Harm due to Solar Radiation Management Geoengineering." *Ethics, Policy and Environment* 17 no. 2:157–74, DOI:10.1080/21550085.2014.927962.

Szerszynski, Bronislaw, Matthew Kearnes, Phil Macnaghten, Richard Owen, and Jack Stilgoe. 2013. "Why Solar Radiation Management Geoengineering and Democracy Won't Mix." *Environment and Planning A* 45:2809–16, DOI:10.1068/a45649.

Talbot, Bernard. 1980. "Introduction to Recombinant DNA Research, Development and Evolution of the NIH Guidelines, and Proposed Legislation." *The University of Toledo Law Review* 12 no. 4:804–14, https://heinonline.org/HOL/P?h=hein.journals/utol12&i=822.

Teller, Edward, Roderick Hyde, and Lowell Wood. 1997. "Global Warming and Ice Ages: I. Prospects for Physics-Based Modulation of Global Change." UCRL-JC-128715, Lawrence Livermore National Laboratory, https://e-reports-ext.llnl.gov/pdf/231636.pdf.

Tilmes, S., B. M. Sanderson, and B. C. O'Neill. 2016. "Climate Impacts of Geoengineering in a Delayed Mitigation Scenario." *Geophysical Research Letters* 43 no. 15: 8222–9, DOI:10.1002/2016GL070122.

Tingley, Dustin, and Gernot Wagner. 2017. "Solar Geoengineering and the Chemtrails Conspiracy on Social Media." *Palgrave Communications* 3 no. 1: article 12, DOI:10.1057/s41599-017-0014-3.

Tol, Richard S. J. 2016. "Distributional Implications of Geoengineering." In *Climate Justice and Geoengineering: Ethics and Policy in the Atmospheric Anthropocene*, edited by Christopher J. Preston, 189–200. London: Rowman & Littlefield.

Trachtman, Joel P. 2008. *The Economic Structure of International Law*. Cambridge, MA: Harvard University Press.

Tribe, Laurence H. 1974. "Ways Not to Think about Plastic Trees: New Foundations for Environmental Law." *Yale Law Journal* 83 no. 7:1315–48, https://heinonline.org/HOL/P?h=hein.journals/ylr83&i=1327.
United Nations. 2018. "Secretary-General's Remarks on Climate Change [as Delivered]." www.un.org/sg/en/content/sg/statement/2018-09-10/secretary-generals-remarks-climate-change-delivered.
—n.d., accessed March 14, 2019. "Have Your Say." http://data.myworld2015.org/.
United Nations Division for Ocean Affairs and the Law of the Sea, Office of Legal Affairs. 2010. *The Law of the Sea: Marine Scientific Research: A Revised Guide to the Implementation of the Relevant Provisions of the United Nations Convention on the Law of the Sea*. New York: United Nations, www.un.org/Depts/los/doalos_publications/publicationstexts/msr_guide%202010_final.pdf.
United Nations Educational, Scientific and Cultural Organization. 2010. "Geoengineering: The Way Forward?" www.unesco.org/new/en/natural-sciences/about-us/single-view/news/geoengineering_the_way_forward/.
United Nations Environment. 2018. *The Emissions Gap Report 2018*. Nairobi: United Nations Environment Programme (UNEP), DOI:20.500.11822/26895.
United Nations Framework Convention on Climate Change. n.d., accessed March 14, 2019. "What Do Adaptation to Climate Change and Climate Resilience Mean?" https://unfccc.int/topics/adaptation-and-resilience/the-big-picture/what-do-adaptation-to-climate-change-and-climate-resilience-mean.
Urpelainen, Johannes. 2012. "Geoengineering and Global Warming: A Strategic Perspective." *International Environmental Agreements: Politics, Law and Economics* 12 no. 4:375–89, DOI:10.1007/s10784-012-9167-0.
Valencia, Mark J., and Kazumine Akimoto. 2006. "Guidelines for Navigation and Overflight in the Exclusive Economic Zone." *Marine Policy* 30 no. 6:704–11, DOI:10.1016/j.marpol.2005.11.002.
Van Hooydonk, Eric. 2014. "The Law of Unmanned Merchant Shipping: An Exploration." *The Journal of International Maritime Law* 20 no. 6:403–23, http://www.ericvanhooydonk.be/media/54f3185ce9304.pdf.
van Vuuren, Detlef P., Jae Edmonds, Mikiko Kainuma, Keywan Riahi, Allison Thomson, Kathy Hibbard et al. 2011. "The Representative Concentration Pathways: An Overview." *Climatic Change* 109 no. 1–2:5–31, DOI:10.1007/s10584-011-0148-z.
Vanderheiden, Steve. 2008. *Atmospheric Justice: A Political Theory of Climate Change*. Oxford: Oxford University Press.
Vaughan, Naomi, and Timothy Lenton. 2011. "A Review of Climate Geoengineering Proposals." *Climatic Change* 109 no. 3–4:791–825, DOI:10.1007/s10584-011-0027-7.
Vergne, Jean-Philippe, and Rodolphe Durand. 2010. "The Missing Link between the Theory and Empirics of Path Dependence: Conceptual Clarification, Testability Issue, and Methodological Implications." *Journal of Management Studies* 47 no. 4:736–59, DOI:10.1111/j.1467-6486.2009.00913.x.
Verlaan, Philomène A. 2007. "Experimental Activities That Intentionally Perturb the Marine Environment: Implications for the Marine Environmental Protection and Marine Scientific Research Provisions of the 1982 United Nations Convention on the Law of the Sea." *Marine Policy* 31 no. 2:210–16, DOI:10.1016/j.marpol.2006.07.004.
—2009. "Geo-Engineering, the Law of the Sea, and Climate Change." *Carbon & Climate Law Review* 3 no. 4:446–58, DOI:10.21552/CCLR/2009/4/115.

—2012. "Marine Scientific Research: Its Potential Contribution to Achieving Responsible High Seas Governance." *The International Journal of Marine and Coastal Law* 27 no. 4:805–12, DOI:10.1163/15718085-12341260.

Victor, David G. 2008. "On the Regulation of Geoengineering." *Oxford Review of Economic Policy* 24 no. 2:322–36, DOI:10.1093/oxrep/grn018.

—2011. *Global Warming Gridlock: Creating More Effective Strategies for Protecting the Planet.* Cambridge: Cambridge University Press.

—2019. "Governing the Deployment of Geoengineering: Institutions, Preparedness, and the Problem of Rogue Actors." In *Governance of the Deployment of Solar Geoengineering*, 41–4. Harvard Project on Climate Agreements. https://www.belfercenter.org/index.php/publication/governance-deployment-solar-geoengineering.

Viikari, Lotta. 2008. *The Environmental Element in Space Law: Assessing the Present and Charting the Future.* Leiden: Martinus Nijhoff.

Virgoe, John. 2009. "International Governance of a Possible Geoengineering Intervention to Combat Climate Change." *Climatic Change* 95 no. 1:103–19, DOI:10.1007/s10584-008-9523-9.

Visschers, Vivianne H. M., Jing Shi, Michael Siegrist, and Joseph Arvai. 2017. "Beliefs and Values Explain International Differences in Perception of Solar Radiation Management: Insights from a Cross-Country Survey." *Climatic Change* 142 no. 3–4:531–44, DOI:10.1007/s10584-017-1970-8.

Volokh, Eugene. 2003. "The Mechanisms of the Slippery Slope." *Harvard Law Review* 116 no. 4:1026–137, https://heinonline.org/HOL/P?h=hein.journals/jfpp17&i=61.

Wagner, Gernot, and Martin L. Weitzman. 2012. "Playing God." *Foreign Policy*, October 24, 2012, https://foreignpolicy.com/2012/10/24/playing-god/.

Walker, George K., ed. 2011. *Definitions for the Law of the Sea: Terms Not Defined by the 1982 Convention.* Leiden: Martinus Nijhoff.

Waters, Colin N., Jan Zalasiewicz, Colin Summerhayes, Ian J. Fairchild, Neil L. Rose, Neil J. Loader et al. 2018. "Global Boundary Stratotype Section and Point (GSSP) for the Anthropocene Series: Where and How to Look for Potential Candidates." *Earth-Science Reviews* 178:379–429, DOI:10.1016/j.earscirev.2017.12.016.

Weili, Weng, and Chen Ying. 2018. "A Chinese Perspective on Solar Geoengineering." In *Geoengineering Our Climate? Ethics, Politics and Governance*, edited by Jason J. Blackstock and Sean Low, 155–8. London: Earthscan.

Weitzman, Martin L. 2015. "A Voting Architecture for the Governance of Free-Driver Externalities, with Application to Geoengineering." *The Scandinavian Journal of Economics* 117 no. 4:1049–68, DOI:10.1111/sjoe.12120.

Wibeck, Victoria, Anders Hansson, and Jonas Anshelm. 2015. "Questioning the Technological Fix to Climate Change: Lay Sense-Making of Geoengineering in Sweden." *Energy Research & Social Science* 7:23–30, DOI:10.1016/j.erss.2015.03.001.

Wibeck, Victoria, Anders Hansson, Jonas Anshelm, Shinichiro Asayama, Lisa Dilling, Pamela M. Feetham et al. 2017. "Making Sense of Climate Engineering: A Focus Group Study of Lay Publics in Four Countries." *Climatic Change* 145 no. 1–2:1–14, DOI:10.1007/s10584-017-2067-0.

Wigley, T. M. L. 2006. "A Combined Mitigation/Geoengineering Approach to Climate Stabilization." *Science* 314 no. 5798:452–4, DOI:10.1126/science.1131728.

Williamson, P., and R. Bodle. 2016. *Update on Climate Geoengineering in Relation to the Convention on Biological Diversity: Potential Impacts and Regulatory Framework.* Montreal: Secretariat of the Convention on Biological Diversity, www.cbd.int/doc/publications/cbd-ts-84-en.pdf.

Wilson, James Q. 1980. "The Politics of Regulation." In *The Politics of Regulation*, edited by James Q. Wilson, 357–94. New York: Basic Books.

Winickoff, David E., Jane A. Flegal, and Asfawossen Asrat. 2015. "Engaging the Global South on Climate Engineering Research." *Nature Climate Change* 5 no. 7:627–34, DOI:10.1038/nclimate2632.

Wong, Pak-Hang. 2016. "Consenting to Geoengineering." *Philosophy & Technology* 26 no. 2:173–88, DOI:10.1007/s13347-015-0203-1.

Wong, Pak-Hang, Tom Douglas, and Julian Savulescu. 2014. "Compensation for Geoengineering Harms and No-Fault Climate Change Compensation," *Climate Geoengineering Governance Working Paper* 8, http://geoengineering-governance-research.org/perch/resources/workingpaper8wongdouglassavulescucompensationfinal-.pdf.

Wood, Robert, Thomas Ackerman, Philip Rasch, and Kelly Wanser. 2017. "Could Geoengineering Research Help Answer One of the Biggest Questions in Climate Science?" *Earth's Future* 5 no. 7:659–63, DOI:10.1002/2017EF000601.

World Bank, The. n.d., accessed March 14, 2019. "Agriculture, Forestry, and Fishing, Value Added." https://data.worldbank.org/indicator/NV.AGR.TOTL.CD.

World Medical Association. 2018. "WMO Declaration of Helsinki: Ethical Principles for Medical Research Involving Human Subjects." Accessed March 15, 2019. www.wma.net/policies-post/wma-declaration-of-helsinki-ethical-principles-for-medical-research-involving-human-subjects/.

World Meteorological Organization. 2014. "WMO Statement on Geoengineering (Draft)." Accessed March 15, 2019. www.wcrp-climate.org/JSC35/documents/WMO%20Statement%20on%20Geoengineering%202.pdf.

—2016. *WMO Operating Plan 2016–2019.* Accessed March 15, 2019. www.wmo.int/pages/about/documents/WMOOP2016-2019_October2016version.pdf.

—2018. *Executive Summary: Scientific Assessment of Ozone Depletion: 2018.* World Meteorological Organization Global Ozone Research and Monitoring Project 58. http://conf.montreal-protocol.org/meeting/mop/mop30/presession/Background-Documents/SAP-2018-Assessment-ES-October2018.pdf.

Wright, Malcolm J., Damon A. H. Teagle, and Pamela M. Feetham. 2014. "A Quantitative Evaluation of the Public Response to Climate Engineering." *Nature Climate Change* 4 no. 2:106–10, DOI:10.1038/nclimate2087.

Xia, L., A. Robock, S. Tilmes, and R. R. Neely, III. 2016. "Stratospheric Sulfate Geoengineering Could Enhance the Terrestrial Photosynthesis Rate." *Atmospheric Chemistry and Physics* 16 no. 3:1479–89, DOI:10.5194/acp-16-1479-2016.

Zaelke, Durwood, and James Cameron. 1989. "Global Warming and Climate Change: An Overview of the International Legal Process." *American University Journal of International Law and Policy* 5 no. 2:249–90, https://heinonline.org/HOL/P?h=hein.journals/amuilr5&i=263.

Zalasiewicz, Jan, Colin N. Waters, Colin P. Summerhayes, Alexander P. Wolfe, Anthony D. Barnosky, Alejandro Cearreta et al. 2017. "The Working Group on the Anthropocene: Summary of Evidence and Interim Recommendations." *Anthropocene* 19:55–60, DOI:10.1016/j.ancene.2017.09.001.

Zelli, Fariborz, Ina Möller, and Harro van Asselt. 2017. "Institutional Complexity and Private Authority in Global Climate Governance: The Cases of Climate Engineering, REDD+ and Short-Lived Climate Pollutants." *Environmental Politics* 26 no. 4:669–93, DOI:10.1080/09644016.2017.1319020.

Zürn, Michael, and Stefan Schäfer. 2013. "The Paradox of Climate Engineering." *Global Policy* 4 no. 3:266–77, DOI:10.1111/gpol.12004.

Index

2°C warming target. *See* two degree warming target

Aarhus Convention on Access to Information, Public Participation in Decision-Making and Access to Justice in Environmental Matters, 112, 135–7, 155–6
  Kiev Protocol on Pollutant Release and Transfer Registers, 77, 135–6
Abate, Randall, 139
abatement of emissions, 1–4, 25, 28, 33, 50, 89–90, 95, 105, 158, 166, 168, 197–9, 206, 211–13, 216, 222–3, *See also* adaptation, relationship with emissions abatement
  displacement by solar geoengineering, 2, 4, 18, 29, 32–53, 65, 68–9, 210, 212–13
  increase by solar geoengineering, 69–70
  problem structure, 17, 56–8, 61–2, 197
academic freedom. *See* human rights
access to information. *See* sharing of data, information, and results and transparency
accountability, 153, 156, 157, 163, 214
acid rain, 27–8, 97–100, 140, 141
Act to Prevent Pollution from Ships. *See* United States and US law
adaptation, 1–3, 11, 16, 29, 34, 41–4, 49–52, 65, 76–7, 92–6, 105, 113, 168, 197–9, 206, 212–13, 219, 222–3
  relationship with emissions abatement, 32–4, 41
adaptive governance, 158, 199, 203, 210
Adelman, Sam, 106
adverse selection, 36, 190
aerosols, tropospheric, 9, 18, 20, 21, 97
African Charter on Human and Peoples' Rights, 102, 104
aggregate effort. *See* problem structure
Agreement on Trade-Related Aspects of Intellectual Property Rights, 171

agriculture, 11–12, 25, 93, 95, 129, 165–6, 168, 218, 219
aircraft, 20, 121, 132–3, 139, 141–2, 186, 202
airspace, 92, 132–3
America. *See* United States and US law
American Association for the Advancement of Science, 18
American Convention on Human Rights, 102
American Geophysical Union, 175
American Meteorological Society, 83
Antarctic Treaty and its Madrid Protocol, 126–8, 148, 187
Antarctica, 10, 72, 76, 114, 126–8, 138–9, 143–4, 148, 187, 210
Anthropocene, 11, 221
areas beyond national jurisdiction, 71, 82, 114, 138–9, 143, 185
Asilomar conference and principles. *See* principles for geoengineering
assessment. *See* environmental impact assessment
assessment of results, 47, 64, 83–4, 96, 99, 106, 127, 157, 162, 169, 202–7, 217
attribution of impacts, 9, 43, 89, 105, 158, 180, 185, 191–5
Australia, 21, 22
authority to implement solar geoengineering. *See* implementation of solar geoengineering, challenges and issues. *See also* compensation for harm
aviation. *See* aircraft

Barrett, Scott, 56–8, 60
Basel Protocol on Liability and Compensation for Damage Resulting from Transboundary Movements of Hazardous Wastes and their Disposal, 187
biodiversity, 4, 11, 89, 114, 129–31
biotechnology, 152, 186, 210
Bipartisan Policy Center, 158

260

Bodansky, Daniel, 60, 87, 213
Boyle, Alan, 188
Brazil, 82
Budyko, Mikhail, 17

C2G2. *See* Carnegie Climate Geoengineering Governance Initiative
Caldeira, Ken, 18
Canada, 38, 99, 133, 187
capacity building, 202, 211, 213–14
carbon dioxide removal. *See* negative emissions technologies
Carnegie Climate Geoengineering Governance Initiative, 19, 205, 207
Charter of Fundamental Rights of the European Union. *See* European Union
Charter of the United Nations. *See* United Nations
Chicago Convention on International Civil Aviation, 132–3
China, 1, 60, 64
cirrus cloud thinning, 20, 22–4, 121, 132, 140, 147, 216
Clean Air Act. *See* United States and US law
climate sensitivity, 10, 43
CLRTAP. *See* Convention on Long-Range Transboundary Air Pollution
code of conduct, 4, 151–4, 160–1, 203
collective action. *See* problem structure
Collingridge dilemma, 3, 199
commercial actors. *See* nonstate actors
common but differentiated responsibilities. *See* principles of international law
common concern of humankind. *See* principles of international law
communitarian governance. *See* nonstate governance
community of shared fate, 153, 162, 200, 204
compensation for harm, 2, 5, 29, 64–5, 90–1, 111, 113, 161, 164–77
 aircraft, 186
 bankruptcy problem, 190
 baseline for, 181–2, 191
 causation, 147, 180, 190, 192, 193
 compensable damages, 179–80, 190
 defenses, 147, 182–3, 186, 190, 192
 from implementation, 191–5, 217
 from outdoor research, 188–91, 201, 203, 212, 214
 from space activities, 128–9, 186–7
 from termination, 194
 fund, 91, 186, 190–1, 193–5, 212, 214
 genetically modified organisms, 186
 identities of injurers and victims, 180–3
 in principles for geoengineering, 155–61

 insurance, 91, 184–5, 186, 189, 192, 193–4, 201
 justice, 184, 186, 194–5
 justice, access to, 185, 187–8
 liability, 46, 88–91, 120–1, 128–9, 146, 156–61, 178–81, 183–94
 negligence, 146, 183, 188, 192
 nuclear power, 186, 212
 oil, maritime transport of, 186
 polluter pays principle, 78, 185
 reasons for, 183–5, 188
 states' rejection of, 187–8
 under customary international law, 88–91, 179–80, 185–6
 United Nations Convention on the Law of the Sea, under the, 120–1, 187
 US law for weather modification, under, 146
 US tort law, under, 146–7
 vicarious liability, 91, 181, 190
Conference on Environment and Development. *See* Rio Declaration on Environment and Development
Conference on the Human Environment. *See* Stockholm Declaration on the Human Environment
Conferences of the Parties. *See* Convention on Biological Diversity, United Nations Framework Convention on Climate Change
conflict. *See* implementation of solar geoengineering, challenges and issues, disagreement and tension
Congress. *See* United States and US law
consent, 62, 119, 121, 125
 to experimentation, 110–11
 to harm, 182, 191
 to international law, states', 72, 183, 189, 191
 to solar geoengineering, 155, 157, 161, 217
consultation with the potentially affected, 62, 75–6, 82, 86–7, 98, 125, 130, 134–5, 155, 157, 189, 201, 211, 216, 219
Contreras, Jorge, 170, 174
Convention for the Prevention of Pollution from Ships, 116, 148
Convention on Biological Diversity, 129–31, 159
 Conferences of the Parties, 82, 130–1, 160, 206
Convention on Civil Liability for Damage Resulting from Activities Dangerous to the Environment, 188
Convention on Human Rights and Biomedicine, 110

Convention on International Liability for Damage Caused by Space Objects. *See* Treaty on Principles Governing the Activities of States in the Exploration and Use of Outer Space, Including the Moon and Other Celestial Bodies
Convention on International Trade in Endangered Species of Wild Fauna and Flora, 144
Convention on Long-Range Transboundary Air Pollution, 97–100, 115
  Gothenburg Protocol, 99
  Helsinki Protocol, 99
  Oslo Protocol, 77, 88, 99
Convention on Nature Protection and Wildlife Preservation in the Western Hemisphere, 144
Convention on the Elimination of All Forms of Discrimination against Women, 104
Convention on the Prohibition of Military or Any Other Hostile Use of Environmental Modification Techniques, 131–2, 201, 213, 219
  Understandings Regarding the Convention, 132
Convention on the Rights of Persons with Disabilities, 104
Convention on the Rights of the Child, 104
cooperation. *See* principles of international law
coordination, 79, 81–3, 154, 158, 174–6, 201, 204, 209, 212, 215, 218
cost of solar geoengineering, 20–3, 28, 58–61, 77–8, 94, 164–7, 201
  sharing of, 58, 60, 64, 212, 216
Costa Rica / Nicaragua. See International Court of Justice
Council of Europe, 102, 110
countermeasures in international law, 89–90
counter-solar geoengineering, 29, 42, 59, 61, 64, 215, 219
Craik, Neil, 106
Crutzen, Paul, 18, 221
customary international law. *See* compensation for harm, International Law Commission, prevention of transboundary harm, and state responsibility

Davies, Gareth, 67, 194
Declaration of Helsinki, 110, 204
Declaration on Permanent Sovereignty over Natural Resources, 85
Declaration on the Human Environment. *See* Stockholm Declaration on the Human Environment
Defenders of Wildlife v. Lujan. See United States and US law

democracy, 48, 60, 67, 110–11, 169, 181, 196, 217
deployment. *See* implementation of solar geoengineering, challenges and issues
developing countries, 1, 10, 13–14, 16, 19, 47, 61, 65, 69, 85, 94, 109, 209, 211, 213–14, 223
development, 10, 37, 79–81, 86, 93, 94, 105, 129, 154, 182
  right to, 80
  sustainable, 80–1, 95, 219
disagreement and tension. *See* implementation of solar geoengineering, challenges and issues
disclosure. *See* transparency
discounting the future, 41, 45
dispute resolution, 73, 84, 102, 114, 119, 136, 185, 212–15, 218
due diligence, 75, 86–8, 94, 100, 117, 181, 182–4, 190, 219
dumping, 77, 121–6, 147–8
duty of care. *See* due diligence

economic value of solar geoengineering, 25–6
emergencies, 38, 113, 139–40, 219
  contingency plans, 86, 130, 205
Endangered Species Act. *See* United States and US law
enforcement of international law, 13, 57, 73–4, 103, 119, 132, 163, 191, 200
England. *See* United Kingdom
ENMOD. *See* Convention on the Prohibition of Military or Any Other Hostile Use of Environmental Modification Techniques
environmental advocacy organizations. *See* nonstate actors
Environmental Defense Fund, 158
Environmental Defense Fund, Inc. v. Massey. See United States and US law
environmental impact assessment, 4, 81–2, 86, 94, 100, 116–17, 133–5, 156, 160, 200–1, 202, 205, 211, 215, 216, 219
Environmental Protection Agency. *See* United States and US law
environmentalism, 17, 51–2, 221–2
equity. *See* principles of international law
erga omnes obligations, 76
Espoo Convention on Environmental Impact Assessment in a Transboundary Context, 133–5
  Protocol on Strategic Environmental Assessment, 134–5
ethics, 2–3, 33–5, 40, 48, 109–11, 130, 178, 204
European Convention on Human Rights, 102
European Geosciences Union, 175

European Union, 1, 18, 62, 72, 73, 83, 133–6, 204
Charter of Fundamental Rights, 107
exclusive economic zone, 116, 121, 122, 123, 138, 147, 148
expertise, 79, 161, 203
externalities, 55–6, 192, 194
extreme weather events, 9, 11, 28, 65, 192, 195, 199, 217

Fish and Wildlife Service. See United States and US law
Fleurke, Floor, 77
Foley, Rider, 106
Forum for Climate Engineering Assessment, 19
fossil fuels, 9, 13, 37, 105, 141, 166–7, 170, 173
Fragnière, Augustin, 159
France, 103
free driver. See problem structure
free rider. See problem structure
Frosch, Robert A., 17
funders, 109, 138, 152, 163, 167, 175, 201, 206–7

Gabčíkovo-Nagymaros Project. See International Court of Justice
Gardiner, Stephen, 159
Gates, Bill, 164–5
General Assembly. See United Nations
Geoengineering Research Evaluation Act. See United States and US law
Geoengineering Research Governance Project, 160–1
Germany, 39, 65, 158
global governance, 56, 67, 74
global public good. See problem structure
Goeschl, Timo, 68
Gore, Al, 33
Gothenburg Protocol. See Convention on Long-Range Transboundary Air Pollution
Great Barrier Reef, 21, 22
Guterres, António, 1

Hale, Benjamin, 34–5, 52
Hartford Fire Insurance Co. v. California. See United States and US law
Harvard University, 167, 172
Heinrich Böll Foundation, 158–9
Helsinki Protocol. See Convention on Long-Range Transboundary Air Pollution
high seas, 72, 114, 119, 123, 133, 138, 143–5, 165
HNS Convention. See International Convention on Liability and Compensation for Damage in Connection with the Carriage of Hazardous and Noxious Substances by Sea

Horton, Joshua, 60, 180, 184–5
hostile or military use of solar geoengineering. See implementation of solar geoengineering, challenges and issues
House of Commons. See United Kingdom
Hubert, Anna-Maria, 160
Hulme, Mike, 220
human rights, 4, 71, 101–13, 135, 155, 198
  academic freedom, 107
  of vulnerable people and groups, 111
  procedural, 104, 111–12
  research subjects, 109–11, 152
  scientists' material interests, 108
  substantive, 112–13
  to enjoy the benefits of scientific progress, 108–9
Human Rights Council. See United Nations
Hunt, Hugh, 170, 174
hydrological cycle, 8–11, 23–6, 69, 75, 100, 217

impact assessment. See environmental impact assessment
implementation of solar geoengineering, challenges and issues
  as civil disobedience, 61, 166
  authority, 2, 28, 62–3, 156–7
  by a nonstate actor, 5, 59–60, 143, 164–7, 209
  disagreement and tension, 28, 59–60, 63–6, 180, 197
  hostile or military use, 28, 64, 65, 131–2, 202, 207, 215, 219
  premature, 197, 205–6, 210, 214
  termination, 2, 20, 40, 67, 87, 180, 183, 194, 218
  uni- or minilateral, 2, 17, 28–9, 58–62, 76, 136, 165, 191, 211, 216
India, 60
influence by interests, 29, 48, 156, 166–7, 168–9, 174
information asymmetry, 35–7, 46, 55
informed consent. See consent
Institute for Advanced Sustainability Studies, 160
Institute of Nuclear Power Operations, 153
insurance. See compensation for harm
Integrated Assessment of Geoengineering Proposals, 38
intellectual property, 5, 54, 135, 156, 161, 164, 167, 169–77, 189, 196, 201, 207
  and human rights, 108
  compulsory licensing and march-in, 172–3, 176, 207
  current patenting activity, 170–1
  defensive patenting and publication, 170, 173–4
  limitations on, 154–5
  patent pool, 173
  pledge community, 174–6

intellectual property (cont.)
  prizes, 176
  research commons for, 174–7, 211
  trade secrets, 170, 172–3, 175–6
intentionality of solar geoengineering, 59, 141, 145, 181, 198–9, 201, 208
Inter-American Court of Human Rights, 104
intergenerational issues, 37, 41, 43–7, 68, 80–1, 93–4, 111, 132, 196, 206, 223
Intergovernmental Panel on Climate Change, 10–11, 15, 23, 33, 34, 83–4, 95, 202, 209
international and intergovernmental organizations, 4, 7, 57, 71, 73–5, 79, 116, 117–18, 120, 122, 151, 160, 175, 189, 191–2, 209
International Atomic Energy Agency, 209
  Statute, 212
International Civil Aviation Organization, 133
International Convention on Liability and Compensation for Damage in Connection with the Carriage of Hazardous and Noxious Substances by Sea, 188
International Council for Science, 82, 83, 175
International Court of Justice, 137, 192
  Costa Rica / Nicaragua, 85, 89
  Gabčíkovo-Nagymaros Project, 90
  Nuclear Weapons (Advisory Opinion), 85
  Pulp Mills, 85
International Covenant on Civil and Political Rights, 102–4, 110, 112, 113
International Covenant on Economic, Social and Cultural Rights, 102, 103–4, 107, 108, 109
International Ethical Guidelines for Health-Related Research Involving Humans, 110, 204
International Law Commission, 84, 187
  Draft Articles on Prevention of Transboundary Harm from Hazardous Activities, 86–7
  Draft Articles on Responsibility of States for Internationally Wrongful Acts, 88–90
  Draft Guidelines on the Protection of the Atmosphere, 92, 100
  Draft Principles on the Allocation of Loss in the Case of Transboundary Harm Arising out of Hazardous Activities, 90–1, 180, 185, 190
International Maritime Organization, 83, 115, 123, 207
International Monetary Fund, 218
International Tribunal for the Law of the Sea, 100, 115
  Responsibilities and Obligations of States Sponsoring Persons and Entities with Respect to Activities in the Area (Advisory Opinion), 87

Japan, 39
Johnson, Eddie Bernice, 148
justice, 5, 19, 184, 186–8, 194–5, 222
  access to, 105, 112, 136, 185, 187–8

Keith, David, 18, 34, 201
Kiev Protocol on Pollutant Release and Transfer Registers. *See* Aarhus Convention on Access to Information, Public Participation in Decision-Making and Access to Justice in Environmental Matters
Kruger, Tim, 161
Kyoto Protocol. *See* United Nations Framework Convention on Climate Change

Lasswell, Harold, 54
Lawrence Livermore National Laboratory, 18
legitimacy, 2, 28, 56, 59, 62, 67, 111, 153, 155, 158–61, 163, 178, 191, 200, 202–7, 210, 213–14, 216
legitimate scientific research, 82, 125–6, 175, 205, 206
liability. *See* compensation for harm
Lin, Albert, 37, 45, 48
lock-in and slippery slope, 29–30, 48, 156, 168, 171, 173, 202–5, 209–10
London Convention and London Protocol, 77–8, 121–6, 147
  Marine Geoengineering Amendment to the London Protocol, 124–6, 132, 201, 206
London Protocol. *See* London Convention and London Protocol
Lugano Convention. *See* Convention on Civil Liability for Damage Resulting from Activities Dangerous to the Environment
Lujan v. Defenders of Wildlife. *See* United States and US law

Madrid Protocol. *See* Antarctic Treaty
marine cloud brightening, 1, 19, 22–4, 121–2, 139, 170, 173, 199
Marine Cloud Brightening Project, 21
marine geoengineering, 124–6, 132, *See also* oceans, solar geoengineering in, on, or above
Marine Geoengineering Amendment to the London Protocol. *See* London Convention and London Protocol
Marine Mammal Protection Act. *See* United States and US law
Marine Protection, Research and Sanctuaries Act. *See* United States and US law
marine scientific research, 114, 117–20, 122–3, 125–6
MARPOL. *See* Convention for the Prevention of Pollution from Ships

McLaren, Duncan, 40–1
McNerney, Jerry, 148
meta-regulation. *See* nonstate governance
microbubbles, 121, 125
Migratory Bird Treaty Act. *See* United States and US law
military involvement in solar geoengineering, 207, 211
military use of solar geoengineering. *See* hostile or military use of solar geoengineering
Millard-Ball, Adam, 69
mitigation. *See* abatement of emissions
models, climate, 2, 6, 20, 22–6, 40, 52, 56, 61, 72, 76, 84, 89, 109, 113, 142, 167, 174, 182, 203
monetary policy, 217–19
monitoring, environmental, 20, 64, 98, 99, 111, 116, 125, 127, 129, 142, 158–60, 167, 172, 203, 205, 211
Montreal Protocol. *See* Vienna Convention for the Protection of the Ozone Layer
moral hazard, 34–7, 46, 52, 190, *See also* abatement of emissions, displacement by solar geoengineering
moratorium, 159, 206–7, 212, 214
Moreno-Cruz, Juan, 69
Morrow, David, 32
Morton, Oliver, 53, 61
Muir, John, 221
mutual restraint. *See* problem structure

National Academies of Sciences, Engineering, and Medicine, 17–19, 22, 34, 148
National Marine Fisheries Service. *See* United States and US law
National Oceanic and Atmospheric Administration. *See* United States and US law
National Weather Modification Policy Act. *See* United States and US law
Natural Environment Research Council. *See* United Kingdom
necessity in international law, 90, 182
negative emissions technologies, 1–2, 6, 14–19, 25, 32–4, 42–4, 49, 50, 51, 154, 206, 216, 222–3
ocean fertilization, 15, 125, 162, 165, 204–5
problem structure, 197–9
negligence. *See* compensation for harm, negligence
New Zealand, 39
nonbinding multilateral agreements, 4, 73, 74, 75, 79–83, 104, 111, 198
nonproliferation, 58, 211–14
nonstate actors, 5, 54, 74, 88, 102, 106, 109, 114, 136, 144, 160, 164–77, 181, 185, 189–90, *See also* funders
commercial actors, 2, 154–5, 162, 167–74, 175–6

environmental advocacy organizations, 19, 51–2, 135–6, 151–2, 162
implementation of solar geoengineering by. *See* implementation of solar geoengineering, challenges and issues
professional societies, 109, 151, 163, 204
publishers, scientific, 109, 163, 175, 203–4
nonstate governance, 3, 4–5, 74, 109, 150–63, 198, 200, 202–5
communitarian governance, 153, 162
meta-regulation, 151–2, 204
private regulation, 151, 204
self-regulation, 151–3, 157, 163, 204
notification, 75, 76, 81–2, 86, 103, 106, 116, 125, 130, 134–5, 145, 155, 202, 215, 216, 219
nuclear power, 153, 162, 186, 211–13, *See also* compensation for harm
nuclear weapons, 58, 60, 211, 220
Nuclear Weapons (Advisory Opinion). *See* International Court of Justice

Obama, Barack, 140
ocean acidification, 27, 40, 41
Ocean Dumping Act. *See* United States and US law
ocean fertilization. *See* negative emissions technologies
oceans, solar geoengineering in, on, or above, 17, 22, 114–26, 147–8. *See also* marine cloud brightening
Office of Science and Technology Policy. *See* United States and US law
Olson, Robert, 168
Oslo Protocol. *See* Convention on Long-Range Transboundary Air Pollution
outer space, 72, 92, 114, 128–9, 133, 138, 143, 144, *See also* compensation for harm, space-based solar geoengineering
Outer Space Treaty. *See* Treaty on Principles Governing the Activities of States in the Exploration and Use of Outer Space, Including the Moon and Other Celestial Bodies
Oxford principles. *See* principles for geoengineering
ozone, 9, 18, 27, 76, 78, 83, 92, 96–7, 117, 141–2, 180

Paris Agreement. *See* United Nations Framework Convention on Climate Change
Parson, Edward, 48–9, 201, 208
participation in the governance of solar geoengineering, 49, 60–1, 172, 174, 175, 189, 190–2

Pasztor, Janos, 19
patents. *See* intellectual property
peer review, 125–6, 175, 205
Permanent Court of Arbitration, 137
Pielke, Roger, Jr., 14
Pinchot, Gifford, 221
plants. *See* agriculture
polluter pays. *See* principles of international law
pollution, solar geoengineering regulated as, 72, 78, 98, 115–17, 120–6, 133, 136, 139–142, 148, 202
precaution. *See* principles of international law
premature implementation of solar geoengineering. *See* implementation of solar geoengineering, challenges and issues
prevention of transboundary harm, customary international law of, 75, 80, 84–9, 91, 93, 112, 129–30, 161, 209, 216, 219, *See also* International Law Commission, Draft Articles on Prevention of Transboundary Harm from Hazardous Activities
principal-agent problem, 55, 181
principles for geoengineering, 4, 150, 154–9, 161–3, 178, 203–4
    Asilomar conference and principles, 19, 157–8
    Oxford principles, 154–62, 169, 172, 203
    Tollgate Principles, 159
principles of international law, 4, 73–9, 81, 86, 93, 100, 185, 187
    common but differentiated responsibilities, 75–6, 78–9, 81, 93
    common concern of humankind, the environment as, 76–7, 90, 93, 155
    cooperation, international, 13, 28, 42, 47, 50, 52, 57–8, 60–1, 67, 71, 74–5, 80–3, 86, 94–5, 99, 103–5, 107–9, 113, 115–18, 127–30, 132–3, 144, 152, 160, 176, 185, 187, 195, 199, 202–3, 209, 211–14, 218–20
    equity, 5, 19, 41, 44, 81, 86, 93, 100, 111, 119, 129, 152, 185, 186, 193–5
    polluter pays, 78, 81, 185, *See also* compensation for harm
    precaution, 77–8, 81, 93, 124, 129
private regulation. *See* nonstate governance
prizes. *See* intellectual property
problem structure, 4, 55–8, *See also* abatement of emissions, negative emissions technologies
    aggregate effort, 56–8
    collective action, 13, 15–17, 42, 55–7, 61, 74, 166, 169, 183, 192, 197, 211, 223
    free driver, 42, 57, 59, 69
    free rider, 13, 55–7, 166, 190
    mutual restraint, 58, 214, 219
    public good, 76, 165, 189
    single best effort, 57, 70

procedural rights. *See* Aarhus Convention on Access to Information, Public Participation in Decision-Making and Access to Justice in Environmental Matters, human rights, public participation, transparency
procurement, 5, 167–8
professional societies. *See* nonstate actors
Protocol on Civil Liability and Compensation for Damage Caused by the Transboundary Effects of Industrial Accidents on Transboundary Waters, 187
Provisions for Co-operation between States in Weather Modification. *See* United Nations Environment (Programme)
public engagement, 158, 161, 162, 200, 205, 207–8, 212
public good. *See* problem structure
public opinion and perception, 14, 37–40, 45–6, 65–7, 157, 162, 168, 199–200, 205, 212
public participation, 104–5, 108–9, 112, 134–6, 155–6, 157, 160, 202–3, 205
publishers, scientific. *See* nonstate actors
Pulp Mills. *See* International Court of Justice

Quaas, Martin, 68

rationality, states', 5–6, 41, 45–6, 54–5, 66
Rayner, Steve, 33, 52
Rees, Joseph, 153
regional solar geoengineering, 126, 209
regulatory capture, 153
Reichwein, David, 87, 160
rent-seeking, 168
reputation, 58, 74, 79, 151, 153, 162, 200, 204, 215
research commons. *See* intellectual property
research of solar geoengineering. *See also* compensation for harm
    governing large-scale research, 208–14
    governing small-scale research, 200–8
Resolution on the Right to Exploit Freely Natural Wealth and Resources, 85
Research subjects. *See* human rights
Responsibilities and Obligations of States Sponsoring Persons and Entities with Respect to Activities in the Area (Advisory Opinion). *See* International Tribunal for the Law of the Sea
Rhode Island, 139
Ricke, Katherine, 69
right to exploit natural resources, 72, 80, 85–6, 93, 115, 129
rights. *See* human rights

Rio Declaration on Environment and
    Development, 77, 78, 80–1, 85, 104, 112, 129,
    155, 156
risk
    aversion, 41, 43, 44, 63
    compensation, 36–7, 47, *See also* abatement of
      emissions, displacement by solar
      geoengineering
    perception, 222
    risk-risk trade-off, 2, 197
Rome Convention on Damage Caused by Foreign
    Aircraft to Third Parties on the Surface, 186
Royal Society, 2, 18–19, 38, 150, 154, 158–61, 175,
    178, 185
Russia and the Soviet Union, 17, 60, 99,
    123, 187

Sarnoff, Joshua, 170, 174
Saudi Arabia, 82
Schelling, Thomas, 17, 58, 220
scientific research in law, 114, 117–20, 122–3, 125–8
SCoPEx. *See* Stratospheric Controlled
    Perturbation Experiment
Security Council. *See* United Nations
self-regulation. *See* nonstate governance
sharing of data, information, and results, 14, 56, 71,
    75, 94, 130, 162, 175, 189, 193, 207, 209, 211, 218,
    See also transparency
ships, 21, 116, 118–20, 123, 139, 141, 148, 165,
    173, 202
    unmanned, 122
side payments, 58, 217
single best effort. *See* problem structure
slippery slope. *See* lock-in and slippery slope
social license to operate, 153, 162, 200
Solar Geoengineering Organization, 209–14,
    217–18
Solar Radiation Management Governance
    Initiative, 19, 158–9, 202, 205
sovereignty, 59, 72–3, 85, 132–3, 194, 215, *See also*
    right to exploit natural resources
Soviet Union. *See* Russia and the Soviet Union
Space Liability Convention. *See* Treaty on
    Principles Governing the Activities of States
    in the Exploration and Use of Outer Space,
    Including the Moon and Other Celestial
    Bodies
space-based solar geoengineering, 2, 17, 20–3, 23,
    128–9, 170, 186–7, *See also* compensation for
    harm
species, endangered, 4, 116, 127, 143–5
SPICE. *See* Stratospheric Particle Injection for
    Climate Engineering
state rationality. *See* rationality, states'

state responsibility, customary international law of,
    88–90, 120, 179, 185, 188, 191, *See also*
    International Law Commission: Draft
    Articles on Responsibility of States for
    Internationally Wrongful Acts
Steele v. Bulova Watch Co., Inc., *See* United States
    and US law
Stilgoe, Jack, 168
Stockholm Declaration on the Human
    Environment, 79–82, 85, 104
stratospheric aerosol injection, 1, 17, 23–4, 27–9, 61,
    67, 83, 88, 96–100, 117, 121–2, 132, 139–42, 147,
    155, 170, 173, 180, 216
Stratospheric Controlled Perturbation
    Experiment, 21, 27, 167
Stratospheric Particle Injection for Climate
    Engineering, 19, 21, 110, 161, 174
sulfur, 18, 21, 27–8, 88, 96–100, 117, 133, 140–2
Supreme Court. *See* United States and US law
surface-based solar geoengineering, 22, 170
sustainable development. *See* development
Sustainable Development Goals, 81
Svoboda, Toby, 106
Sweden, 38–9
Switzerland, 82
Szerszynski, Bronislaw, 67

technocracy, 217
technology transfer, 71, 94–6, 97, 108, 121,
    171, 213
tension between climate change and solar
    geoengineering, 72, 77, 98, 116–17, 121, 126,
    127, 158
termination of solar geoengineering.
    See implementation of solar geoengineering,
    challenges and issues
territorial sea, 118
Tol, Richard, 194
Tollgate Principles. *See* principles for
    geoengineering
transboundary harm. *See* prevention of
    transboundary harm, customary international
    law of
transparency, 47, 81, 83, 96, 112, 130–1, 135, 153–4,
    155–6, 157–60, 162–3, 169–71, 172, 174, 203, 211,
    215, *See also* sharing of data, information, and
    results
Treaty on Principles Governing the Activities of
    States in the Exploration and Use of Outer
    Space, Including the Moon and Other
    Celestial Bodies, 128–9, 186–7
    Convention on International Liability for
    Damage Caused by Space Objects,
    186–7

Index

tropospheric aerosols. *See* aerosols, tropospheric
two degree warming target, 12, 14–15, 34, 63, 81, 95, 96, 166

UNECE. *See* United Nations, Economic Commission for Europe
uni- or minilateral implementation of solar geoengineering. *See* implementation of solar geoengineering, challenges and issues
United Kingdom, 21, 38–9, 65, 99, 124, 154, 156, 161
  House of Commons, 154–6, 161
  Natural Environment Research Council, 38
United Nations, 1, 14, 79–85, 115, 125, 128, 133, 174, 208, 214–16
  Charter, 75, 86
  Economic Commission for Europe, 98, 133–5
  General Assembly, 62–3, 80, 85, 136, 208, 218
  Human Rights Council, 105, 108
  Security Council, 63, 72, 136–7, 208, 211, 220
United Nations Convention on the Law of the Sea, 78, 114–24, 126, 187, 193, 218–19, *See also* compensation for harm
United Nations Education, Social and Cultural Organization, 82–3, 207
  Universal Declaration on Bioethics and Human Rights, 110
United Nations Environment (Programme), 81–3, 96, 207, 213, 218
  Provisions for Co-operation between States in Weather Modification, 82–3
United Nations Framework Convention on Climate Change, 14–15, 33, 62–3, 75–9, 89–90, 92–7, 112, 212–13
  Conferences of the Parties, 14, 33, 62–3, 95, 208
  Kyoto Protocol, 14, 33, 89, 95
  Paris Agreement, 12, 14–15, 33, 34, 79, 81, 89, 95–6, 101, 105, 187, 208, 213, 219
United States and US law, 1, 4, 17, 21, 27, 33–34, 38–9, 60, 62, 64, 83, 95, 99, 102–3, 115, 123, 129, 138–49, 152–3, 158, 162, *See also* compensation for harm
  Act to Prevent Pollution from Ships, 148
  Clean Air Act, 140–3, 148
  Congress, 140, 143
  Defenders of Wildlife v. Lujan, 145
  Endangered Species Act, 138, 143–5
  Environmental Defense Fund, Inc. v. Massey, 139
  Environmental Protection Agency, 140–2, 147
  extraterritorial application, 138–9, 142–5
  Fish and Wildlife Service, 143
  Geoengineering Research Evaluation Act, 148
  Hartford Fire Insurance Co. v. California, 139
  liability for harm, 146–7
  Lujan v. Defenders of Wildlife, 145
  Marine Mammal Protection Act, 145
  Marine Protection, Research and Sanctuaries Act (Ocean Dumping Act), 147–8
  Migratory Bird Treaty Act, 145
  National Environmental Policy Act, 138–9
  National Marine Fisheries Service, 143
  National Oceanic and Atmospheric Administration, 145
  National Weather Modification Policy Act, 146
  Office of Science and Technology Policy, 148
  Secretary of Commerce, 145–6
  Steele v. Bulova Watch Co., Inc., 139
  Supreme Court, 140, 144
  United States v. Aluminum Co. of America (Alcoa), 139
  Weather Modification Reporting Act, 145
  White House, 34, 148
United States v. Aluminum Co. of America (Alcoa). *See* United States and US law
Universal Declaration of Human Rights, 102, 108
Universal Declaration on Bioethics and Human Rights. *See* United Nations Education, Social and Cultural Organization
University of Calgary, 160
University of Oxford, 160

Verlaan, Philomene, 117
vessels. *See* ships
Victor, David, 58, 150, 164, 202
Vienna Convention for the Protection of the Ozone Layer, 96–7, 141
  Montreal Protocol on Substances that Deplete the Ozone Layer, 96–7, 141

Wagner, Gernot, 57
weather. *See* extreme weather events
weather modification, 4, 82–3, 145–6, 202, *See also* Convention on the Prohibition of Military or Any Other Hostile Use of Environmental Modification Techniques, United States and US law
Weather Modification Reporting Act. *See* United States and US law
Weitzman, Martin, 57, 69
welfarism, 5, 54, 184–5, 188
White House. *See* United States and US law
Wood, Lowell, 18
Working Group on Coupled Modeling, 174
World Academy of Sciences, the, 158
World Climate Research Programme, 82–3, 174
World Medical Association, 152
World Meteorological Organization, 82–3, 207, 213, 218
World Trade Organization, 171

For EU product safety concerns, contact us at Calle de José Abascal, 56–1°, 28003 Madrid, Spain or eugpsr@cambridge.org.

www.ingramcontent.com/pod-product-compliance
Lightning Source LLC
LaVergne TN
LVHW051915060526
838200LV00004B/157